Medical Parasitology

Abhay R. Satoskar, MD, PhD
Ohio State University
Columbus, Ohio, USA

Gary L. Simon, MD, PhD
The George Washington University
Washington DC, USA

Peter J. Hotez, MD, PhD
The George Washington University
Washington DC, USA

Moriya Tsuji, MD, PhD
The Rockefeller University
New York, New York, USA

LANDES BIOSCIENCE
Austin, Texas
U.S.A.

VADEMECUM
Parasitology
LANDES BIOSCIENCE
Austin, Texas USA

Please address all inquiries to the Publisher:
Landes Bioscience, 1002 West Avenue, Austin, Texas 78701, USA
Phone: 512/ 637 6050; FAX: 512/ 637 6079

ISBN: 978-1-57059-695-7

Library of Congress Cataloging-in-Publication Data

Medical parasitology / [edited by] Abhay R. Satoskar ... [et al.].
p. ; cm.
Includes bibliographical references and index.
ISBN 978-1-57059-695-7
1. Medical parasitology. I. Satoskar, Abhay R.
[DNLM: 1. Parasitic Diseases. WC 695 M489 2009]
QR251.M426 2009
616.9'6--dc22

2009035449

While the authors, editors, sponsor and publisher believe that drug selection and dosage and the specifications and usage of equipment and devices, as set forth in this book, are in accord with current recommendations and practice at the time of publication, they make no warranty, expressed or implied, with respect to material described in this book. In view of the ongoing research, equipment development, changes in governmental regulations and the rapid accumulation of information relating to the biomedical sciences, the reader is urged to carefully review and evaluate the information provided herein.

Dedications

To Anjali, Sanika and Monika for their support —Abhay R. Satoskar

To Vicki, Jason and Jessica for their support —Gary L. Simon

To Ann, Matthew, Emily, Rachel, and Daniel —Peter J. Hotez

To Yukiko for her invaluable support —Moriya Tsuji

About the Editors...

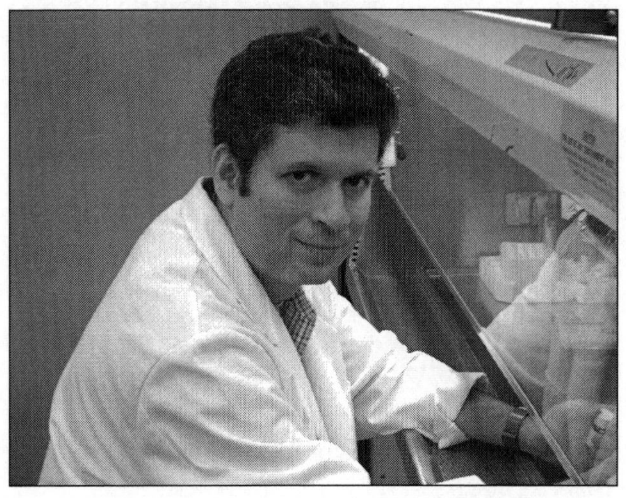

ABHAY R. SATOSKAR is Associate Professor of Microbiology at the Ohio State University, Columbus. Main research interests include parasitology and development of immunotherapeutic strategies for treating parasitic diseases. He is a member of numerous national and international scientific organizations including American Association of Immunologists and American Society of Tropical Medicine and Hygiene. He has served as a consultant for several organizations including NIH (USA), National Research Foundation (South Africa), Wellcome Trust (UK) and Sheikh Hamadan Foundation (UAE). He holds visiting faculty appointments in institutions in India and Mexico. Abhay Satoskar received his medical degree (MB, BS and MD) from Seth G. S. Medical College and King Edward VII Memorial Hospital affiliated to University of Bombay, India. He received his doctoral degree (PhD) from University of Strathclyde, Glasgow.

About the Editors...

GARY L. SIMON is the Walter G. Ross Professor of Medicine and Director of the Division of Infectious Diseases at The George Washington University School of Medicine. He is also Vice-Chairman of the Department of Medicine. Dr. Simon is also Professor of Microbiology, Tropical Medicine and Immunology and Professor of Biochemistry and Molecular Biology. His research interests are in the diagnosis and treatment of HIV infection and its complications. He is especially interested in the interaction between HIV and diseases of sub-Saharan Africa, notably tuberculosis.

Dr. Simon is a native of Brooklyn, New York, but grew up in the Washington, DC metropolitan area. He obtained his undergraduate degree in chemistry from the University of Maryland and a PhD degree in physical chemistry from the University of Wisconsin. He returned to the University of Maryland where he received his MD degree and did his internal medicine residency. He did his infectious disease training at Tufts-New England Medical Center in Boston.

About the Editors...

PETER J. HOTEZ is Distinguished Research Professor and the Walter G. Ross Professor and Chair of the Department of Microbiology, Immunology and Tropical Medicine at The George Washington University, where his major research and academic interest is in the area of vaccine development for neglected tropical diseases and their control. Prof. Hotez is also the President of the Sabin Vaccine Institute, a non-profit medical research and advocacy organization. Through the Institute, Dr. Hotez founded the Human Hookworm Vaccine Initiative, a product development partnership supported by the Bill and Melinda Gates Foundation, to develop a recombinant vaccine for human hookworm disease, and the Global Network for Tropical Neglected Diseases Control, a new partnership formed to facilitate the control of neglected tropical diseases in developing countries. He is also the Founding Editor-in-Chief of *PLoS Tropical Neglected Diseases.*

Dr. Hotez is a native of Hartford, Connecticut. He obtained his BA degree in Molecular Biophysics *Phi Beta Kappa* from Yale University (1980) and his MD and PhD from the medical scientist-training program at Weill Cornell Medical College and The Rockefeller University.

About the Editors...

MORIYA TSUJI is Aaron Diamond Associate Professor and Staff Investigator, HIV and Malaria Vaccine Program at the Aaron Diamond AIDS Research Center, The Rockefeller University, New York. He is also Adjunct Associate Professor in the Department of Medical Parasitology at New York University School of Medicine. He is a member of various national and international scientific organizations, including Faculty of 1000 Biology, United States-Israel Binational Science Foundation, the Center for Scientific Review at the National Institute of Health of the United States Department of Health and Human Services, the Science Programme at the Wellcome Trust of the United Kingdom, the French Microbiology Program at the French Ministry of Research and New Technologies, and the Board of Experts for the Italian Ministry for University and Research. He is also an editorial board member of the journal *Virology: Research and Treatment*. His major research interests are (i) recombinant viral vaccines against microbial infections, (ii) identification of novel glycolipid-based adjuvants for HIV and malaria vaccines, and (iii) the protective role of CD1 molecules in HIV/malaria infection. Moriya Tsuji received his MD in 1983 from The Jikei University School of Medicine, Tokyo, Japan, and in 1987 earned his PhD in Immunology from the University of Tokyo, Faculty of Medicine.

Contents

Section III. Cestodes

Section IV. Protozoans

Editors

Abhay R. Satoskar, MD, PhD
Department of Microbiology
and
Department of Molecular Virology, Immunology
and Medical Genetics
Ohio State University
Columbus, Ohio, USA
Email: satoskar.2@osu.edu
Chapters 20, 21, 25

Gary L. Simon, MD, PhD
Department of Medicine
and
Department of Microbiology, Immunology
and Tropical Medicine
and
Department of Biochemistry and Molecular Biology
Division of Infectious Diseases
The George Washington University
Washington DC, USA
Email: gsimon@mfa.gwu.edu
Chapters 5

Peter J. Hotez, MD, PhD
Department of Microbiology, Immunology
and Tropical Medicine
The George Washington University
Washington DC, USA
Email: mtmpjh@gwumc.edu
Chapter 16

Moriya Tsuji, MD, PhD
HIV and Malaria Vaccine Program
The Aaron Diamond AIDS Research Center
The Rockefeller University
New York, New York, USA
Email: mtsuji@adarc.org
Chapters 26, 32

Contributors

Wafa Alnassir, MD
Department of Medicine
Division of Infectious Diseases
University Hospitals of Cleveland
Cleveland, Ohio, USA
Email: wafanassirali@yahoo.com
Chapter 18

Subash Babu, PhD
Helminth Immunology Section
Laboratory of Parasitic Diseases
National Institutes of Health
Bethesda, Maryland, USA
Email: sbabu@niaid.nih.gov
Chapter 12

Guy Caljon, PhD
Unit of Cellular and Molecular Immunology
Department of Molecular and Cellular Interactions
VIB, Vrije Universiteit Brussel
Brussels, Belgium
Email: gucaljon@vub.ac.be
Chapter 23

Matthew W. Carroll, MD
Division of Infectious Diseases
The George Washington University School of Medicine
Washington DC, USA
Email: mcarroll@gwu.edu
Chapter 6

Christopher M. Cirino, DO, MPH
Division of Infectious Diseases
The George Washington University School of Medicine
Washington DC, USA
Email: ccirino710@hotmail.com
Chapter 7

Allen B. Clarkson, Jr, PhD
Department of Medical Parasitology
New York University School of Medicine
New York, New York, USA
Email: clarka01@med.nyu.edu
Chapter 31

John Cmar, MD
Department of Medicine
Divisions of Infectious Diseases and Internal Medicine
Sinai Hospital of Baltimore
Baltimore, Maryland, USA
Email: doc.operon@gmail.com
Chapter 13

Hannah Cummings, BS
Department of Microbiology
Ohio State University
Columbus, Ohio, USA
Email: cummings.123@osu.edu
Chapters 20, 21

Erin Elizabeth Dainty, MD
Department of Obstetrics and Gynecology
University of Pennsylvania
Philadelphia, Pennsylvania, USA
Email: erin.dainty@uphs.upenn.edu
Chapter 11

Janine R. Danko, MD, MPH
Department of Infectious Diseases
Uniformed Services University of the Health Sciences
Naval Medical Research Center
Bethesda, Maryland, USA
Email: janine.danko@med.navy.mil
Chapter 1

John R. David, MD
Department of Immunology
and Infectious Diseases
Harvard School of Public Health
Boston, Massachusetts, USA
Email: jdavid@hsph.harvard.edu
Chapter 25

Patrick De Baetselier, PhD
Unit of Cellular and Molecular
Immunology
Department of Molecular and Cellular
Interactions
VIB, Vrije Universiteit Brussel
Brussels, Belgium
Email: pdebaets@vub.ac.be
Chapter 23

David J. Diemert, MD
Human Hookworm Vaccine Initiative
Albert B. Sabin Vaccine Institute
Washington DC, USA
Email: david.diemert@sabin.org
Chapter 4

Daniel J. Eichinger, PhD
Department of Medical Parasitology
New York University
School of Medicine
New York, New York, USA
Email: eichid01@med.nyu.edu
Chapter 28

David M. Engman, MD, PhD
Departments of Pathology
and Microbiology-Immunology
Northwestern University
Chicago, Illinois, USA
Email: d-engman@northwestern.edu
Chapter 22

Cynthia Livingstone Gibert, MD
Department of Medicine
Division of Infectious Diseases
The George Washington University
Washington VA Medical Center
Washington DC, USA
Email: cynthia.gibert@med.va.gov
Chapter 11

Murliya Gowda, MD
Infectious Disease Consultants (IDC)
Fairfax, Virginia, USA
Email: pgowda2000@yahoo.com
Chapter 8

Thaddeus K. Graczyk, MSc, PhD
Department of Environmental
Health Sciences
Division of Environmental
Health Engineering
Johns Hopkins Bloomberg
School of Public Health
Baltimore, Maryland, USA
Email: tgraczyk@jhsph.edu
Chapter 16

Ambar Haleem, MD
Department of Internal Medicine
University of Iowa
Iowa City, Iowa, USA
Email: ambar-haleem@uiowa.edu
Chapter 24

Raymond M. Johnson, MD, PhD
Department of Medicine
Indiana University School of Medicine
Indianapolis, Indiana, USA
Email: raymjohn@iupui.edu
Chapter 30

Kevin C. Kain, MD, FRCPC
Department of Medicine
University of Toronto
Department of Global Health
McLaughlin Center for Molecular Medicine
and
Center for Travel and Tropical Medicine
Toronto General Hospital
Toronto, Ontario, Canada
Email: kevin.kain@uhn.on.ca
Chapter 32

Vijay Khiani, MD
Department of Medicine
University Hospitals of Cleveland
Cleveland, Ohio, USA
Email: vijay.khiani@gmail.com
Chapter 19

Charles H. King, MD, FACP
Center for Global Health and Diseases
Case Western Reserve University School of Medicine
Cleveland, Ohio, USA
Email: chk@cwru.edu
Chapters 17-19

Ann M. Labriola, MD
Department of Medicine
Division of Infectious Diseases
The George Washington University
Washington VA Medical Center
Washington DC, USA
Email: ann.labriola@va.gov
Chapter 10

Claudio M. Lezama-Davila, PhD
Department of Microbiology
and
Department of Molecular Virology, Immunology and Medical Genetics
Ohio State University
Columbus, Ohio, USA
Email: lezama-davila.1@osu.edu
Chapter 25

Angelike Liappis, MD
Departments of Medicine and Microbiology, Immunology and Tropical Medicine
Division of Infectious Diseases
The George Washington University
Washington DC, USA
Email: mtmapl@gwumc.edu
Chapter 15

Stefan Magez, PhD
Unit of Cellular and Molecular Immunology
Department of Molecular and Cellular Interactions
VIB, Vrije Universiteit Brussel
Brussels, Belgium
Email: stemagez@vub.ac.be
Chapter 23

Bradford S. McGwire, MD, PhD
Division of Infectious Diseases
and
Center for Microbial Interface Biology
Ohio State University
Columbus, Ohio, USA
Email: brad.mcgwire@osumc.edu
Chapter 22

Salim Melari, PhD
Department of Biochemistry
Fels Institute for Cancer Research
and Molecular Biology
Temple University School of Medicine
Philadelphia, Pennsylvania, USA
Email: salim.merali@temple.edu
Chapter 31

Rohit Modak, MD, MBA
Division of Infectious Diseases
The George Washington University
Medical Center
Washington DC, USA
Email: Rohitmodak@yahoo.com
Chapter 2

Thomas B. Nutman, MD
Helminth Immunology Section
Laboratory of Parasitic Diseases
National Institutes of Health
Bethesda, Maryland, USA
Email: tnutman@niaid.nih.gov
Chapter 12

David M. Parenti, MD, MSc
Department of Medicine
and
Department of Microbiology,
Immunology and Tropical Medicine
Division of Infectious Diseases
The George Washington University
Washington DC, USA
Email: dparenti@mfa.gwu.edu
Chapter 9

Michelle Paulson, MD
National Institute of Allergy
and Infectious Diseases
National Institutes of Health
Bethesda, Maryland, USA
Email: paulsonm@niaid.nih.gov
Chapter 14

Afsoon D. Roberts, MD
Department of Medicine
and
Department of Microbiology,
Immunology and Tropical Medicine
Division of Infectious Diseases
The George Washington University
School of Medicine
Washington DC, USA
Email: aroberts@mfa.gwu.edu
Chapter 3

Miriam Rodriguez-Sosa, PhD
Unidad de Biomedicina
FES-Iztacala
Universidad Nacional Autómonia
de México
México
Email: rodriguezm@campus.iztacala.
unam.mx
Chapter 21

Edsel Maurice T. Salvana, MD
Department of Medicine
Division of Infectious Diseases
University Hospitals of Cleveland
Cleveland, Ohio, USA
Email: edsel.salvana@case.edu
Chapter 17

Photini Sinnis, MD
Department of Medicine
and
Department of Medical Parasitology
New York University School of
Medicine
New York, New York, USA
Email: photini.sinnis@med.nyu.edu
Chapter 27

Sam R. Telford, III, SD, MS
Department of Biomedical Sciences
Infectious Diseases
Tufts University School
of Veterinary Medicine
Grafton, Massachusetts, USA
Email: sam.telford@tufts.edu
Chapter 33

Luis I. Terrazas, PhD
Unidad de Biomedicina
FES-Iztacala
Universidad Nacional Autónoma
de México
México
Email: literrazas@campus.iztacala.
unam.mx
Chapter 20

Sandhya Vasan, MD
The Aaron Diamond AIDS
Research Center
The Rockefeller University
New York, New York, USA
Email: svasan@adarc.org
Chapter 26

Mary E. Wilson, MD, PhD
Departments of Internal Medicine,
Microbiology and Epidemiology
Iowa City VA Medical Center
University of Iowa
Iowa City, Iowa, USA
Email: mary-wilson@uiowa.edu
Chapter 24

Sharon H. Wu, MS
Department of Microbiology,
Immunology and Tropical Medicine
The George Washington University
Washington DC, USA
Email: sharonwu@gwu.edu
Chapter 16

Gerasimos J. Zaharatos, MD
Division of Infectious Diseases,
Department of Medicine
and
Department of Microbiology
Jewish General Hospital
McGill University
Montreal, Quebec, Canada
Email: gerasimos.zaharatos@mcgill.ca
Chapter 29

Preface

Infections caused by parasites are still a major global health problem. Although parasitic infections are responsible for a significant morbidity and mortality in the developing countries, they are also prevalent in the developed countries. Early diagnosis and treatment of a parasitic infection is not only critical for preventing morbidity and mortality individually but also for reducing the risk of spread of infection in the community. This concise book gives an overview of critical facts for clinical and laboratory diagnosis, treatment and prevention of parasitic diseases which are common in humans and which are most likely to be encountered in a clinical practice. This book is a perfect companion for primary care physicians, residents, nurse practitioners, medical students, paramedics, other public health care personnel and as well as travelers. The editors would like to thank all the authors for their expertise and their outstanding contributions. We would also like to thank Dr. Ronald Landes and all other staff of Landes Bioscience who has worked tirelessly to make this magnificent book possible.

Abhay R. Satoskar, MD, PhD
Gary Simon, MD, PhD
Moriya Tsuji, MD, PhD
Peter J. Hotez, MD, PhD

Section I

Nematodes

Chapter 1

Enterobiasis

Janine R. Danko

Background

Enterobius vermicularis, commonly referred to as pinworm, has the largest geographical distribution of any helminth. Discovered by Linnaeus in 1758, it was originally named *Oxyuris vermicularis* and the disease was referred to as *oxyuriasis* for many years. It is believed to be the oldest parasite described and was recently discovered in ancient Egyptian mummified human remains as well as in DNA samples from ancient human coprolite remains from North and South America.

Enterobius is one of the most prevalent nematodes in the United States and in Western Europe. At one time, in the United States there are an estimated 42 million infected individuals. It is found worldwide in both temperate and tropical areas. Prevalence is highest among the 5-10 year-old age group and infection is uncommon in children less than two years old. Enterobiasis has been reported in every socioeconomic level; however spread is much more likely within families of infected individuals, or in institutions such as child care centers, orphanages, hospitals and mental institutions. Humans are the only natural host for the parasite.

Infection is facilitated by factors including overcrowding, wearing soiled clothing, lack of adequate bathing and poor hand hygiene, especially among young school-aged children. Infestation follows ingestion of eggs which usually reach the mouth on soiled hands or contaminated food. Transmission occurs via direct anus to mouth spread from an infected person or via airborne eggs that are in the environment such as contaminated clothing or bed linen. The migration of worms out of the gastrointestinal tract to the anus can cause local perianal irritation and pruritus. Scratching leads to contamination of fingers, especially under fingernails and contributes to autoinfection. Finger sucking and nail biting may be sources of recurrent infection in children. Spread within families is common. *E. vermicularis* may be transmitted through sexual activity, especially via oral and anal sex.

When swallowed via contaminated hands, food or water, the eggs hatch releasing larvae (Fig. 1.1). The larvae develop in the upper small intestine and mature in 5 to 6 weeks without undergoing any further migration into other body cavities (i.e., lungs). Both male and female forms exist. The smaller male is 2-5 mm in length and 0.3 mm in diameter whereas the female is 8-13 mm long and up to 0.6 mm in diameter (Fig. 1.2). Copulation occurs in the distal small bowel and the adult females settle in the large intestine where they can survive for up to 13 weeks (males live for approximately 7 weeks). The adult female can produce approximately 11,000 eggs. A gravid female can migrate out through the anus to lay her eggs. This phenomenon usually occurs at night and is thought to be secondary to the drop in host body

Medical Parasitology, edited by Abhay R. Satoskar, Gary L. Simon, Peter J. Hotez and Moriya Tsuji.

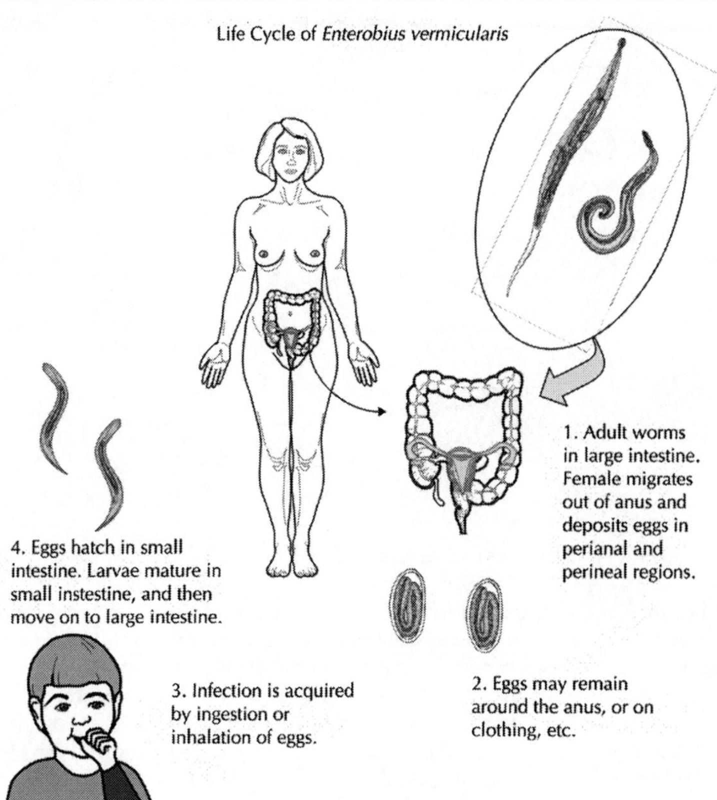

Figure 1.1. Life-cycle of *Enterobius vermicularis*. Reproduced from: Nappi AJ, Vass E, eds. Parasites of Medical Importance. Austin: Landes Bioscience, 2002:84.

temperature at this time. The eggs embyonate and become infective within 6 hours of deposition. In cool, humid climates the larvae can remain infective for nearly 2 weeks, but under warm, dry conditions, they begin to lose their infectivity within 2 days. Most infected persons harbor a few to several hundred adult worms.

Disease Signs and Symptoms

The majority of enterobiasis cases are asymptomatic; however the most common symptom is perianal or perineal pruritus. This varies from mild itching to acute pain. Symptoms tend to be most troublesome at night and, as a result, infected individuals often report sleep disturbances, restlessness and insomnia. The most common complication of infection is secondary bacterial infection of excoriated skin. Folliculitis has been seen in adults with enterobiasis.

Gravid female worms can migrate from the anus into the female genital tract. Vaginal infections can lead to vulvitis, serous discharge and pelvic pain. There are

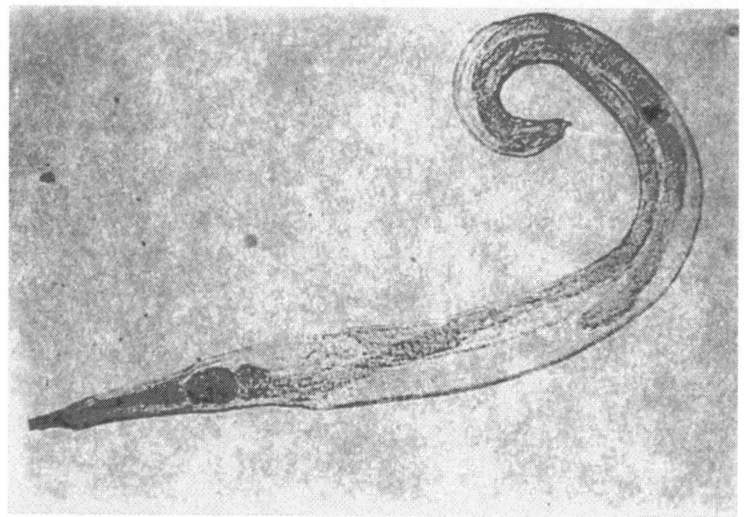

Figure 1.2. *Enterobius vermicularis.*

numerous reports of granulomas in the vaginal wall, uterus, ovary and pelvic peritoneum caused by *E. vermicularis* dead worms or eggs. Pre-pubertal and adolescent girls with *E. vermicularis* infection can develop vulvovaginitis. Scratching may lead to introital colonization with colonic bacteria and thus may increase susceptibility to urinary tract infections.

Although ectopic lesions due to *E. vermicularis* are rare, pinworms can also migrate to other internal organs, such as the appendix, the prostate gland, lungs or liver, the latter being a result of egg embolization from the colon via the portal venous system. Within the colonic mucosa or submucosa granulomas can be uncomfortable and may mimic other diseases such as carcinoma of the colon or Crohn's disease. *E. vermicularis* has been found in the lumen of uninflamed appendices in patients who have been operated on for acute appendicitis. Although eosinophilic colitis has been described with enterobiasis, eosinophilia is uncommon in infected individuals.

Diagnosis

The diagnosis of *E. vermicularis* infestation rests on the recognition of dead adult worms or the characteristic ova. In the perianal region, the adult female worm may be visualized as a small white "piece of thread". The most successful diagnostic method is the "Scotch tape" or "cellophane tape" method (Fig. 1.3). This is best done immediately after arising in the morning before the individual defecates or bathes. The buttocks are spread and a small piece of transparent or cellulose acetate tape is pressed against the anal or perianal skin several times. The strip is then transferred to a microscope slide with the adhesive side down. The worms are white and transparent and the skin is transversely striated. The egg is also colorless, measures 50-54 × 20-27 mm and has a characteristic shape,

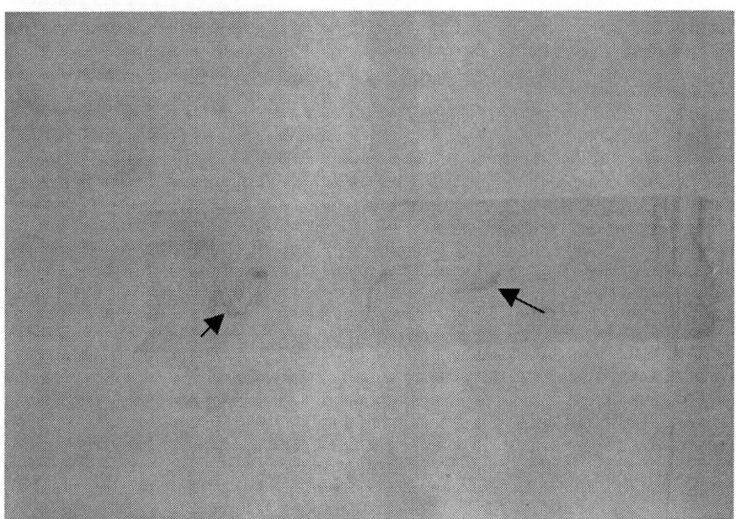

Figure 1.3. *Enterobius vermicularis* captured on scotch tape.

flattened on one side. Examination of a single specimen detects approximately 50% of infections; when this is done on three consecutive mornings sensitivity rises to 90%. Parija et al. found a higher sensitivity if lacto-phenol cotton blue stain was used in detecting eggs after the tape test was performed. Six consecutive negative swabs on separate days are necessary to exclude the diagnosis. Stool examination for eggs is usually not helpful, as only 5-15% of infected persons will have positive results. Rarely, *E. vermicularis* eggs have been found in cervical specimens (done for routine Papanicolaou smears), in the urine sediment, or the worms have been seen during colonoscopy. Serologic tests specific for *E. vermicularis* are not available.

Treatment

E. vermicularis is susceptible to several anthelmintic therapies, with a cure rate of >90%. Mebendazole (100 mg), albendazole (400 mg), or pyrantel pamoate (11 mg/kg of base) given as a single dose and then repeated after 14 days are all effective regimens. Mebendazole or albendazole are preferred because they have relatively few side effects. Their mode of action involves inhibition of the microtubule function in adult worms and glycogen depletion. For children less than 2, 200 mg should be administered. Although equally effective, pyrantel pamoate is associated with more side effects including gastrointestinal distress, neurotoxicity and transient increases in liver enzymes. Both mebendazole and albendazole are category C drugs, thus contraindicated in pregnancy although an Israeli study by Diav-Citrin et al of 192 pregnant women exposed to mebendazole, failed to reveal an increase in the number of malformations or spontaneous abortions compared to the general population. Persons with eosinophilic colitis should be treated for three successive days with mebendazole (100 mg twice daily). Experience with mebendazole or albendazole with ectopic enterobiasis is limited; persons who present with pelvic pain, those who have salpingitis, tuboovarian abscesses or painful perianal granulomas or signs or symptoms of appendicitis often proceed to surgery. In most reported cases, the antiparasitic agent is given after surgery when the diagnosis of pinworm has been established. Conservative therapy with local or systemic antibiotics is usually appropriate for perianal abscesses due to enterobiasis. Ivermectin has efficacy against pinworm but is generally not used for this indication and is not approved for enterobiasis in the United States. Overall, prognosis with treatment is excellent. Because pinworm is easily spread throughout households, the entire family of the infected person should be treated. All bedding and clothing should be thoroughly washed. The same rule should be applied to institutions when an outbreak of pinworm is discovered.

Prevention and Prophylaxis

There are no effective prevention or prophylaxis strategies available. Although mass screening campaigns and remediation for parasite infection is costly, treatment of pinworm infection improves the quality of life for children. The medications, coupled with improvements in sanitation, especially in rural areas can provide a cost-effective way at treating this nematode infection. Measures to prevent reinfection and spread including clipping fingernails, bathing regularly and frequent hand washing, especially after bowel movements. Routine laundering of clothes and linen is adequate to disinfect them. House cleaning should include vacuuming around beds, curtains and other potentially contaminated areas to eliminate other environmental eggs if possible. Health education about route of infection, especially autoinfection and these prevention tactics should always be incorporated into any treatment strategy.

Disclaimer

The views expressed in this chapter are those of the author and do not necessarily reflect the official policy or position of the Department of the US Navy, the Department of Defense or the US Government.

I am a military service member (or employee of the US Government). This work was prepared as part of my official duties. Title 17 USC §105 provides that 'Copyright protection under this title is not available for any work of the United States Government.' Title 17 USC §101 defines a US Government work as a work prepared by a military service member or employee of the US Government as part of that person's official duties. —Janine R. Danko

Suggested Reading

1. Al-Rufaie HK, Rix GH, Perez Clemente MP et al. Pinworms and postmenopausal bleeding. J Clin Path 1998; 51:401-2.
2. Arca MJ, Gates RL, Groner JL et al. Clinical manifestations of appendiceal pinworms in children: an institutional experience and a review of the literature. Pediatr Surg Int 2004; 20:372-5.
3. Beaver PC, Kriz JJ, Lau TJ. Pulmonary nodule caused by Enterobium vermicularis. Am J Trop Med Hyg 1973; 22:711-13.
4. Bundy D, Cooper E. In: Strickland GT, ed. Hunter's Tropical Medicine and Emerging Infectious Diseases, 8th Edition. Philadelphia: W.B. Saunders Company, 2000.

5. Diav-Citrin O, Shechtman S, Arnon J et al. Pregnancy outcome after gestational exposure to mebendazole: a prospective controlled cohort study. Am J Obstet Gynecol 2003; 188:282-5.
6. Fernandez-Flores A, Dajil S. Enterobiasis mimicking Crohn's disease. Indian J Gastroenterol 2004; 23:149-50.
7. Georgiev VS. Chemotherapy of enterobiasis. Exp Opin Pharmacother 2001; 2:267-75.
8. Goncalves ML, Araujo A, Ferreira LF. Human intestinal parasites in the past: New findings and a review. Mem Inst Oswaldo Cruz 2003; 98:103-18.
9. Herrstrom P, Fristrom A, Karlsson A et al. Enterobius vermicularis and finger sucking in young Swedish children. Scand. J Prim Healthcare 1997; 115:146-8.
10. Little MD, Cuello CJ, D'Allessandra A . Granuloma of the liver due to Enterobius vermicularis: report of a case. Am J Trop Med Hyg 1973; 22:567-9.
11. Liu LX, Chi J, Upton MP. Eosinophilic colitis associated with larvae of the pinworm Enterobius vermicularis. Lancet 1995; 346:410-12.
12. Neva FA, Brown HW. Basic Clinical P, 6th Edition. Norwalk: Appleton and Lange, 1994.
13. Parija SC, Sheeladevi C, Shivaprakash MR et al. Evaluation of lacto-phenol cotton blue stain for detection of eggs of Enterobius vermicularis in perianal surface samples. Trop Doctor 2001; 31:214-5.
14. Petro M, Iavu K, Minocha A. Unusual endoscopic and microscopic view of E. vermicularis: a case report with a review of the literature. South Med Jrnl 2005; 98:927-9.
15. Smolyakov R, Talalay B, Yanai-Inbar I et al. Enterobius vermicularis infection of the female genital tract: a report of three cases and review of the literature. Eur J Obstet Gynecol Reproduct Biol 2003; 107:220-2.
16. Sung J, Lin R, Huang L et al. Pinworm control and risk factors of pinworm infection among primary-school chdilren in Taiwan. Am J Trop Med Hyg 2001; 65:558-62.
17. Tornieporth NG, Disko R, Brandis A et al. Ectopic enterobiasis: a case report and review. J Infect 1992; 24:87-90.
18. Wagner ED, Eby WC. Pinworm prevalence in California elementary school children and diagnostic methods. Am J Trop Med Hyg 1983; 32:998-1001.

CHAPTER 2

Trichuriasis

Rohit Modak

Background

Trichuris trichiura is an intestinal nematode affecting an estimated 795 million persons worldwide. Also known as whipworm due to its characteristic shape, *Trichuris* can be classified as a soil-transmitted helminth because its life cycle mandates embryonic development of its eggs or larvae in the soil. It is the second most common nematode found in humans, behind *Ascaris*.

Trichuriasis is more common in areas with tropical weather such as Asia, Sub-Sarahan Africa and the Americas, particularly in impoverished regions of the Caribbean. It is also more common in poor rural communities and areas that lack proper sanitary facilities with easily contaminated food and water. A large number of individuals who are infected actually harbor fewer than 20 worms and are asymptomatic; those with a larger burden of infection (greater than 200 worms) are most likely to develop clinical disease. School age children tend to be most heavily infected.

There is no reservoir host for *Trichuris*. Transmission occurs when contaminated soil reaches the food, drink, or hands of a person and is subsequently ingested. Therefore, poor sanitary conditions is a major risk factor. It is noteworthy that patients are often coinfected with other soil-transmitted helminths like *Ascaris* and hookworm due to similar transmission modalities.

Life Cycle

Adult female worms shed between 3,000 to 20,000 eggs per day, which are passed with the stool. In the soil, the eggs develop into a 2-cell stage, an advance cleavage stage and then embryonate. It is the embryonated egg that is actually infectious. Environmental factors such as high humidity and warm temperature quicken the development of the embryo. This helps explain the geographic predilection for tropical environments. Under optimal conditions, embryonic development occurs between 15-30 days. Infection begins when these embryonated eggs are ingested.

The eggs first hatch in the small intestine and release larvae that penetrate the columnar epithelium and situate themselves just above the lamina propria. After four molts, an immature adult emerges and is passively carried to the large intestine. Here, it re-embeds itself into the colonic columnar cells, usually in the cecum and ascending colon. Heavier burdens of infection spread to the transverse colon and rectum. The worm creates a syncytial tunnel between the mouths of crypts; it is here that the narrow anterior portion is threaded into the mucosa and its thicker

Medical Parasitology, edited by Abhay R. Satoskar, Gary L. Simon, Peter J. Hotez and Moriya Tsuji.

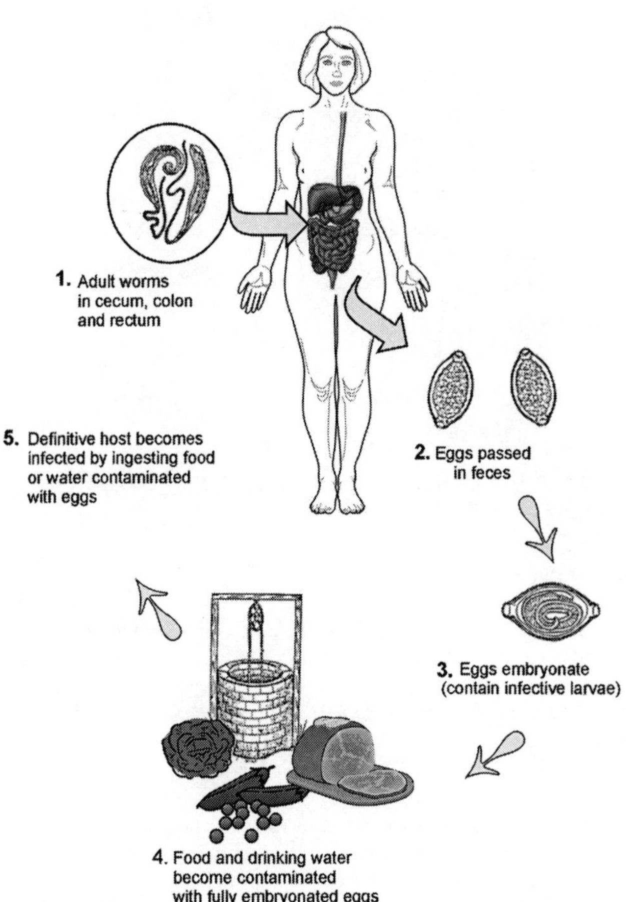

Figure 2.1. Life Cycle of Trichuris Trichura. Reproduced from: Nappi AJ, Vass E, eds. Parasites of Medical Importance. Austin: Landes Bioscience, 2002:73.

posterior end protrudes into the lumen, allowing its eggs to escape. Maturation and mating occur here as well.

The pinkish gray adult worm is approximately 30-50 mm in length, with the female generally being slightly larger than the male. The nutritional requirements of *Trichuris* are unclear; unlike hookworm however, it does not appear that *Trichuris* is dependent on its host's blood. Eggs are first detectable in the feces of those infected about 60-90 days following ingestion of the embryonated eggs. The life span of an adult worm is about one to three years. Unlike *Ascaris* and hookworm, there is no migratory phase through the lung.

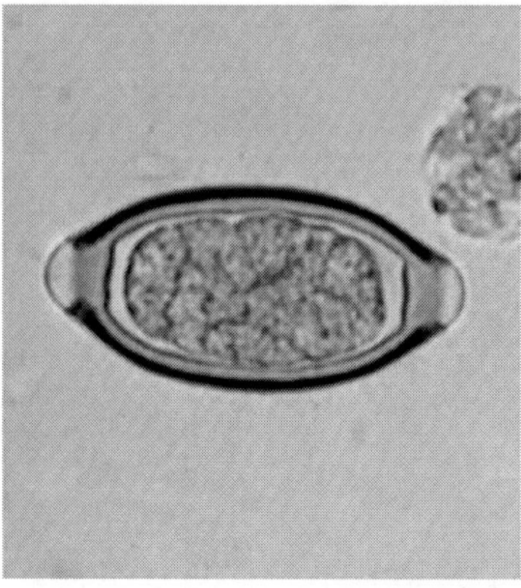

Figure 2.2. Egg of *Trichuris trichiura*. Reproduced from: Centers for Disease Control and Prevention (CDC) (http://www.dpd.cdc.gov/DPDx/).

Disease Signs and Symptoms

Frequently, infection with *Trichuris* is asymptomatic or results only in peripheral eosinophilia. Clinical disease most often occurs in children, as it is this population that tends to be most heavily-infected and presents as *Trichuris* colitis. In fact, this is the most common and major disease entity associated with infection. Acutely, some patients will develop *Trichuris* dysentery syndrome, characterized by abdominal pain and diarrhea with blood and mucus. With severe dysentery, children develop weight loss and become emaciated. Anemia is common and results from both mucosal bleeding secondary to capillary damage and chronic inflammation. The anemia of trichuriasis is not as severe as that seen with hookworm. *Trichuris* infection of the rectum can lead to mucosal swelling. In that case, tenesmus is common and if prolonged can lead to rectal prolapse, especially in children. Adult worms can be seen on the prolapsed mucosa.

Chronic trichuriasis often mimics inflammatory bowel disease. Physical symptoms include chronic malnutrition, short stature and finger clubbing. These symptoms are often alleviated with appropriate anthelminthic treatment. Rapid growth spurts have been reported in children following deworming with an anthelminthic agent. Deficits in the cognitive and intellectual development of children have also been reported in association with trichuriasis.

Host Response

Infection with *Trichuris* results in a low-grade inflammatory response that is characterized by eosinophilic infiltration of the submucosa. There is an active humoral immune response to *Trichuris* infection, but it is not fully protective. Like hookworm infections, anthelminthic therapy in endemic areas provides only transient relief and reexposure to contaminated soil leads to reinfection. The T-cell immune response to *Trichuris* infection is primarily a Th2 response. This suggests that trichuriasis, like other nematode infections, has modest immunomodulatory effects.

Diagnosis

Infection can be diagnosed by microscopic identification of *Trichuris* eggs in feces. The eggs are quite characteristic, with a barrel or lemon shape, thick shell and a clear plug at each end.

Because the level of egg output is high (200 eggs/g feces per worm pair), a simple fecal smear is usually sufficient for diagnosis. However in light infections, a concentration procedure is recommended.

Trichuriasis can also be diagnosed by identifying the worm itself on the mucosa of a prolapsed rectum or during colonoscopy. The female of the species is generally longer, while the male has a more rounded appearance.

Because of the frequency of co-infections, a search for other protozoa, specifically *Ascaris* and hookworm should be considered. Charcot-Leyden crystals in the stool in the absence of eggs in the stool should lead to further stool examinations for *T. trichuria*. Although inflammatory bowel disease is often in the differential, the sedimentation rate (ESR) is generally not elevated in trichuriasis and the degree of inflammation evident on colonoscopic examination is much less than that seen with Crohn's disease or ulcerative colitis.

Treatment

Benzimidazoles are the drugs of choice in treating trichuriasis. Their anthelminthic activity is primarily due to their ability to inhibit microtubule polymerization by binding to beta-tubulin, a protein unique to invertebrates. A single dose of albendazole has been suggested for treatment; however, despite the appeal of adequate single dose therapy, clinical studies have shown a cure rate of less than 25 percent. Longer duration of therapy, resulting in higher cure rates, is recommended for heavier burdens of infection. High cure rates are difficult to establish because of the constant re-expsoure to the organisms. Mebendazole at a dose of 100 mg twice daily for three days is also effective, with cure rates of almost 90 percent. In some countries, pyrantel-oxantel is used for treatment, with the oxantel component having activity against *Trichuris* and the pyrantel component having activity against *Ascaris* and hookworm.

Albendazole and mebendazole are generally well tolerated when given at doses used to treat trichuriasis, even in pediatric populations. Adverse effects include transient abdominal pain, diarrhea, nausea and dizziness. With long-term use, reported toxicities include bone marrow suppression, alopecia and hepatotoxicity. Both drugs are not recommended in pregnancy, as they have been shown to be teratogenic and embryotoxic in laboratory rats. However, albendazole should be considered in pregnant women in the second or third trimester when the potential benefit outweighs the risks to the fetus. Although these drugs have not been studied in very young children, the World Health Organization (WHO) has recommended that both agents may be used for treatment in patients as young as 12 months, albeit at reduced dosages.

Prevention and Prophylaxis

Drinking clean water, properly cleaning and cooking food, hand washing and wearing shoes are the most effective means of preventing soil-transmitted helminth infections. Adequately sanitizing areas in which trichuriasis is prevalent is extremely problematic; these communities often lack the resources needed for such a substantial undertaking.

Direct exposure to sunlight for greater than 12 hours or temperatures exceeding 40 degrees C in excess of 1 hour kills the embryo within the egg, but under optimal conditions of moisture and shade in the warm tropical and subtropical soil, *Trichuris* eggs can remain viable for months. There is relative resistance to chemical disinfectants and eggs can survive for prolonged periods even in treated sewage. Therefore, proper disposal of sewage is vital to control this infection. In areas of the world where human feces is used as fertilizer, this is practically impossible.

Because the prevalence of trichuriasis has been estimated to be up to 80% in some communities and can frequently be asymptomatic, the WHO advocates empiric treatment of soil-transmitted helminths by administering anthelminthic drugs to populations at risk. Specifically, WHO recommends periodic treatment of school-aged children, the population in whom the burden of infection is greatest. The goal of therapy is to maintain the individual worm burden at a level less than that needed to cause significant morbidity or mortality. This strategy has been used successfully in preventing and reversing malnutrition, iron-deficiency anemia, stunted growth and poor school performance. This is in large part due to the efficacy and broad spectrum activity of a single dose of anthelminthic drugs like albendazole. Because reinfection is a common problem as long as poor sanitary conditions remain, it is proposed that single dose therapy be given at regular intervals (1-3 times per year). The WHO hopes that by 2010, 75% of all school-aged children at risk for heavy infection will have received treatment.

Major challenges to controlling the infection include continued poor sanitary conditions. Additionally, the use of benzimidazole drugs at regular intervals may lead to the emergence of drug resistance. Resistance has been documented in livestock and suspected in humans. Since the single dose regimen is not ideal (although the most feasible), continued monitoring and screening is necessary.

Suggested Reading

1. Adams VJ, Lombard CJ, Dhansay MA et al. Efficacy of albendazole against the whipworm Trichuris trichiura: a randomised, controlled trial. S Afr Med J 2004; 94:972-6.
2. Albonico M, Crompton DW, Savioli L. Control strategies for intestinal nematode infections. Adv Parasito 1999; 42:277-341.
3. Albonico M, Bickle Q, Haji HJ et al. Evaluation of the efficacy of pyrantel-oxantel for the treatment of soil-transmitted nematode infections. Trans R Soc Trop Med Hyg 2002; 96:685-90.
4. Belding D. Textbook of Parasitology. New York: Appleton-Century-Crofts, 1965:397-8.
5. Cooper ES. Bundy DAP: Trichuris is not trivial. Parasitol Today 1988; 4:301-5.
6. De Silva N. Impact of mass chemotherapy on the morbidity due to soil-transmitted nematodes. Acta Tropica 2003; 86:197-214.
7. de Silva NR, Brooker S, Hotez PJ et al. Soil-transmitted helminth infections: updating the global picture. Trends Parasitol 2003; 19:547-51.

8. Despommier DD, Gwadz RW, Hotez PJ et al, eds. Parasitic Diseases, 5th Edition. New York: Apple Trees Productions, LLC, 2005:110-5.
9. Division of Parasitic Diseases, National Centers for Infectious Diseases, Centers for Disease Control and Prevention, Atlanta, GA: [http://www.dpd.cdc.gov/DPDx/].
10. Geerts S, Gryseels B. Anthelminthic resistance in human helminthes: a review. Trop Med Int Health 2001; 6:915-21.
11. Gilman RH, Chong YH, Davis C et al. The adverse consequences of heavy Trichuris infection. Trans R Soc Trop Med Hyg 1983; 77:432-8.
12. Lin AT, Lin HH, Chen CL. Colonoscopic diagnosis of whip-worm infection. Hepato-Gastroenterol 1998; 45:2105-9.
13. Legesse M, Erko B, Medhin G. Comparative efficacy of albendazole and three brands of mebendazole in the treatment of ascariasis and trichuriasis. East Afr Med J 2004; 81:134-8.
14. MacDonald TT, Choy MY, Spencer J et al. Histopathology and immunohistochemistry of the caecum in children with the Trichuris dysentery syndrome. J Clin Pathol 1991; 44:194-9.
15. Maguire J. Intestinal Nematodes. Mandell, Douglas and Bennett's Principles and Practice of Infectious Diseases, 6th Edition. Philadelphia: Elsevier, 2005:3263-4.
16. Montresor A, Awasthi S, Crompton DWT. Use of benzimadazoles in children younger than 24 months for the treatment of soil-transmitted helminthiases. Acta Tropica 2003; 86:223-32.
17. Sirivichayakul C, Pojjaroen-Anant C, Wisetsing P et al. The effectiveness of 3, 5, or 7 days of albendazole for the treatment of Trichuris trichiura infection. Ann Trop Med Parasitol 2003; 97:847-53.

2

Chapter 3

Ascariasis

Afsoon D. Roberts

Introduction

Ascariasis, a soil-transmitted infection, is the most common human helminthic infection. Current estimates indicate that more than 1.4 billion people are infected worldwide. In the United States, there are an estimated 4 million people infected, primarily in the southeastern states and among immigrants. The etiologic agent, *Ascaris lumbricoides,* an intestinal roundworm, is the largest nematode to infect humans. The adult worm lives in the small intestine and can grow to a length of more than 30 cm. The female worms are larger than the males. Important factors associated with an increased prevalence of disease include socio-economic status, defecation practices and cultural differences relating to personal and food hygiene as well as housing and sewage systems. Most infections are subclinical; more severe complications occur in children who tend to suffer from the highest worm burdens.

Epidemiology and Transmission

There are a number of factors that contribute to the high frequency of infection with *Ascaris lumbricoides.* These include its ubiquitous distribution, the high number of eggs produced by the fecund female parasite and the hardy nature of the eggs which enables them to survive unfavorable conditions. The eggs can survive in the absence of oxygen, live for 2 years at 5-10° C and be unaffected by dessication for 2 to 3 weeks. In favorable conditions of moist, sandy soil, they can survive for up to 6 years, even in freezing winter conditions. The greatest prevalence of disease is in tropical regions, where environmental conditions support year round transmission of infection. In dry climates, transmission is seasonal and occurs most frequently during the rainy months.

Ascariasis is transmitted primarily by ingestion of contaminated food or water. Although infection occurs in all age groups, it is most common in preschoolers and young children. Sub-optimal sanitation is an important factor, leading to increased soil and water contamination. In the United States, improvements in sanitation and waste management have led to a dramatic reduction in the prevalence of disease.

Recently, patterns of variation in the ribosomal RNA of *Ascaris* worms isolated in North America were compared to those of worms and pigs from other worldwide locations. Although repeats of specific restriction sites were found in most parasites from humans and pigs in North America, they were rarely found in parasites from elsewhere. This evidence suggests that perhaps human infections in North America may be related to *Ascaris suum.*

Medical Parasitology, edited by Abhay R. Satoskar, Gary L. Simon, Peter J. Hotez and Moriya Tsuji.

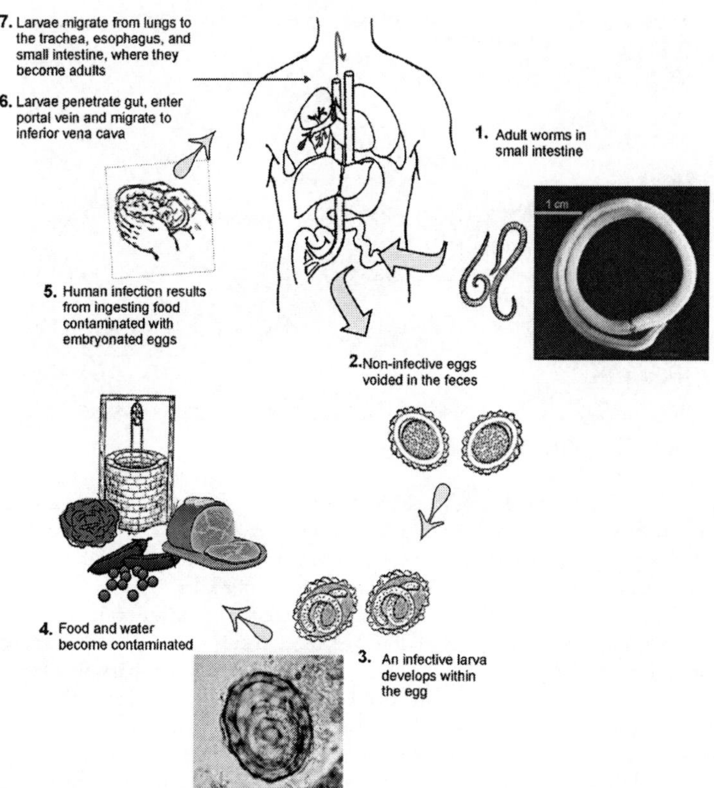

Figure 3.1. Life Cycle of *Ascaris lumbricoides*. Reproduced from: Nappi AJ, Vass E, eds. Parasites of Medical Importance. Austin: Landes Bioscience, 2002:82.

Life Cycle

The adult ascaris worms reside in the lumen of the small intestine where they feed on predigested food (Fig. 3.1). Their life span ranges from 10 to 24 months. The adult worms are covered with a tough shell composed of collagens and lipids. This outer covering helps protect them from being digested by intestinal hydrolases. They also produce protease inhibitors that help to prevent digestion by the host.

The adult female worm can produce 200,000 eggs per day (Fig. 3.2). The eggs that pass out of the adult worm are fertilized, but not embryonated. Once the eggs exit the host via feces, embryonation occurs in the soil and the embryonated eggs are subsequently ingested. There is a mucopolysaccharide on the surface that

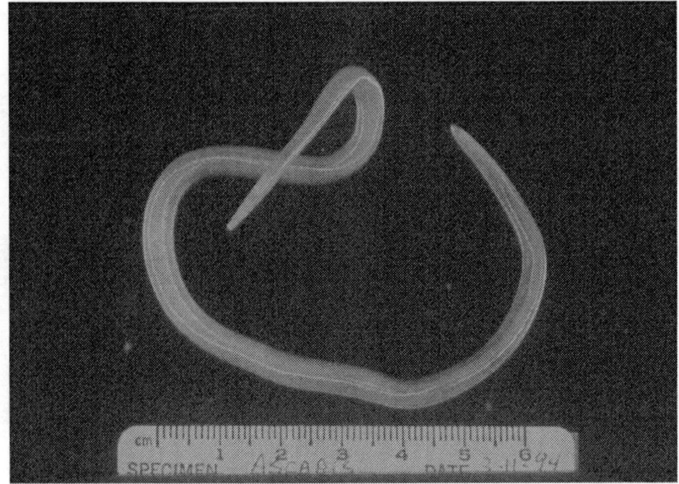

Figure 3.2. *Ascaris lumbricoides.*

promotes adhesion of the eggs to environmental surfaces. Within the embryonated egg, the first stage larva develops into the second stage larva. This second stage larva is stimulated to hatch by the presence of both the alkaline conditions in the small intestine and the solubilization of its outer layer by bile salts.

The hatched parasite that now resides in the lumen of the intestine penetrates the intestinal wall and is carried to the liver through the portal circulation. It then travels via the blood stream to the heart and lungs by the pulmonary circulation. The larva molts twice, enlarges and breaks into the alveoli of the lung. They then pass up through the bronchi and into the trachea, are swallowed and reach the small intestine once again. Within the small intestine, the parasites molt twice more and mature into adult worms. The adult worms mate, although egg production may precede mating.

Clinical Manifestations

Although most individuals infected with *Ascaris lumbricoides* are essentially asymptomatic, the burden of symptomatic infection is relatively high as a result of the high prevalence of infection on a worldwide basis. Symptomatic disease is usually related to either the larval migration stage and manifests as pulmonary disease, or to the intestinal stage of the adult worm.

The pulmonary manifestations of ascariasis occur during transpulmonary migration of the organisms and are directly related to the concentration of larvae. Thus, symptoms are more pronounced with higher burdens of migratory worms. The transpulmonary migration of helminth larvae is responsible for the development of a transient eosinophilic pneumonitis characteristic of Loeffler's syndrome with peripheral eosinophilia, eosinophilic infiltrates and elevated serum IgE concentrations. Symptoms usually develop 9-12 days after ingestion of the eggs, while the larvae reside in the lung. Affected individuals often develop bronchospasm, dyspnea and wheezing. Fever, a persistent, nonproductive cough and, at times chest pain, can also occur. Hepatomegaly may also be present. In some areas of the world such as Saudi Arabia where transmission of infection is related to the time of the year, seasonal pneumonitis has been described.

The diagnosis of *Ascaris*-related pneumonitis is suspected in the correct clinical setting by the presence of infiltrates on chest X-ray which tend to be migratory and usually completely clear after several weeks. The pulmonary infiltrates are usually round, several millimeters to centimeters in size, bilateral and diffuse.

Among the more serious complications of *Ascaris* infection is intestinal obstruction. This occurs when a large number of worms are present in the small intestine and is usually seen in children with heavy worm burdens. These patients present with nausea, vomiting, colicky abdominal pain and abdominal distention. In this condition worms may be passed via vomitus or stools. In endemic areas, 5-35% of all cases of intestinal obstruction can be attributable to ascariasis. The adult worms can also perforate the intestine leading to peritonitis. *Ascaris* infection can be complicated by intussusception, appendicitis and appendicular perforation due to worms entering the appendix.

A potenital consequence of the intestinal phase of the infection relates to the effect it may have on the nutritional health of the host. Children heavily infected with *Ascaris* have been shown to exhibit impaired digestion and absorption of proteins and steatorrhea. Heavy infections have been associated with stunted growth and a reduction in cognitive function. However, the role of *Ascaris* in these deficiencies is not clearly defined. Some of these studies were done in developing countries where additional nutritional factors cannot be excluded. There is also a high incidence of co-infection with other parasites that can affect growth and nutritional status. Interestingly, a controlled study done in the southern United States failed to demonstrate significant differences in the nutritional status of *Ascaris* infected and uninfected individuals.

Hepatobiliary and Pancreatic Symptoms

Hepatobiliary symptoms have been reported in patients with Ascariasis and are due to the migration of adult worms into the biliary tree. Affected individuals can experience biliary colic, jaundice, ascending cholangitis, acalculous cholecystitis and perforation of the bile duct. Pancreatitis may develop as a result of an obstruction of the pancreatic duct. Hepatic abscesses have also been reported. Sandouk et al studied 300 patients in Syria who had biliary or pancreatic involvement. Ninety-eight percent of the patients presented with abdominal pain, 16% developed ascending cholangitis, 4% developed pancreatitis and 1% developed obstructive jaundice. Both ultrasonography, as well as endoscopic retrograde cholangiopancreatography (ERCP) have been used as diagnostic tools for biliary or pancreatic ascariasis. In Sandouk's study extraction of the worms endoscopically resulted in resolution of symptoms.

Diagnosis

The diagnosis of ascariasis is made through microscopic examination of stool specimens. *Ascaris* eggs are easily recognized, although if very few eggs are present the diagnosis may be easily missed (Fig. 3.3). Techniques for concentrating the stool

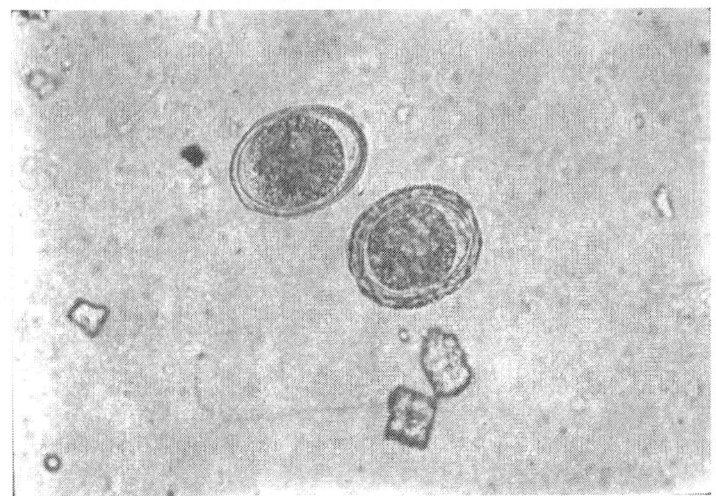

Figure 3.3. *Ascaris lumbricoides* egg.

specimen will increase the yield of diagnosis through microscopy. Occasionally an adult worm is passed via rectum. Eosinophilia may be present, especially during the larval migration through the lungs. In very heavily infected individuals a plain X-ray of the abdomen may sometimes reveal a mass of worms.

Treatment

Both albendazole and mebendazole are effective therapies for ascariasis. Mebendazole can be prescribed as 100 mg BID for 3 days or 500 mg as a single dose. The adverse effects of the drug include gastrointestinal symptoms, headache and rarely leukopenia. Albendazole is prescribed as a single dose of 400 mg. Albendazole's side effect profile is similar to mebendazole.

The drug piperazine citrate is an alternative therapeutic option, but it is not widely available and has been withdrawn from the market in some developed countries as other less toxic and more effective therapy is available. However, in cases of intestinal or biliary obstruction it can be quite useful as it paralyses the worms, allowing them to be expelled by peristalsis. It is dosed as 50-75 mg/kg QD, up to a maximum of 3.5 g for 2 days. It can be administered as piperazine syrup via a naso-gastric tube.

Finally, pyrantel pamoate can be used at a single dose of 11 mg/kg, up to a maximum dose of 1 g. This drug can be used in pregnancy. The side effects of pyrantel pamoate include headache, fever, rash and gastrointestinal symptoms. It has been reported to be up to 90% effective in treating the infection.

These medications are all active against the adult worm and are not active against larval stage. Thus, reevaluation of infected individuals is recommended following therapy. Family members should also be screened as infection is common among other members of a household. Treatment does not protect against reinfection.

3

Prevention

Given the high prevalence of infection with *Ascaris lumbricoides* and the potential health and educational benefits of treating the infection in children, the World Health Organization (WHO) has recommended global deworming measures aimed at school children. The goal of recent helminth control programs has been to recommend periodic mass treatment where the prevalence of infection in school aged children is greater than 50%. The current goal is to treat infected individual 2 to 3 times a year with either mebendazole or albendazole. Integrated control programs combining medical treatment with improvements in sanitation and health education are needed for effective long-term control.

Suggested Reading

1. Ali M, Khan AN. Sonography of hepatobiliary ascariasis. J Clin Ultrasound 1996; 24:235-41.
2. Anderson TJ. Ascaris infections in human from North America: Molecular evidence for cross infection. Parasitology 1995; 110:215-29.
3. Bean WJ. Recognition of ascariasis by routine chest or abdomen roentgenograms. Am J Roentgenol Rad Ther Nucl Med 1965; 94:379.
4. Blumenthal DS, Schultz AG. Incidence of intestinal obstruction in children infected by Ascaris lumbricoides. Am J Trop Med Hyg 1974; 24:801.
5. Blumenthal DS, Schultz MG. Effect of Ascaris infection on nutritional status in children. Am J Trop Med Hyg 1976; 25:682.
6. Chevarria AP, Schwartzwelder JC et al. Mebendazole, an effective broad spectrum anti-heminthic. Am J Trop Med Hyg 1973; 22:592-5.
7. Crampton DWT, Nesheim MC, Pawlowski ZS, eds. Ascariasis and Its Public Health Significance. London: Taylor and Francis, 1985.
8. De Silva NR, Guyatt HL, Bundy DA. Morbidity and mortality due to Ascaris-induced intestinal obstruction. Trans R Soc Trop Med Hyg 1997; 91:31-6.
9. DeSilva NR, Chan MS, Bundy DA. Morbidity and mortality due to ascariasis: Re-estimation and sensitivity analysis of global numbers at risk. Trop Med Int Health 1997; 2:519-28.
10. Despommier DD, Gwadz RW, Hotez PJ et al, eds. Parasitic Diseases, 5th edition. New York: Apple Tree Productions, 2005:115-20.
11. Gelpi AP, Mustafa A. Seosonal pneumonitis with eosinophilia: A study of larval ascariasis in Saudi Arabia. Am J Trop Med Hyg 1967; 16:646.
12. Jones JE. Parasites in Kentucky: the past seven decades. J KY Med Assoc 1983; 81:621.
13. Khuroo MS. Ascariasis. Gastroenterol Clin North Am 1996; 25:553-77.
14. Khuroo MS. Hepato-biliary and pancreatic ascariasis. Indian J Gastroenterol 2001; 20:28.
15. Loeffler W. Transient lung infiltrations with blood eosinophilia. Int Arch Allergy Appl Immunol 1956; 8:54.
16. Mandell GL, Bennett JE, Dolin R, eds. Principles and Practices of Infectious Disease, 5th edition. Philadelphia: Churchill Livingstone, 2000:2941.
17. Norhayati M, Oothuman P, Azizi O et al. Efficacy of single dose albendazole on the prevalence and intensity of infection of soil-transmitted helminths in Orang Asli children in Malaysia. Southeast Asian J Trop Med Public Health 1997; 28:563.
18. O'Lorcain, Holland CU. The public health importance of Ascaris lumbricoides. Parasitology 2000; 121:S51-71.
19. Phills JA, Harold AJ, Whiteman GV et al. Pulmonary infiltrates, asthma, eosinophilia due to Ascaris suum infestation in man. N Engl J Med 1972; 286:965.

20. Reeder MM. The radiographic and ultrasound evaluation of ascariasis of the gastrointestinal, biliary and respiratory tract. Semin Roentgenol 1998; 33:57.
21. Sandou F, Haffar S, Zada M et al. Pancreatic-biliary ascariasis: Experience of 300 cases. Am J Gastroenterol 1997; 92:2264-7.
22. Sinniah B. Daily egg production of Ascaris lumbricoides: The distribution of eggs in the feces and the variability of egg counts. Parasitology 1982; 84:167.
23. Stephenson LS. The contribution of Ascaris lumbricoides to malnutrition in children. Parasitolgy 1980; 81:221-33.
24. Warren KS, Mahmoud AA. Algorithems in the diagnosis and management of exotic diseases, xxii ascariasis and toxocariaisis. J Infec Dis 1977; 135:868.
25. WHO Health of school children: Treatment of intestinal helminths and schistosomiasis (WHO/Schisto/95.112; WHO/CDS/95.1). World Health Organisation 1995.

CHAPTER 4

Hookworm

David J. Diemert

Introduction

Human hookworm infection is a soil-transmitted helminth infection caused primarily by the nematode parasites *Necator americanus* and *Ancylostoma duodenale*. It is one of the most important parasitic infections worldwide, ranking second only to malaria in terms of its impact on child and maternal health. An estimated 576 million people are chronically infected with hookworm and another 3.2 billion are at risk, with the largest number of afflicted individuals living in impoverished rural areas of sub-Saharan Africa, southeast Asia and tropical regions of the Americas. *N. americanus* is the most widespread hookworm globally, whereas *A. duodenale* is more geographically restricted in distribution. Although hookworm infection does not directly account for substantial mortality, its greater health impact is in the form of chronic anemia and protein malnutrition as well as impaired physical and intellectual development in children.

Humans may also be incidentally infected by the zoonotic hookworms *Ancylostoma caninum*, *Ancylostoma braziliensis* and *Uncinaria stenocephala*, which can cause self-limited dermatological lesions in the form of cutaneous larva migrans. Additionally, *Ancylostoma ceylanicum*, normally a hookworm infecting cats, has been reported to cause hookworm disease in humans especially in Asia, whereas *A. caninum* has been implicated as a cause of eosinophilic enteritis in Australia.

Life Cycle

Hookworm transmission occurs when third-stage infective filariform larvae come into contact with skin (Fig. 4.1). Hookworm larvae have the ability to actively penetrate the cutaneous tissues, most often those of the hands, feet, arms and legs due to exposure and usually through hair follicles or abraded skin. Following skin penetration, the larvae enter subcutaneous venules and lymphatics to gain access to the host's afferent circulation. Ultimately, they enter the pulmonary capillaries where they penetrate into the alveolar spaces, ascend the brachial tree to the trachea, traverse the epiglottis into the pharynx and are swallowed into the gastrointestinal tract. Larvae undergo two molts in the lumen of the intestine before developing into egg-laying adults approximately five to nine weeks after skin penetration. Although generally one centimeter in length, adult worms exhibit considerable variation in size and female worms are usually larger than males (Fig. 4.2).

Adult *Necator* and *Ancylostoma* hookworms parasitize the proximal portion of the human small intestine where they can live for several years, although differences

Medical Parasitology, edited by Abhay R. Satoskar, Gary L. Simon, Peter J. Hotez and Moriya Tsuji.

4

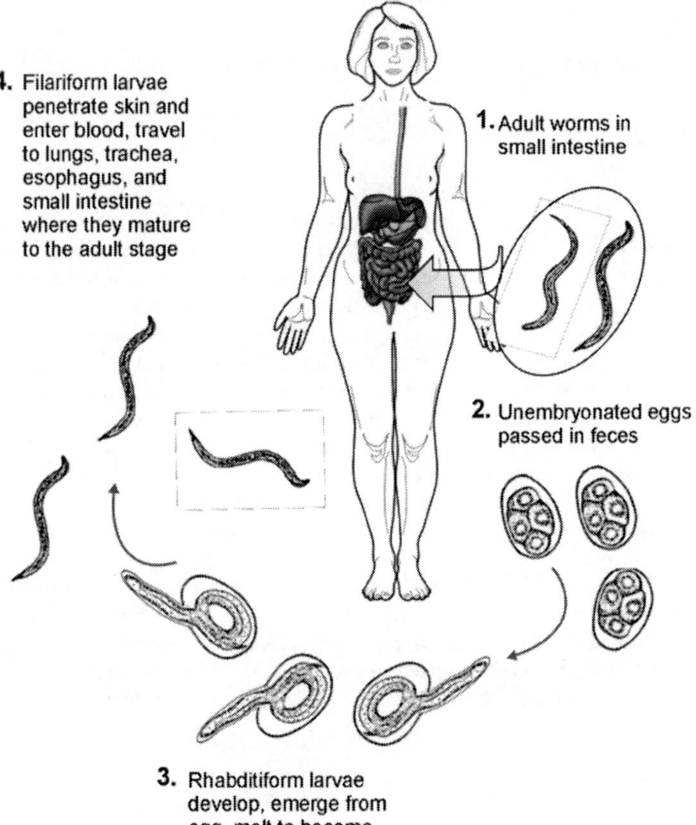

Figure 4.1. Life cycle of the hookworm, *Necator americanus*. Reproduced from: Nappi AJ, Vass E, eds. Parasites of Medical Importance. Austin: Landes Bioscience, 2002:80.

exist between the life spans of the two species: *A. duodenale* survive for on average one year in the human intestine whereas *N. americanus* generally live for three to five years (Fig. 4.3). Adult hookworms attach onto the mucosa of the small intestine by means of cutting teeth in the case of *A. duodenale* or a rounded cutting plate in the case of *N. americanus*. After attachment, digestive enzymes are secreted that enable the parasite to burrow into the tissues of the submucosa where they derive nourishment from eating villous tissue and sucking blood into their digestive tracts. Hemoglobinases within the hookworm digestive canal enable digestion of human hemoglobin, which is a primary nutrient source of the parasite.

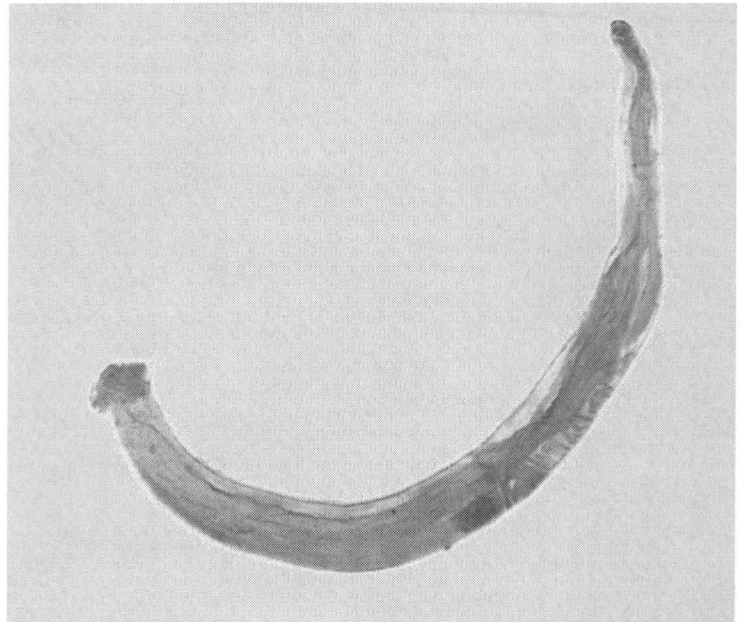

Figure 4.2. Adult male *Ancylostoma duodenale* hookworm. Reproduced with permission from: Despommier DD, Gwadz RW, Hotez PJ, Knirsch CA. Parasitic Diseases. New York: Apple Trees Productions, 2005:121.

Humans are considered the only major definitive host for these two parasites and there are no intermediate or reservoir hosts; in addition, hookworms do not reproduce within the host. After mating in the host intestinal tract, each female adult worm produces thousands of eggs per day which then exit the body in feces. *A. duodenale* female worms lay approximately 28,000 eggs daily, while the output from *N. americanus* worms is considerably less, averaging around 10,000 a day. *N. americanus* and *A. duodenale* hookworm eggs hatch in warm, moist soil, giving rise to rhabditiform larvae that grow and develop, feeding on organic material and bacteria. After about seven days, the larvae cease feeding and molt twice to become infective third-stage filariform larvae. Third-stage larvae are nonfeeding but motile organisms that seek out higher ground such as the tips of grass blades to increase the chance of contact with human skin and thereby complete the life cycle. Filariform larvae can survive for up to approximately two weeks if an appropriate host is not encountered.

A. duodenale larvae can also be orally infective and have been conjectured to infect infants during breast feeding.

Epidemiology and Burden of Disease

Human hookworm infections are widely distributed throughout the tropics and sub-tropics (Fig. 4.4). *N. americanus* is the most prevalent hookworm worldwide, with the highest rates of infection in sub-Saharan Africa, the tropical

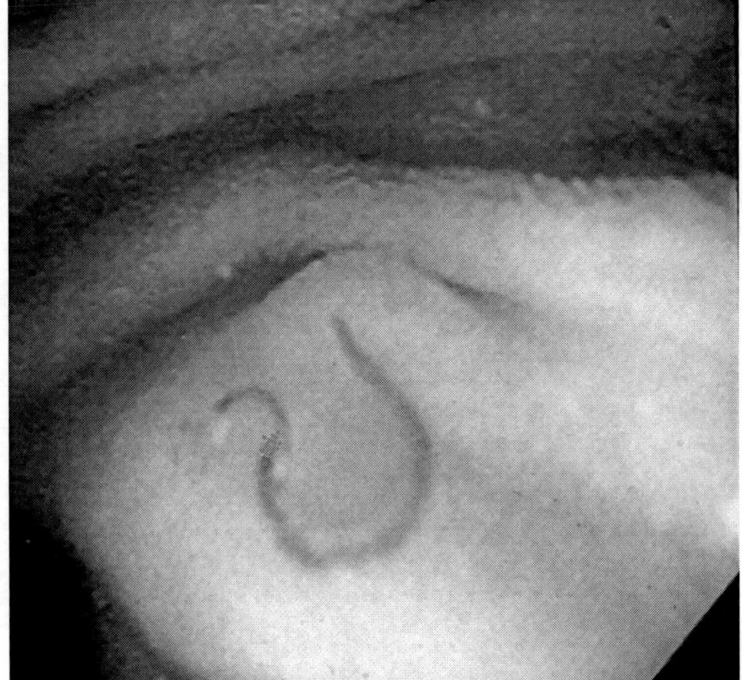

Figure 4.3. Adult hookworm, diagnosed by endoscopy. Reproduced with permission from: Despommier DD, Gwadz RW, Hotez PJ, Knirsch CA. Parasitic Diseases. New York: Apple Trees Productions, 2005:123.

regions of the Americas, south China and southeast Asia, whereas *A. duodenale* is more focally endemic in parts of India, China, sub-Saharan Africa, North Africa and a few regions of the Americas. Climate is an important determinant of hookworm transmission, with adequate moisture and warm temperature essential for larval development in the soil. An equally important determinant of infection is poverty and the associated lack of sanitation and supply of clean water. In such conditions, other helminth species are frequently co-endemic, with emerging evidence that individuals infected with multiple different types of helminths (most commonly the triad of *Ascaris lumbricoides*, hookworm and *Trichuris trichiura*) are predisposed to developing even heavier intensity infections than those who harbor single-species infections. Because morbidity from hookworm infections and the rate of transmission are directly related to the number of worms harbored within the host, the intensity of infection is the primary epidemiological parameter used to describe hookworm infection as measured by the number of eggs per gram of feces.

While prevalence in endemic areas increases markedly with age in young children and reaches a plateau by around an age of ten years, intensity of infection

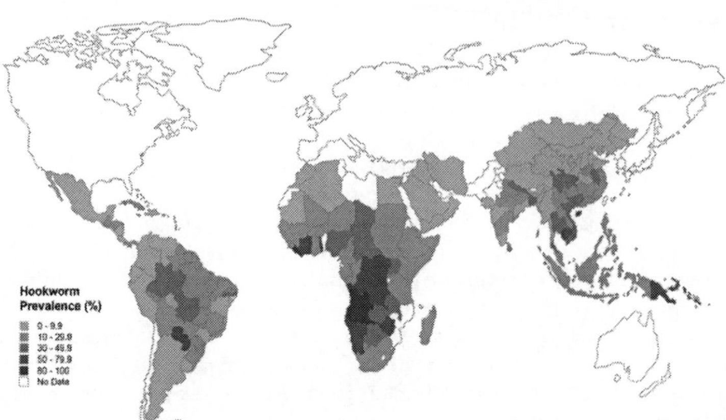

Figure 4.4. Prevalence of hookworm infection worldwide. From: Hotez et al, Public Library of Science.

rises at a slower rate during childhood, reaching a plateau by around 20 years and then increasing again from age 60 years onward. Controversy persists whether such age-dependency reflects changes in exposure, acquired immunity, or a combination of both. Although heavy hookworm infections also occur in childhood, it is common for prevalence and intensity to remain high into adulthood, even among the elderly. Hookworm infections are often referred to as "overdispersed" in endemic communities, such that the majority of worms are harbored by a minority of individuals in an endemic area. There is also evidence of familial and household aggregation of hookworm infection, although the relative importance of genetics over the shared household environment is debated.

Although difficult to ascertain, it is estimated that worldwide approximately 65,000 deaths occur annually due to hookworm infection. However, hookworm causes far more disability than death. The global burden of hookworm infections is typically assessed by estimating the number of Disability Adjusted Life Years (DALYs) lost. According to the World Health Organization, hookworm accounts for the loss of 22 million DALYs annually, which is almost two-thirds that of malaria or measles. This estimate reflects the long-term consequences of hookworm-associated malnutrition, anemia and delays in cognitive development, especially in children.

Recent data support the high disease burden estimates of hookworm infection and highlight its importance as a maternal and child health threat. For example, studies in Africa and Asia show that between one-third and one-half of moderate to severe anemia among pregnant women can be attributable to hookworm and recent evidence from interventional studies further suggest that administration of anthelmintics antenatally can substantially improve maternal hemoglobin as well as birth weight and neonatal and infant survival. In addition to pregnant women, hookworm contributes to moderate and severe anemia among both preschool and primary school-aged children.

Clinical Manifestations

The clinical features of hookworm infection can be separated into the acute manifestations associated with larval migration through the skin and other tissues and the acute and chronic manifestations resulting from parasitism of the gastrointestinal tract by adult worms.

Migrating hookworm larvae provoke reactions in many of the tissues through which they pass, including several cutaneous syndromes that result from skin-penetrating larvae. Repeated exposure to *N. americanus* and *A. duodenale* filariform larvae can result in a hypersensitivity reaction known as "ground itch", a pruritic local erythematous and papular rash that appears most commonly on the hands and feet. In contrast, when zoonotic hookworm larvae (typically *A. braziliensis, A. caninum* or *U. stenocephala*) penetrate the skin, usually after direct contact between skin and soil or sandy beaches contaminated with animal feces, they produce cutaneous larva migrans, most often on the feet, buttocks and abdomen. Since these zoonotic larvae are unable to complete their life cycle in the human host, they eventually die after causing a typical clinical syndrome of erythematous linear tracts with a serpiginous appearance and intense pruritus. These tracts can elongate by several centimeters a day; larvae can migrate for up to one year, but the lesions usually heal spontaneously within weeks to months although secondary pyogenic infection may occur at these sites as well as those of ground itch.

One to two weeks following skin invasion, hookworm larvae travel through the vasculature and enter the lungs, where they can uncommonly result in pneumonitis. The pulmonary symptoms that may develop are usually mild and transient, consisting of a dry cough, sore throat, wheezing and slight fever. The pulmonary symptoms are more pronounced and of longer duration with *A. duodenale* than with *N. americanus* infection. Acute symptomatic disease may also result from oral ingestion of *A. duodenale* larvae, referred to as the Wakana syndrome, which is characterized by nausea, vomiting, pharyngeal irritation, cough, dyspnea and hoarseness.

In hookworm infection, the appearance of eosinophilia coincides with the development of adult hookworms in the intestine. The major pathology of hookworm infection, however, results from the intestinal blood loss that results from adult parasite invasion and attachment to the mucosa and submucosa of the small intestine. Usually only moderate and high intensity hookworm infections in the gastrointestinal tract produce clinical manifestations, with the highest intensity infections occurring most often in children, although even in low intensity infections, initial symptoms may include dyspepsia, nausea and epigastric distress. *A. duodenale* may also result in acute enteritis with uncontrollable diarrhea and foul stools that may last indefinitely.

In general, the precise numerical threshold at which worms cause disease has not been established since this is highly dependent on the underlying nutritional status of the host. Chronic hookworm disease occurs when the blood loss due to infection exceeds the nutritional reserves of the host, thus resulting in iron-deficiency anemia. It has been estimated that the presence of more than 40 adult worms in the small intestine is sufficient to reduce host hemoglobin levels below 11 g per deciliter, although the exact number depends on several factors including the species of hookworm and host iron reserves. Worm-for-worm, *A. duodenale* causes more blood loss than *N. americanus*: whereas each *N. americanus* worm produces a daily blood loss of 0.03 to 0.1 ml, the corresponding figure for *A. duodenale* is between 0.15 and 0.26 ml.

The clinical manifestations of chronic hookworm disease resemble those of iron deficiency anemia due to other etiologies, while the protein loss from heavy hookworm infection can result in hypoproteinemia and anasarca. The anemia and protein malnutrition that results from chronic intestinal parasitism cause long-term impairments in childhood physical, intellectual and cognitive development. As iron deficiency anemia develops and worsens, an infected individual may have weakness, palpitations, fainting, dizziness, dyspnea, mental apathy and headache. Uncommonly, there may be constipation or diarrhea with occult blood in the stools or frank melena, especially in children; there may also be an urge to eat soil (pica). Overwhelming hookworm infection may cause listlessness, coma and even death, especially in infants under one year of age. Because children and women of reproductive age have reduced iron reserves, they are considered populations that are at particular risk for hookworm disease. As noted above, the severe iron deficiency anemia that may arise from hookworm disease during pregnancy can result in adverse consequences for the mother, her unborn fetus and the neonate.

Zoonotic infection with the dog hookworm *A. caninum* has been reported as a cause of eosinophilic enteritis in Australia, although considering the ubiquitous worldwide distribution of this parasite, unrecognized clinical disease due to this worm may be more widespread than previously recognized. Reported clinical features of this syndrome include abdominal pain, diarrhea, abdominal bloating, weight loss and rectal bleeding. *A. ceylanicum*, while normally a hookworm that parasitizes cats, has been reported as a cause of chronic intestinal infection in humans living in Asia.

The most common manifestations of hookworm infection seen by the average healthcare practitioner in developed countries are cutaneous larva migrans in returning travelers and chronic intestinal infection with resulting anemia and peripheral eosinophilia in immigrants and long-term expatriate residents or military personnel returning from long-term postings in endemic areas.

Diagnosis

Diagnosis of established hookworm infections is made primarily by means of microscopic identification of characteristic eggs in the stool (Fig. 4.5). In an infected human, a single adult female hookworm will produce thousands of eggs per day. Because hookworm infections will often not present with specific signs and symptoms, the clinician typically requires some index of suspicion, such as local epidemiology or country of origin or travel, to request a fecal examination for ova and parasites.

Hookworm eggs are colorless and have a single thin hyaline shell with blunted ends, ranging in size from 55-75 μm by 36-40 μm. Several sensitive egg concentration techniques, such as the formalin-ethyl acetate sedimentation method, can be used to detect even light infections. Where concentration procedures are not available, a direct wet mount examination of the specimen is adequate for detecting moderate to heavy infections. A single stool sample is often sufficient to diagnose

4

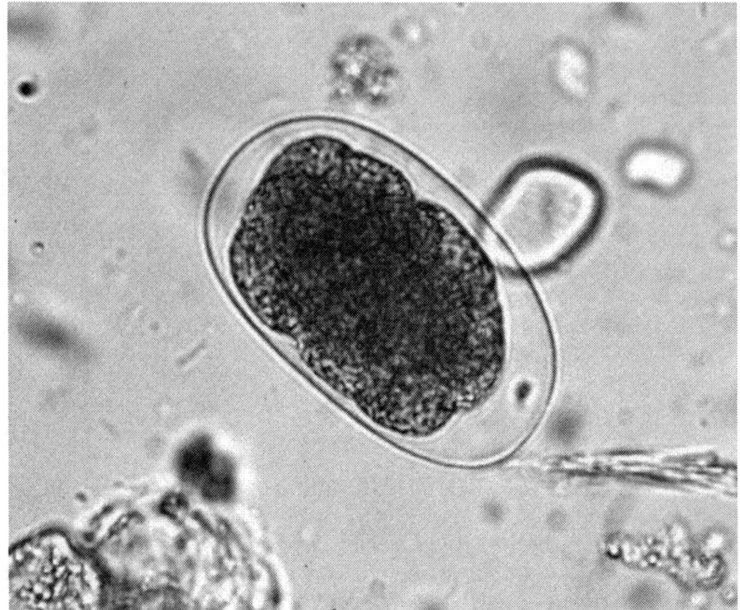

Figure 4.5. Fertilized, embryonated hookworm egg. Reproduced with permission from: Despommier DD, Gwadz RW, Hotez PJ, Knirsch CA. Parasitic Diseases. New York: Apple Trees Productions, 2005:123.

hookworm infection, with diagnostic yield not appreciably increased by examining further specimens. Although examination of the eggs cannot distinguish between *N. americanus* and *A. duodenale*, this is not clinically relevant. Differentiation between the two species can be made by either rearing filariform larvae from a fecal sample smeared on a moist filter paper strip for five to seven days (the Harada-Mori technique) or by recovering the adult worms following treatment and examining for such features as the mouthparts.

Besides microscopic examination of feces, eosinophilia is a common finding in persistent infection and also during the phase of migration of larvae through the lungs. A chest radiograph will usually be negative during the pulmonary phase of larval migration, although sputum examination may reveal erythrocytes, eosinophils and rarely migrating larvae.

Dermatological infection with the zoonotic hookworms *A. caninum*, *A. braziliensis* and *U. stenocephala* is primarily diagnosed clinically, although eosinophilia and elevated serum IgE occur in between 20% to 40% of patients with cutaneous larva migrans. Since larvae do not migrate to the gastrointestinal tract to develop into adult worms, fecal examination will be negative in these cases.

Treatment

The goal of treatment for *N. americanus* and *A. duodenale* infections is to eliminate adult worms from the gastrointestinal tract. The most common drugs used for the treatment of hookworm infections worldwide are members of the benzimidazole anthelmintic class of drugs, of which mebendazole and albendazole are the two principle members. The benzimidazoles bind to hookworm β-tubulin and inhibit microtubule polymerization, which causes death of adult worms through a process that can take several days. Although both albendazole and mebendazole are considered broad-spectrum anthelmintic agents, there are important therapeutic differences that affect their use in clinical practice. For hookworm, a single 500 mg dose of mebendazole often achieves a low cure rate, with a higher efficacy with a single, 400 mg dose of albendazole, although multiple doses of the benzimidazoles are often required.

Systemic toxicity, such as hepatotoxicity and bone marrow suppression, is rare for the benzidimazoles in the doses used to treat hookworm infections. However, transient abdominal pain, diarrhea, nausea, dizziness and headache can commonly occur. Uncommon adverse reactions include alopecia, elevated hepatic transaminases and rarely, leukopenia. Because the benzidimazoles are embryotoxic and teratogenic in pregnant rats, there are concerns regarding their use in children younger than one year of age and during pregnancy. Overall, the experience with these drugs in children under the age of six years is limited, although data from published studies indicate that they are probably safe. Both pyrantel pamoate and levamisole are considered alternative drugs for the treatment of hookworm and they are administered by body weight.

Cutaneous larva migrans should be treated empirically with either albendazole 400 mg daily for three days, or ivermectin 200 mcg/kg/d for 1-2 days; topical thiabendazole (10% to 15%) has also been used to treat this manifestation of hookworm infection in the past. Mebendazole is poorly absorbed from the gastrointestinal tract and therefore does not attain high tissue levels necessary to kill larvae in the skin and treat this condition.

Prevention and Control

In the past, efforts at hookworm control in endemic areas have focused on behaviour modification, such as encouraging the use of proper footwear and improving the disposal of human feces. Unfortunately, these efforts largely met with failure due to such reasons as the continued high rate of occupational exposure to hookworm during agricultural work, the ability of *N. americanus* larvae to penetrate the skin of any part of the body and the use of human feces as fertilizer for crops. Instead, recent control efforts have focused on the use of anthelmintic chemotherapy as a useful tool for large-scale morbidity reduction in endemic communities throughout the world. Annual mass administration of benzidimazoles to school-aged children reduces and maintains the adult worm burden below the threshold associated with disease. The benefits of regular deworming in this age group include improvements in iron stores, growth and physical fitness, cognitive performance and school attendance. In preschool children, studies have demonstrated improved nutritional indicators such as wasting malnutrition, stunting and appetite. Treated children experience improved scores on motor and language testing in their early development. In addition to children, pregnant women and their newborns in endemic areas, if treated once or twice during pregnancy, achieve significant improvement of maternal anaemia and benefit from a reduction of

4

low-birth weight and infant mortality at six months. In areas where hookworm infections are endemic, anthelmintic treatment is recommended during pregnancy except in the first trimester.

However, periodic anthelmintic administration is not expected to lead to complete elimination of hookworm. Reasons include the variable efficacy of benzimidazoles against hookworm, rapid reinfection following treatment especially in areas of high transmission where this can occur within four to 12 months, the diminished efficacy of the benzimidazoles with frequent and repeated use and the possibility drug resistance may develop in time, as has been observed in veterinary medicine. Development of sensitive tools for the early detection of anthelmintic resistance is underway, with special attention being given to in vitro tests and molecular biology techniques that could be adaptable to field conditions. Since no new anthelmintic drugs will be entering the marketplace any time soon, the efficacy of currently available products must be preserved. Furthermore, these concerns have prompted interest in developing alternative tools for hookworm control, such as antihookworm vaccines. Vaccination to prevent high intensity hookworm infection would alleviate much of the global public health impact of this widespread infection.

Suggested Reading

1. Bethony J, Brooker S, Albonico M et al. Soil-transmitted helminth infections: ascariasis, trichuriasis and hookworm. This is a thorough review of the soil-transmitted helminth infections, including hookworm. Lancet 2006; 367:1521-32.
2. Brooker S, Bethony J, Hotez PJ. Human hookworm infection in the 21st Century. Adv Parasitol 2004; 58:197-288.
3. Hotez P, Brooker S, Bethony J et al. Hookworm Infection. N Engl J Med 2004; 351:799-807. An excellent review.
4. Chan MS, Bradley M, Bundy DA. Transmission patterns and the epidemiology of hookworm infection. Int J Epidemiol 1997; 26:1392-1400.
5. Brooker S, Bethony JM, Rodrigues L et al. Epidemiological, immunological and practical considerations in developing and evaluating a human hookworm vaccine. Expert Rev Vaccines 2005; 4:35-50.
6. Jelinek T, Maiwald H, Nothdurft HD et al. Cutaneous larva migrans in travelers: synopsis of histories, symptoms and treatment of 98 patients. Clin Infect Dis 1994; 19:1062-6.
7. Crompton DW. The public health importance of hookworm disease. Parasitology 2000; 121:S39-S50.

CHAPTER 5

Strongyloidiasis

Gary L. Simon

Background, Epidemiology

Strongyloidiasis is an intestinal infection caused by a parasitic nematode that has a widespread distribution throughout the tropics and subtropics. The number of individuals infected with this nematode is unknown, but estimates range from 30 million to 100 million. Infection in the United States is relatively uncommon and the available prevalence data dates back more than two decades. Among the populations that have been shown to be at increased risk are immigrants from developing countries, veterans who have served in endemic areas, especially among those who were prisoners of war, and residents of Appalachia. Epidemiologic studies in rural Kentucky have revealed prevalence rates of 3-4%. A much lower prevalence is found in urban centers in the Southeast. Evidence of strongyloidiasis was noted in only 0.4% of stool samples in Charleston, West Virginia and New Orleans, Louisiana. On the other hand, a large study in New York City found a prevalence of 1%, presumably due to the large Central American immigrant population. Asymptomatic infection is common in Latin America, Southeast Asia and sub-Saharan Africa. Although less common, infection still occurs in Europe, most notably in the southern and eastern portions of the continent.

Strongyloides may persist in the host for decades. Evidence of infection for more than 40 years has been documented among British soldiers who were prisoners of war in the Far East during World War II. There is no endemic focus of strongyloides in the United Kingdom so that the original infection must have occurred during their incarceration in Thailand or Burma. Among 2,072 former prisoners who were studied between 1968 and 2002, 12% had strongyloides infection. There was a strong association between being held in captivity along the infamous Thai-Burma railway and the presence of strongyloides in stool specimens. Among 248 individuals with strongyloides, 70% had the typical larva currens rash and 68% had eosinophilia.

Strongyloides infection is not limited to human hosts. Dogs and nonhuman primates are susceptible to infection and this may play an important role in transmission. There have been a number of outbreaks of strongyloidiasis among animal handlers. Human-to-human transmission has also been described among homosexual men and in day care centers and mental institutions. Most infections, however, are related to exposure to soil that has been contaminated with fecal material that contained *Strongyloides stercoralis* larvae.

Medical Parasitology, edited by Abhay R. Satoskar, Gary L. Simon, Peter J. Hotez and Moriya Tsuji.

Causative Agent

The vast majority of infections are caused by *S. stercoralis*. There is another species, *S. fuelleborni,* which infects infants in Papua, New Guinea and sub-Saharan Africa. These children develop a "swollen belly syndrome" that is characterized by generalized edema and respiratory distress. There is a significant mortality associated with this condition. In a study done in Democratic Republic of Congo, Africa, 34% of 76 children less than 7 months of age were infected with this organism and, in one mother, the organism was found in breast milk suggesting transmammary passage as a means of infection.

Life Cycle

Strongyloides stercoralis exists in both a free living form in the soil and as an intestinal parasite (Fig. 5.1). The parasitic females are 2.2 mm in length, semi-transparent and colorless and lie embedded within the mucosal epithelium of the proximal small intestine where they deposit their eggs. A single female worm will produce up to 50 eggs per day. There is no parasitic male and reproduction is by parthenogenesis.

The embryonated eggs hatch within the mucosa and emerge into the lumen of the small bowel as noninfectious rhabditiform larvae (Fig. 5.2). The rhabditiform larvae are excreted in the stool and, in a warm, humid environment, mature and become free-living adult male and female worms. The adult worms mate and the female produces embryonated eggs (Fig. 5.3) that, after several molts, ultimately become

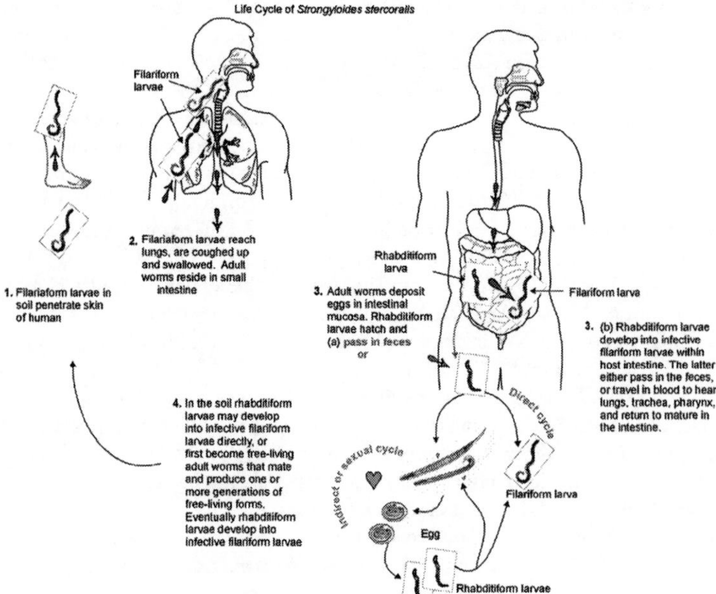

Figure 5.1. Life Cycle of Strongyloides. Adapted from Nappi AJ, Vass E, eds. Parasites of Medical Importance. Austin: Landes Bioscience, 2002:77.

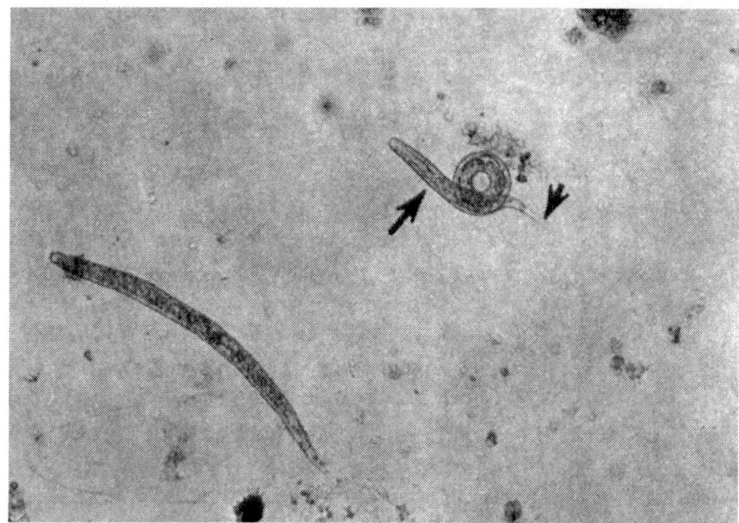

Figure 5.2. Rhabditiform larvae of *S. stercoralis*.

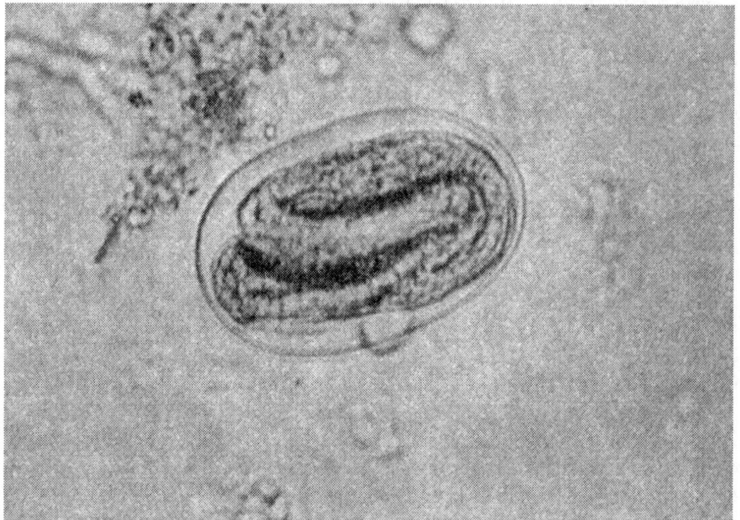

Figure 5.3. Ova of *S. stercoralis*.

infectious filariform larvae (Fig. 5.4). Alternatively, there may be several free-living cycles which precede the development of the filariform larvae. The filariform larvae are long and slender whereas the rhabditiform larvae are shorter and thicker. The filariform larvae penetrate intact skin leading to the parasitic phase of infection.

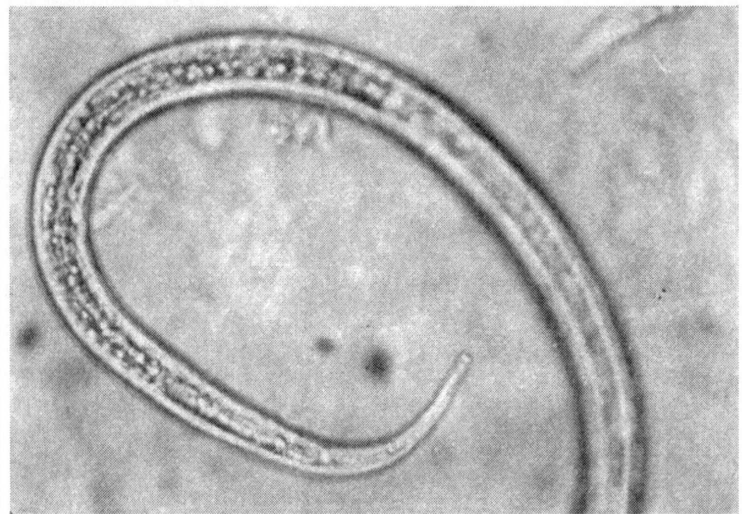

Figure 5.4. Posterior end of filariform larva of *S. stercoralis*.

After penetrating the skin, the filariform larvae migrate to the right side of the heart, either through the lymphatics or the venules and then to the lungs via the pulmonary circulation. In the lungs the larvae penetrate the alveoli, migrate up the tracheobronchial tree and are swallowed. In the proximal small intestine the larvae molt, become adult female worms and lodge in submucosal tissues. Egg production begins in 25-30 days after initially penetrating the skin.

Autoinfection, Hyperinfection

One of the more unusual features of strongyloidiasis is the concept of autoinfection. In the distal colon or moist environment of the perianal region rhabditiform larvae can undergo transformation to filariform larvae. These infectious forms may penetrate the colonic mucosa or perianal skin and "reinfect" the host. This process of autoinfection may be quite common and accounts for the persistence of this organism in the host for decades. In fact, as a result of this autoinfection process, *S. stercoralis* can actually increase its numbers in the same individual without additional environmental exposures.

A more severe form of autoinfection is the hyperinfection syndrome in which large numbers of rhabditiform larvae transform to filariform larvae, penetrate the colonic mucosa and cause severe disease. This occurs in debilitated, malnourished or immunosuppressed individuals. Administration of prednisone is a common risk factor and cases have been reported following treatment of asthma with glucocorticosteroids. Other risk factors include impaired gut motility, protein-calorie malnutrition, alcoholism, hypochlorhydria, lymphoma, organ transplantation and lepromatous leprosy. Human T-cell lymphotropic virus Type 1 (HTLV-1) has also been associated with severe strongyloides infection. The most severe form of the hyperinfection syndrome is disseminated infection in which larvae are found in other organs including the liver, kidney and central nervous system.

Immune Response

Eosinophils, mast cells and T-cell mediated immune function appear to play a role in resistance to strongyloides. Eosinophilia is common in controlled chronic strongyloides infection whereas eosinophils are absent in patients with the hyperinfection syndrome. Eosinophils, when in close proximity to a helminth, can kill the organism through exocytosis of granular contents, eosinophil major basic protein and eosinophil cationic protein. Eosinophils have IgE receptors on their surface and their cytotoxic effects may be mediated through an IgE-antibody dependant cellular cytotoxicity.

Mast cells have several functions in the host defense against helminths. They promote peristalsis and mucus production which aid in expelling the parasite, they produce IL-4 and they attract eosinophils. Release of mast cell granules is an important gastrointestinal inflammatory stimulus.

The immunosuppressive character of those conditions which predispose to hyperinfection has indicated that T-cell mediated immunity is an important component of the host response to Strongyloides. Thus, although initially predicted to be an AIDS-related disease, it was rather surprising to find very little evidence of excessive numbers of advanced strongyloidiasis among HIV-infected individuals. One possible explanation for these findings may be the distinguishing features of the TH-1 and TH-2 helper lymphocyte response.

TH-1 cells are primarily associated with the cellular immune response and the production of proinflammatory cytokines such as IL-2, IL-12 and γ-interferon. The predominant defect in HIV-infected individuals is in the TH-1 mediated inflammatory response. On the other hand, control of helminthic infections is more a function of the TH-2 response which stimulates the production of cytokines such as IL-4, IL-5 and IL-10. Interleukin-4 and IL-5 promote IgE synthesis; IL-5 regulates eosinophil migration and activation. Both IL-4 and IL-10 are antagonists of the TH-1 response, thus promoting the shift to a TH-2 humoral response. It is the absence of an effective TH-2 response that appears to be one of the major factors in promoting development of the hyperinfection syndrome.

Corticosteroids promote the hyperinfection syndrome through a variety of mechanisms. Corticosteroids inhibit the proliferation of eosinophils, induce apoptosis of eosinophils and T-lymphocytes and inhibit the transcription of IL-4 and IL-5. Corticosteroids may also have direct effects on the worms themselves. They may stimulate female worms to increase larval output and promote molting of rhabditiform larvae into the invasive filariform larvae.

Clinical Findings

The clinical manifestations of uncomplicated strongyloidiasis are cutaneous, pulmonary and gastrointestinal. Following penetration of the skin, there may be a localized, erythematous, papular, pruritic eruption. Migrating larvae may produce a serpiginous urticarial rash, larva currens (Fig. 5.5), that can progress as fast as 10 cm/hr. This is frequently seen on the buttocks, perineum and thighs and may represent autoinfection. Within a few days of initial infection, pulmonary symptoms such as cough, wheezing or shortness of breath may develop as well as fleeting pulmonary infiltrates and eosinophilia. With the development of gastrointestinal involvement there may be epigastric discomfort suggestive of peptic ulcer disease, bloating, nausea and diarrhea. Rarely, more severe infections have been associated

5

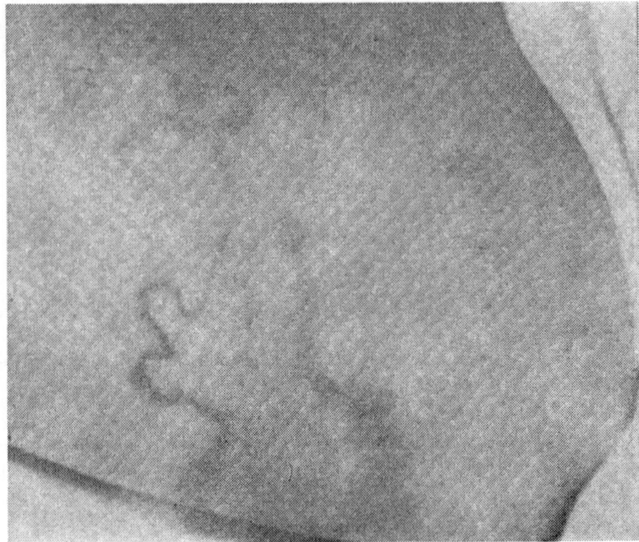

Figure 5.5. Larva currens.

with malabsorption and protein-calorie malnutrition. However, in many chronically infected individuals, symptoms resolve with the appearance of larvae in the stool, although most have persistent eosinophilia.

Radiologic findings are quite variable ranging from no abnormalities to duodenal dilatation and thickened mucosal folds. In severe cases, there may be nodular intraluminal defects secondary to granuloma formation as well as loss of mucosal architecture and luminal narrowing secondary to fibrosis.

The nonspecific symptoms of chronic strongyloidiasis may lead to endoscopic evaluation for consideration of peptic ulcer disease, especially in urban centers. Endoscopic findings consist of mucosal hyperemia and white punctate lesions in the duodenum. Histologic examination reveals larvae and adult worms in the duodenal crypts, submucosa and lamina propria.

The large numbers of organisms in the intestines and lungs in patients with the hyperinfection syndrome result in symptoms that are in sharp contrast to the often relatively benign nature of chronic strongyloidiasis. Abdominal pain, nausea, vomiting and profuse diarrhea are common. An ileus may be present and bowel edema leading to intestinal obstruction has been described. Duodenal ulceration and hemorrhage, perforation and peritonitis are rare complications. Pulmonary findings include cough, hemoptysis, shortness of breath, diffuse pulmonary infiltrates and even respiratory failure. Larvae are often found in the sputum of patients with the hyperinfection syndrome. Patients with the hyperinfection syndrome may present with Gram-negative sepsis as a result of colonic bacteria entering the host either through intestinal perforation resulting from the invasion of filariform larvae, or on the back or in the gut of the larvae. In patients with bacteremia, respiratory failure may ensue as a result of the direct effects of larvae migrating through the lungs, accompanying Gram-negative pneumonia or from the development of sepsis-mediated adult respiratory distress syndrome. Meningitis is frequently seen in patients with disseminated infection. Pneumonia, peritonitis and infection of other organs may also occur in patients with disseminated infection. Most patients with severe hyperinfection or disseminated infection do not survive their illness.

Diagnosis

The diagnosis of strongyloidiasis can be established by identification of the larvae in the stool. This is often quite difficult, even for the experienced parasitologist, because of the low number of larvae produced on a daily basis by an adult worm. Furthermore, release of larvae is intermittent so that even careful examination of a single stool specimen may not reveal the organism. The sensitivity for finding *S. stercoralis* in a single stool specimen from an infected individual has been estimated to be only 30%. Multiple microscopic examinations using large volumes of stool concentrated by a sedimentation method are often necessary. Frequently, the only clue to the diagnosis is the presence of unexplained eosinophilia.

A variety of methods have been developed to increase the likelihood of finding larvae in a stool specimen. One method employs an agar culture on which a stool sample is placed on nutrient agar and then incubated for several days. As the motile larvae crawl over the agar, they carry bacteria with them leaving visible tracks. In one large study, this technique was found to be 96% sensitive in determining the presence of strongyloides. Unfortunately, it is rather laborious and time-consuming and is not usually employed in routine clinical laboratories. Most clinical laboratories utilize a lugol-iodine staining technique of a stool sample that has been subjected to a sedimentation procedure in order to concentrate the larvae.

Examination of specimens obtained directly from the duodenum has been used in the past. Both duodenal aspiration, especially in children and the string test have been utilized with positive results. The latter employs a procedure in which a gelatin capsule with a string attached is swallowed and then retrieved after several hours. The string is then stripped of mucus and examined microscopically for the presence of larvae. In many developed areas, these tests have been replaced by esophagogastroduodenoscopy.

Serologic tests have been developed that can aid in the diagnosis of strongyloidiasis. An enzyme-linked immunoassy is available from the Centers for Disease Control (Atlanta, GA) which has a sensitivity of 95%. Specificity is less as a result of cross-reactivity with other helminth infections. An immediate hypersensitivity type of skin test has been developed which has a sensitivity of 82-100%, but has significant cross-reactivity with filarial infections.

Treatment

The goal of treatment of strongyloidiasis is eradication of the organism. Unlike hookworm, where simply reducing worm burden is efficacious, the process of autoinfection will result in prolonged infection, potentially increasing worm burden over time and the risk of hyperinfection in anyone with residual organisms. The drug of choice for treatment of strongyloidiasis is ivermectin given orally at a dose of 200 µg/kg once daily for 2 days. Thiabendazole is equally effective at a dose of 25 mg/kg twice daily for 3 days, but has a much greater incidence of side effects including nausea, vomiting, foul taste and foul smelling urine. Albendazole has also been used, but is less effective than either thiabendazole or ivermectin, and requires 7 days of treatment.

Hyperinfection syndrome may be treated with either thiabendazole or ivermectin. Therapy should be continued for at least 2 weeks, an autoinfection cycle, after larvae can no longer be detected. Either drug can be administered by nasogastric tube or per rectum in patients unable to tolerate oral medications. If possible, immunosuppressive therapy should be discontinued. In those patients in whom the immunosuppressive condition cannot be altered, it may be prudent to repeat a brief course of therapy every few weeks. Preemptive therapy should be considered in individuals from endemic areas who are going to undergo immunosuppressive therapy, especially if eosinophilia is present.

Suggested Reading

1. Ashford RW, Vince JD, Gratten MJ et al. Strongyloides infection associated with acute infantile disease in Papua, New Guinea. Trans R Soc Trop Med Hyg 1978; 72:554.
2. Barnes PJ. Corticosteroids, IgE and atopy. J Clin Invest; 2001; 107:265-6.
3. Berkmen YM, Rabinowitz J. Gastrointestinal manifestations of the Strongyloidiasis. Amer J Roentg, Rad Ther, Nuc Med 1972; 115:306-11.
4. Brown RC, Girardeau MHF. Trans-mammary passage of Strongyloides spp. Larvae in the human host. Amer J Trop Med Hyg 1977; 26:215-9.
5. Concha R, Harrington, W Jr et al. Intestinal Strongyloidiasis: recognition, management and determinants of outcome. J Clin Gastro 2005; 39:203-11.
6. Despommier DD, Gwadz RW, Hotez PJ et al. Strongyloides stercoralis. In: Parasitic Diseases. New York: Apple Tree Productions, 2005:129-34.
7. Gatti S, Lopes R, Cevini C. Intestinal parasitic infections in an institution for the mentally retarded. Ann Trop Med Parasitol 2000; 94:453-60.
8. Genta RM. Dysregulation of strongyloidiasis: a new hypothesis. Clin Microbiol Rev 1992; 5:345-55.
9. Genta RM. Global Prevalence of Strongyloidiasis: Critical review with epidemiologic insights into the prevention of disseminated disease. Rev Inf Dis 1989; 11:755-67.
10. Georgi JR, Sprinkle CL. A case of human strongyloidiasis apparently contracted from asymptomatic colony dogs. Amer J Trop Med Hyg 1974; 23:899-901.
11. Gill GV, Welch E, Bailey JW et al. Chronic strongyloides stercoralis infection if former british far east prisoners of war. Quar J Med 2004; 97:789-95.
12. Goka AK, Rolston DD, Mathan VI et al. Diagnosis of Strongyloides and hookworm infections: comparison of faecal and duodenal fluid microscopy. Trans R Soc Trop Med Hyg 1990; 84:829-31.
13. Grove DI. Human Strongyloidiasis. Adv Parasitol 1996; 38:251-309.
14. Lindo JF, Conway DJ, Atkins NS et al. Prospective evaluation of enzyme-linked immunosorbent assay and immunoblot methods for the diagnosis of endemic Strongyloides stercoralis infection. Amer J Trop Med Hyg 1994; 51:175-9.
15. Neva FA, Gamm AA, Maxwell C et al. Skin test antigens for immediate hypersensitivity prepared from infective larvae of Strongyloides stercoralis. Amer J Trop Med Hyg 2001; 65:567-72.
16. Overstreet K, Chem J, Wang J et al. Endoscopic and histopathologic findings of Strongyloides stercoralis in a patient with AIDS. Gastrointest Endos 2003; 58:928-31.
17. Salazar SA, Berk SH, Howe D et al. Ivermectin vs. Thiabendazole in the treatment of strongyloidiasis. Infect Med 1994; 50-9.
18. Sato Y, Kobayashi J, Toma H et al. Efficacy of stool examination for detection of Stronyloides infection. Amer J Trop Med Hyg 1995; 53:248-50.
19. Siddiqui AA, Berk SL. Diagnosis of Strongyloides stercoralis infection. Clin Inf Dis 2001; 33:1040-7.
20. Zaha O, Hirata T, Kinjo F et al. Strongyloidiasis—progress in diagnosis and treatment. Intern Med 2000; 39:695-700

Chapter 6

Trichinellosis

Matthew W. Carroll

Background

Epidemiology

Trichinella spp. are tissue nematodes of nearly worldwide distribution. Only the Australian mainland and Puerto Rico remain trichinella free. Members of the genus Trichinella are able to infect a wide range domestic and wild (sylvatic) animals and cause the clinical syndrome trichinellosis, often referred to as trichinosis, in humans. Trichinellosis is rare in the United States; fewer than 100 cases are reported annually. The disease has become more common in Eastern Europe and Asia. This is due to a breakdown in governmental inspection of pork, lack of education regarding the cooking of pork and ongoing feeding of raw garbage to swine. Pigs will also eat rats and mice that may be infected, participate in pig cannibalism and eat fecal matter which can lead to ingestion of live adult worms. On a worldwide basis, pigs represent the most frequently identified vehicle of trichinellosis, although a variety of other animals have been recognized as transmitting this illness. In the United States, nearly one-half of the cases are due to consumption of wild game. Bear meat, cougar and wild boar have been implicated as vehicles of trichinellosis. In Russia, more than 90% of cases were traced to ingestion of bear or wild boar meat. Horse meat is a common vehicle in France and Italy.

Causative Agents

Trichinellosis is most commonly caused by the consumption of raw or undercooked pork products infected with *Trichinella spiralis*. However, there are a number of other *Trichinella spp.* that have been associated with human infection. In North America both *T. murelli* and *T. nativa* have been recognized as causes of sylvatic (wild) trichinellosis. *Trichinella murelli* is most commonly acquired following ingestion of wild bear meat. *Trichinella nativa* tends to have an arctic or subarctic distribution and is found in polar bears, walrus, seals, wolves and sled dogs. Other species of Trichinella have been reported to cause human illness. In Europe, *T. brivoti* has been recognized as a cause of sylvatic disease in wild boar and red fox. There are two Trichinella species that do not encyst in muscle. *T. pseudospiralis* has been described in birds of prey as well as pigs and wild game. *T. paupae*, which is endemic to New Guinea, infects domestic and wild swine. Human disease is common among native hunters in New Guinea who often consume undercooked game meat.

Medical Parasitology, edited by Abhay R. Satoskar, Gary L. Simon, Peter J. Hotez and Moriya Tsuji.

6

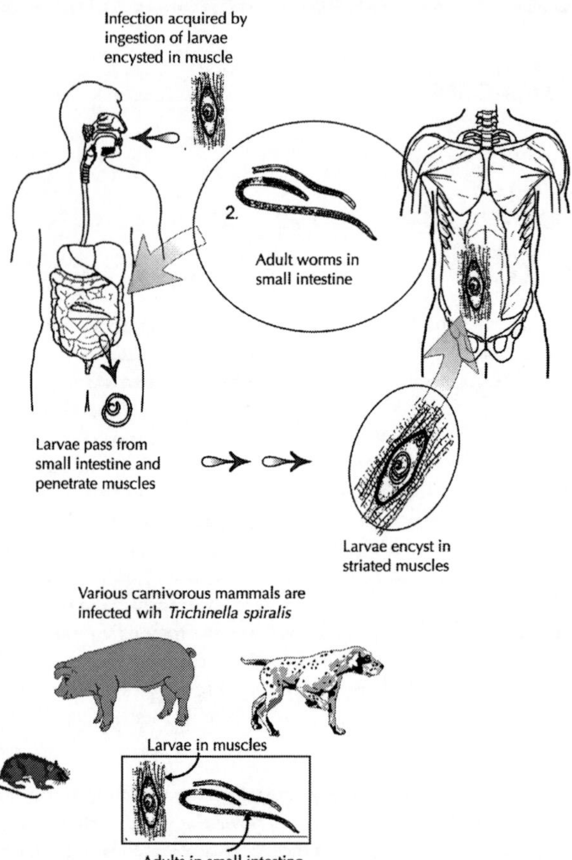

Figure 6.1. Life Cycle of *Trichinella spiralis*. Reproduced from: Nappi AJ, Vass E, eds. Parasites of Medical Importance. Austin: Landes Bioscience, 2002:75.

Life Cycle

Infection with *Trichinella spp*. occurs when raw or undercooked meat containing the nurse-cell larva complex in ingested (Fig. 6.1). Larvae are released by acid-pepsin digestion in the stomach then enter the small intestine where they penetrate the columnar epithelium at the base of a villous. They are obligate intracellular organisms and localize in the intestinal epithelial cells. Within the intestine the larvae molt 4 times during a 30-hour period and develop into adults. The adult female measures 3 × 0.036 mm, the male is smaller, 1.5 × 0.030 mm. After copulation, which occurs within the intestinal epithelium, the female worms expel newborn larvae. This occurs 4-7 days after infection. Females survive 4-16 weeks and, during that time, may give birth to 1,500 larvae.

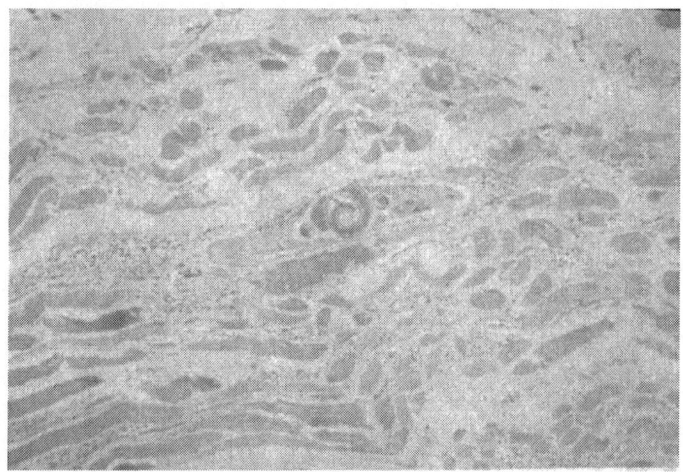

Figure 6.2. Trichinella cysts in muscle.

Newborn larvae measure 0.08 mm long by 7 μm in diameter and possess a unique sword-like stylet. This enables the larvae to cut an entry hole in the lamina propria and enter the mesenteric lymphatics or capillaries. The newborn larvae subsequently make their way to the arterial circulation and disseminate throughout the host. The larvae are capable of entering any cell type but will only survive in striated muscle. Once within the skeletal muscle the Trichinella larva induce the myocyte to transform into a new cell type, a nurse cell, which sustains the life of the larva (Fig. 6.2). During this metamorphosis the muscle cell switches to anaerobic respiration and, for most species of *Trichinella*, develops into a cyst of collagen and hyaline. These larva-containing cysts can persist for many years although most calcify and die within a few months.

Pathogenesis and Host Response

The development of adult worms in the intestinal epithelium leads to an immune-mediated inflammatory response in an effort to expel the organism. IgE is increased and, pathologically, there is an inflammatory infiltrate, rich in eosinophils and mast cells. Eosinophils elaborate major basic protein and toxic oxygen species that help clear the parasite but also cause tissue damage. Mast cells produce a molecule, mast cell protease-1 (MCP-1), that is also toxic to the organism. Knock-out mice that lack the gene encoding MCP-1 have delayed expulsion of the worms from the GI tract as well as increased numbers of encysted larvae. Interleukin-5, which inhibits eosinophil apoptosis, is an integral part of this process. Genetically altered "knockout mice" that lack this gene also have delayed clearance of worms.

As larvae enter cells in various tissues during the parenteral phase of infection, there is widespread inflammation. The resulting pathology is proportionate to the extent of infection. Cell death is frequent and with heavy infections there is edema, proteinuria and organ toxicity including the development of central nervous system disorders and cardiomyopathies.

Trichinella species tend to persist within head and neck muscles including the tongue, laryngeal muscles, masseter muscles, as well as respiratory muscles, especially the intercostals and diaphragm.

Clinical Features

Ingestion of larvae is characterized by abdominal pain, secretory diarrhea, nausea and vomiting during the first week of infection. Occasionally, constipation may be present. Parasite inoculum is a major factor in disease severity and, in many cases, infected individuals may be asymptomatic. On the other hand, in patients with very heavy worm loads, severe enteritis can occur leading to severe dehydration.

The parenteral phase, which occurs approximately one week after infection, is accompanied by systemic symptoms; fever, often 40-41°C, myalgias and muscle tenderness. Petechial hemorrhages may appear in the mucous membranes, conjunctivae and subungal skin and, rarely, an urticarial rash occurs. Periorbital edema is a classic finding that has been reported to occur in 77% of those infected. This can be so severe that the infected individual is unrecognizable. This is thought to be a result of a generalized allergic reaction and typically lasts 1 to 2 weeks. Its resolution is heralded by spontaneous diuresis.

Muscle pain and swelling are the most prominent features of trichinellosis. Muscles become stiff, hard, tender and edematous and often appear hypertrophic. Change in voice is commonly seen with involvement of laryngeal muscles. Pharyngeal, lingual and masseter involvement may make eating and swallowing difficult. Diplopia may occur when the extraocular muscles are affected. These symptoms usually peak 2-3 weeks after infection and then resolve. Laboratory abnormalities include a modest leukocytosis with eosinophila which can be as high as 50%, as well as a mild elevation of creatine kinase and other muscle enzymes.

Other *Trichinella* species may exhibit different symptoms. *Trichinella nativa* tends to produce a prolonged diarrheal illness with a paucity of systemic and myalgic complaints, whereas Thai patients with *T. pseudospiralis* infection may have severe myalgias for up to 4 months.

Involvement of tissues other than muscle can lead to a wide variety of complications. Central nervous system involvement has been reported in 10-25% of patients. Most commonly, patients with neurotrichinosis present with meningoencephalitic signs. Paralysis, delirium, psychosis and peripheral neuropathies have been described.

Cardiac involvement may also occur in patients with trichinellosis. Pericardial effusions were found in nearly 10% of patients in one study. Myocarditis is uncommon, but in patients with severe infection it can be lethal. Electrocardiographic abnormalities may also be found including arrhythmias and heart blocks.

Lethal infection due to Trichinella is uncommon, but deaths have been reported, mostly in patients with severe worm burdens. Besides myocarditis, central nervous system involvement and pneumonia may lead to a fatal outcome. Sepsis and death can occur in patients who develop bacteremia with enteric organisms following penetration of larvae from the GI tract into the bloodstream. A sudden reduction in eosinophils in a patient with severe trichinellosis may herald death or indicate superimposed bacterial infection.

Diagnosis

In the United States trichinellosis is quite rare. Since its symptoms mimic common diseases such as viral gastroenteritis or food poisoning, many clinicians fail to consider it when formulating a differential diagnosis. Sporadic cases often go undetected and thus the true prevalence of this disease is often underestimated.

Trichinellosis should be considered in the presence of fever, myalgias, periorbital edema and eosinophilia; and an epidemiologic history of consumption of raw or undercooked meat. Muscle biopsy is the "gold standard" for the diagnosis of trichinellosis. Typically a 1-4 cubic millimeter piece of muscle is pressed between two slides and viewed under the microscope for the presence of larvae. This is positive in heavy infections. Organisms such as *T. psuedospiralis* and *T. papaue* that do not encyst, as well as other *Trichinella* species if examined prior to encysting, may be missed since they resemble muscle tissue. Alternatively, the muscle can be digested in 1% HCl-1% pepsin at 37° C for 1 hour after which the larvae are released from nurse cells and are much easier to identify. This process often destroys young larvae making it more difficult to diagnose early infection. Routine histological examination may reveal a basophilic reaction in muscle tissue as well as demonstrating the presence of larvae.

Specific immunodiagnostic tests are also available. Several tests designed to look for the presence of larvae group 1 antigens are available. These include indirect hemaggutination, bentonite flocculation, indirect immunoflouresence and, the most sensitive test, an enzyme linked immunoabsorbant assay (ELISA). These tests are quite specific for trichinellosis, but may be insensitive during the first few weeks of infection. Seroconversion usually occurs between the third and fourth week of infection; the ELISA may be positive after 12 days. By day 50 virtually 100% of infected individuals will have IgG antibody directed against the infecting organism. Tests for the detection of larval antigens are also available and indicate active infection. Cross-reactions occur with both antigen and antibody assays, typically in individuals with autoimmune disease or patients infected with parasites that exhibit cross reactivity.

Newer molecular biologic techniques can be used for both diagnosis and to distinguish between species of Trichinella. Polymerase chain reaction (PCR) assays are able to determine the species of Trichinella with as few as one larva. Differences in 5S rRNA can be used to distinguish between *Trichinella spp.* that are morphologically identical.

Treatment

There is no specific anthelminthic therapy that is effective for the treatment of trichinellosis. Benzimidazoles, mebendazole and albendazole, are active only against the adult worm. When used shortly after ingestion, they will reduce worm burden. Albendazole is given at a dose of 400 mg twice daily for 8-14 days. Mebendazole 200-400 mg three times a day for 3 days then 400-500 mg three times a day for ten days is also effective. The initial dose of mebendazole and albendazole is lower to prevent a Mazzoti reaction, an anaphylactic type reaction due to antigenic simulation following the death of the adult worms after treatment.

Treatment of the parenteral stage of the disease is directed towards symptomatic relief. In general, mild disease does not require treatment. Treatment of moderate to severe disease can decrease morbidity and, possibly, mortality. In such patients, corticosteroids have been given to reduce tissue damage. These drugs may be lifesaving in patients with myocarditis or neurotrichinellosis. Prednisolone at doses of 40 to 60 mg per day are typically used. However, this will slow the clearance of adult worms from the gut and should not be given for a prolonged period. Analgesics and antipyretics such as NSAIDs and acetaminophen provide symptomatic relief.

Prevention

6

Effective prevention of trichinellosis requires proper animal husbandry practices as well as adequate cooking of meat. Feeding swine only cooked scraps and controlling the population of infected rodents will help to prevent disease. Freezing pork usually kills *T. spiralis* effectively. However, wild game infected with *T. nativa* are resistant to freezing. Cooking pork to when it is no longer pink (170°F; 77°C) exceeds the thermal death point of Trichinella spp. Game meat is much darker and it is necessary to use a meat thermometer to determine if it has been adequately cooked. Microwave cooking often leaves "cold spots", so this cooking method should not be used with possibly infected meat.

Suggested Reading

1. Centers for Disease Control. Horsemeat associated Trichinosis—France; MMWR1986; 35:291-2,297-8.
2. Bruschi F, Murell KD. New aspects of human Trichinellosis: The impact of new Trichinella Species. Postgrad Med J 2002; 78:15-22.
3. Capo V, Despommier DD. Clinical aspects of infection with Trichinella spp. Clin Microbiol Rev 1996; 9(1):47-54.
4. Despommier DD, Gwadz RW, Hotez PJ et al, eds. Parasitic Diseases, 5th Edition. New York: Apple Tree Production LLC, 2002:135-42.
5. Escalante M, Romarís F, Rodríguez M. Evaluation of Trichinella spiralis group 1 antigens for the serodiagnosis of human trichinellosis. J Clin Microbio 2004; 42:4060-6.
6. Knight PA, Wright SH, Lawrence CE. Delayed expulsion of the nematode T. spiralis in mice lacking the mucosal mast cell-specific granule chymase, mouse mast cell protease-1. J Exp Med 2000; 192:1849-56.
7. Pérez-Martín JE, Serrano FJ, Reina D et al. Sylvatic trichinellosis in Southwestern Spain. J Wildl Dis 2000; 36:531-4.
8. Pozio E, La Rosa G. Trichinella murelli n.sp: eitiologic agent of sylvatic trichinellosis in temprate areas of North America. J Parisitology 2000; 86:134-9.
9. Jongwutiwes S, Chantachum N, Kraivichian P et al. First outbreak of human trichinellosis caused by Trichinella pseudospiralis. Clin Inf Dis 1998; 26:111-5.
10. Ranque S, Faugère B, Pozio E et al. Trichinella pseudospiralis outbreak in France. Emerg Infect Dis 2000; 6:1-7.
11. Rehmet S, Sinn G, Robstad O et al. Trichinellosis Outbreaks—Northrhine-Westfalia, Germany, 1988-1999. MMWR 1999; 48:488-92.
12. Rombout YB, Bosch S, Van Der Giessen JW. Detection and identification of eight trichinella genotypes by reverse line blot hybidization. J Clin Microbiol 2001; 39:642-6.
13. Rotolo R, Garcia R, Habib A et al. Epidemiologic notes and reports common source outbreaks of thicinosis—New York City, Rhode Island. MMWR1982; 31:161-4.
14. Roy S, Lopez A, Schantz P. Trichinellosis survey—United States, 1997-2001; MMWR 52:1-8.
15. Centers for Disease Control and Prevention. Trichinellosis associated with bear meat in New York and Tennessee—2003; MMWR 2004; 53:606-9.
16. Centers for Disease Control and Prevention. Outbreak of Trichinellosis associated with eating cougar jerky—Idaho. MMWR 1996; 45:205-6.
17. Watt G, Saisorn S, Jongsakul K et al. Blinded, placebo controlled trial of antiparasitic drugs for Trichinosis myositis. J Infect Dis 2000; 182:371-4.
18. Zarnke RL, Worley DE, Ver Hoef JM et al. Trichinella sp. in wolves from the interior of Alaska. J Wildl Dis 1999; 35:94-7.

CHAPTER 7

Onchocercosis

Christopher M. Cirino

Background

Onchocercosis (river blindness) is caused by the filarial nematode *Onchocerca volvulus* and transmitted by several species of blackflies in the genus *Simulium*. The disease represents a leading infectious cause of blindness and visual impairment, second to *Chlamydia trachomatis*. An estimated 17.7 million people are infected worldwide, with 270,000 blind and 500,000 with severe visual impairment as a result.

Onchocercosis is distributed primarily in Africa between the tropics of Cancer and Capricorn, while smaller foci exist in Central and Latin America (Guatemala, Mexico, Brazil, Ecuador, Venezuela and Columbia) as well as in the Arabian peninsula (Yemen and Saudi Arabia). Nigeria alone accounts for more than one-third of the global prevalence of disease. In these areas, the disease usually is found along rapid-flowing rivers and streams, where blackflies lay their eggs. Consequently, prior to eradication efforts, whole areas of arable land had been uninhabitable because of the disease.

Life Cycle

The life cycle of *O. volvulus*, consists of developmental stages in the blackfly vector and the definitive human host (Fig. 7.1). When an infected female black fly takes a bloodmeal, infective larvae (larval stage 3 (L3)) penetrate a break in the skin and migrate into connective tissues. After several stages of molting, the larvae develop into adult male (4 cm × 0.2 mm) and female worms (50 cm × 0.4 mm) (macrofilariae, Fig. 7.2) and localize in fibrous subcutaneous nodules, known as onchocercomas. Within these nodules female worms release 1,300-1,900 microfilariae daily for as long as 14 years. Microfilariae (250 to 300 μm in size) migrate into the subcutaneous tissues via the lymphatic vessels. These microfilariae may survive for 6 to 30 months, but the majority do not complete the life cycle. When a blackfly bites an infected human, microfilariae (L1) are ingested with the bloodmeal. They develop into infective larvae (L3) in the fly and migrate to the insect's mouth within a few weeks.

Microfilariae released from onchocercomas migrate through the subcutaneous tissues, eliciting only a minimal inflammatory response. A greater reaction occurs with microfilarial death and degeneration in the tissue, which is exacerbated by onchocercosis chemotherapy. Both B- and T-cell mediated responses occur in response to filarial products with antibody production, involvement of macrophages and recruitment of neutrophils and eosinophils. Additionally, macrophages, dendritic cells and T-cells have been detected in onchocercomata.

Medical Parasitology, edited by Abhay R. Satoskar, Gary L. Simon, Peter J. Hotez and Moriya Tsuji.

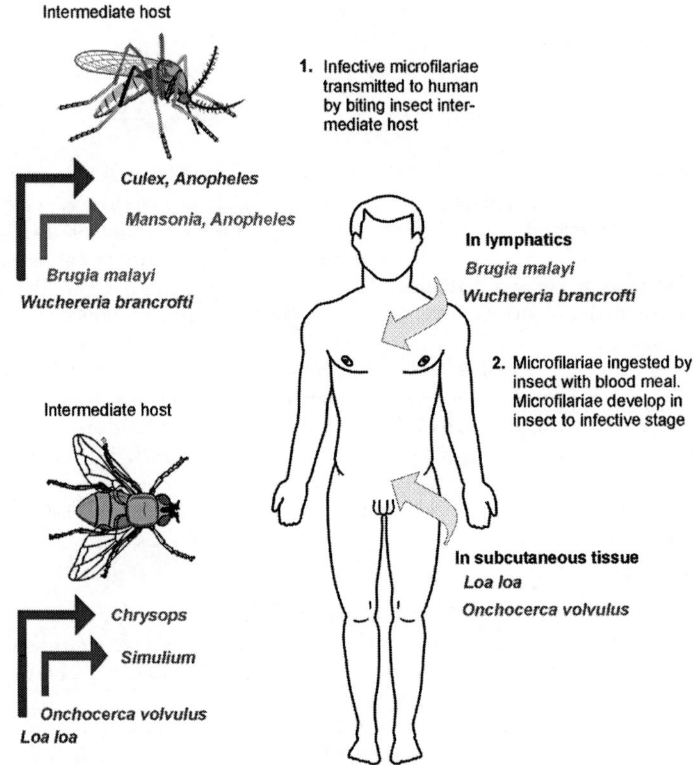

Figure 7.1. Life cycle of *Onchocerca volvulus*. Adapted from: Nappi AJ, Vass E, eds. Parasites of Medical Importance. Austin: Landes Bioscience, 2002:93.

The mechanism by which *O. volvulus* and other filarial nematodes are able to evade the host immune system is under intensive investigation and has been reviewed elsewhere. Three immune manifestations of onchocerciasis have been described: the generalized form, the hyperreactive form and the putatively immune individuals (PI). Implicated are filarial-derived molecules, T-helper subset populations, cytokine and immunoglobulin regulation, host genetic differences between various forms and a host response to the onchocercal endosymbiont *Wolbachia*. An association with high skin and blood worm loads in filarial infections was made with a decreased response of T-helper cell Type 2 (Th2) and increases in the immunosuppressive cytokines interleukin-10 and transforming growth factor-β. Unlike generalized onchocerciasis, the hyperreactive form (sowda) is associated with a heightened T-helper 2 (Th2) which is eosinophil predominant

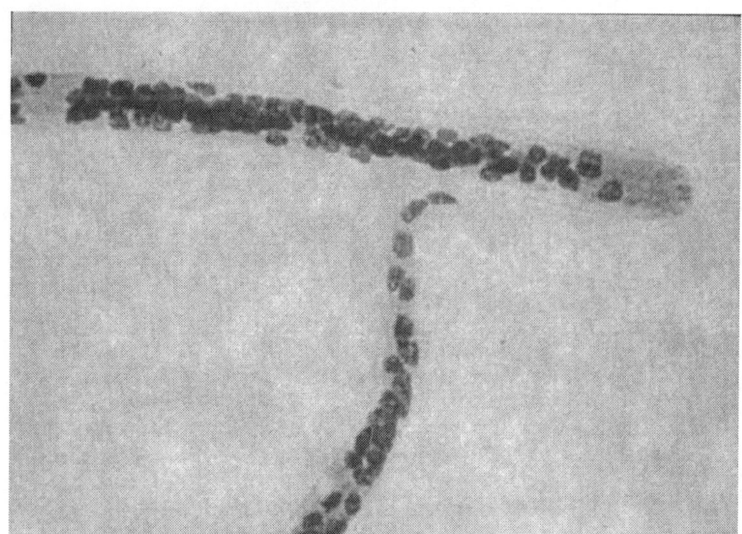

Figure 7.2. *Onchocerca volvulus*. Obtained with kind permission from Dr. Abhay Satoskar.

in the setting of minimal microfilaridermia. Levels of IgG1, IgG3 as well as IgE are elevated in this disease manifestation. Sowda has been found to have a genetic basis with particular major histocompatibility complex alleles and a mutation in the interleukin-13 gene associated with other hyperreactive states (asthma, atopy). Those putatively immune have been found to have higher interferon gamma and interleukin-5 levels and a mixed Th1 and Th2 subset response.

Endobacteria of the genus *Wolbachia* have been identified as essential for worm fertility in many filarial nematodes, including *Onchocerca volvulus*. The release of *Wolbachia* molecules after microfilariae death may contribute directly to the pathogenesis of skin lesions and visual impairment (e.g., keratitis), the severity of posttreatment reactions and the persistence of infection in the host. In a murine model, the inflammatory response following intracorneal injection of filarial extracts was contingent on the presence of *Wolbachia*, with only a minimal response seen from extracts of worms pretreated with doxycycline (*Wolbachia* negative). There is evidence of an innate immune response in the cornea with neutrophil activation mediated by host cell toll-like receptors 2 and 4 (TLR2, TLR4) to an endotoxin-like surface protein and *Wolbachia* surface protein (WSP) of this bacteria. This may contribute to a shift in T-helper subset balance, which allows the parasite to persist in the host through decreased immune activation.

Disease Signs and Symptoms

Largely, the immune response to microfilariae death and degeneration leads to the clinical findings of the disease, including pruritis, acute and chronic dermatitis and visual impairment and blindness. The earliest and most common symptom of onchocerciasis is pruritis, which may be severe. Acute or chronic lymphedema, inguinal lymph node swelling and loss of elasticity resulting in an adenolymphocele, or "hanging groin", have been described in some individuals as a result of microfilariae migration. A classification scheme was developed for the myriad skin manifestations of onchocercosis:

Acute Papular Onchodermatitis (APOD)

APOD presents as a scattered, pruritic, papular eruption, which may be transient and associated with erythema or swelling. Areas affected are predominantly the trunk and limbs. APOD is generally seen in younger populations in hyperendemic areas but may be seen in returning travelers who have acquired the infection.

Chronic Papular Onchodermatitis (CPOD)

CPOD consists of larger, scattered papules over the waist, buttocks and shoulders, which may be hyperpigmented. Pruritis is often severe and the resulting excoriation can predispose to secondary infections. CPOD is the most common manifestation of onchocerciasis in hyperendemic areas.

Lichenified Onchodermatitis (LOD)

Lichenified onchodermatitis represents a more localized hyper-reactive form of CPOD. The lesions are hyperpigmented plaques, which are usually asymmetrical, involving one limb with generally less numbers of microfilariae and onchocercomas. Regional lymphadenopathy is seen in these chronic lesions. LOD is referred to as "sowdah", Arabic for "black", in southern Saudi Arabia and Yemen, where it is endemic.

Atrophy (ATR)

The skin after prolonged infection becomes wrinkled, thin and atrophic, with loss of elasticity and hair. It most commonly occurs on the buttocks and limbs.

Depigmentation (DPM)

Depigmentation is a result of longstanding inflammation and scarring from onchocercosis infection. It is most commonly seen pretibially and has a patchy distribution, which has been referred to as "leopard skin".

Onchocercomas

Onchocercomas are nodules composed of coiled adult female and male worms that average 3 cm in diameter. They are usually found on bony prominences such as the pelvis, chest wall, head or limbs. The variation in location of these nodules between Central and South American and Africa appears to be related to the biting habits of the vector. In the Americas, *Simulium ochraceum* have a tendency to bite on the upper part of the body, whereas in Africa, *Simulium damnosum* typically bite below the waste line. Isolated onchocercomas caused by zoonotic species of *Onchocercoca* (usually *Onchocerca gutturosa*, a cattle nematode) have also been reported.

Visual Impairment and Blindness

The risk of impaired vision and blindness is associated with the proximity of the onchocercoma to the eyes, such as with head nodules, as well as disease burden.

Moreover, blindness appears to be geographically associated with the onchocercal strains in the West African savannahs as opposed to rainforest areas where more severe skin disease is present. Microfilarial migration from the conjunctiva into the cornea likely plays a role in the pathogenesis of ocular disease. Disease involvement in the eye may be present in the anterior and/or posterior segment. Anterior disease causes punctate keratitis, identified as small (0.5 mm), discrete areas of the superficial corneal stroma (snowflakes opacities), which are generally reversible. Other findings such as limbitis, chemosis, conjunctival injection and epiphora may accompany punctate keratitis. Iris and ciliary body involvement may lead to anterior uveitis, iridocyclitis and ultimately sclerosing keratitis, leading to severely impaired vision or complete blindness. With extensive inflammation and synechiae formation, seclusion- and occlusion- pupillae, secondary cataracts and glaucoma may also ensue. Posterior disease is caused typically by microfilariae that have entered the retina along posterior ciliary vessels, leading to chorioretinitis. Posterior disease can result in optic neuritis and lead to optic atrophy as a result of recurrent episodes of acute inflammation.

Other Disease Manifestations

Systemic signs and symptoms have been described in onchocercosis infections. Low body weight and diffuse musculoskeletal pain may occur. Growth arrest in severe disease, known as the Nakalonga syndrome, has been described in western Uganda. Onchocercosis has been linked to infertility, amenorrhea and spontaneous abortion.

Diagnosis

Presumptive diagnosis of onchocercosis usually can be made in a patient presenting with a travel history to an endemic area and the clinical syndrome of pruritis, dermatitis, ocular findings and skin nodules. The standard method of diagnosis of onchocerciasis involves skin snippings to identify microfilaria or the identification of macrofilariae in extirpated onchocercoma. The probability that skin snips will detect microfilaria relates to the parasite burden and the amount of samples taken. Sensitivities range from 80-100%, but lower sensitivities have been noted in areas with long-term ivermectin programs. Skin nodules are more likely to be seen in the setting of greater parasite burden, and the macrofilariae may be directly visualized following removal and examination of a nodule.

Indirect diagnosis may be made through the use of provocation methods, such as the Mazzotti reaction, originally described following the administration of 6 mg of oral diethylcarbamazine (DEC), characterized by fever, pruritis and is potentially sight threatening. More favorable is a topical application of DEC, referred to as the DEC-patch assay, which produces a local reaction with a raised, pustular rash in those harboring *Onchocerca*. The sensitivity of this method ranges from 30% to 92%, with specificities of 80 to 95%. The strength of the reaction was independent of the level of microfilaridermia, suggesting that this is a qualitative method. Such methods may play a greater role in screening endemic areas for recrudescence of infection.

The use of serologic studies and more recently polymerase chain reaction (PCR) methods have received particular attention because these methods are more sensitive and potentially less invasive than skin snips. Using ELISA and various onchocercal antigen "cocktails" have achieved a sensitivity ranging from 70-96% with a specificity of 98-100%. However, these serologic methods are unable to discern active from past infection, as well as cross-reactivity with other parasites.

PCR methods detect the parasite DNA sequence O-150, which is only found in *O. volvulus* and have shown a nearly 100% sensitivity and specificity when used on skin snips. Skin abrasion methods may reduce invasiveness and afford similar sensitivity. Newer PCR methods, such as an antigen detection dipstick, are even less invasive and are useful in the setting of lower parasitemia and disease prevalence. A
7 urine dipstick assay demonstrated a sensitivity and specificity near 100%.

Treatment: Current Therapy and Alternatives

Identification and removal of skin nodules can be curative in patients with a minimal exposure history, but its role is otherwise unclear. Nodulectomy in patients with head onchocercomata and evidence of ocular disease may reduce the likelihood of blindness and is therefore recommended.

Ivermectin

Ivermectin (Mectizan), released in 1987, is the most widely used treatment for onchocercosis. It primarily functions as a microfilaricide. A single dose of ivermectin clears the skin of microfilariae and suppresses further microfilaridermia for several months. However, since the macrofilariae are not very susceptible to ivermectin, the microfilarial burden is eventually replaced and after several months to a year, repeat treatment is necessary. This has formed the crux of onchocercosis mass treatment programs with ivermectin. Ivermectin leads to improvement in skin lesions and pruritis and retards the progression of anterior and posterior eye disease. Because high dose ivermectin has been associated with visual symptoms, it is usually safely dosed at 150 micrograms/kg, and repeated every 6-12 months until asymptomatic. A mild Mazzotti reaction may occur in 10-20% of patients receiving the first doses of ivermectin, but can be more severe in patients with higher microfilarial counts. Most reactions require only supportive care.

Diethylcarbamazine (DEC)

Diethylacarbamazine, a piperazine derivative, has microfilaricide activity against *O. volvulus*, but its use has gone into disfavor, because of the potentially sight-threatening inflammatory reaction that is induced from rapidly killing microfilariae. Ocular complications reported in the treatment of onchocercosis infection with DEC are constriction of visual fields, damage to the optic nerves, chorioretinitis, anterior uveitis and punctate keratitis. Additionally, DEC was shown to induce encephalopathy in patients co-infected with *Loa loa* who had very high microfilarial burdens.

Doxycycline and Wolbachia Endobacteria

Wolbachia species are Rickettsial bacteria that have been identified as "obligatory symbionts" presumably for fertility in many filarial species, in that they are required for all stages of embryogenesis. Targeting *Wolbachia* with doxycycline therapy dosed at 100 mg daily for 6 weeks demonstrated potent activity against embryogenesis of *Onchocerca*. In one study of bovine onchocercosis (*O. ochengi*), it even demonstrated partial macrofilaricide activity. Although doxycycline is contraindicated in some patient populations (children and pregnant or breastfeeding women), it may have a future role as prophylaxis, such as for those leaving an endemic area after a long stay, or aid in eradication efforts through more prolonged clearance of microfilariae.

There is yet to be developed an effective pharmacologic treatment for macrofilariae, which may survive for more than a decade, mandating long-term maintenance of ivermectin treatment programs. Suramin has activity against the macrofilaria of onchocerciasis, but has limited applicability due to its toxicity and intravenous administration route. Ivermectin may actually have some macrofilaricidal properties. Recent studies have shown it may exhibit macrofilarial activity, particularly if given at 3-monthly dosing regimens. Doxycycline's activity against *Wolbachia* endosymbionts in *Onchocerca* may play a potential role as a macrofilaride, as demonstrated in a bovine models.

Although an effective vaccine has yet to be discovered, animal models have suggested that effective immunity can be mounted against stage 3 larvae (L3) and microfilariae (Mf).

Prevention and Prophylaxis

Strategies toward eradication of onchocerciasis focus on the interruption of the onchocercal life cycle, whether through large scale vector control or microfilaricide treatment. The Onchocerciasis Control Programme (OCP) began in 1974 and was comprised of vector control in seven (eventually 11) West African countries with hyperendemic onchocercosis. These measures were intensified to include weekly, aerial larvicide spraying in a wide area of 700,000 km^2. Infection control models which focused on the identification of the Community Microfilarial Load (CMFL) through skin snip sampling indicated that the OCP was successful, allaying concerns that migration of black flies from outside areas would hamper efforts. A micro simulation model formed in collaboration with the Erasmus University of Rotterdam, ONCHOSIM, predicted that a 14 year span of a satisfactory vector control program would control onchocerciasis infection without recrudescence. Vector control was discontinued after 14 years in 1989 and epidemiologic studies almost a decade later demonstrated that the onchocercosis threat was largely eradicated except for small residual foci. The OCP concluded in 2002, after more than 25 years of activity. As testimony to its success, ocular disease from onchocercosis, widespread prior to the OCP, has not been observed.

With the donation of ivermectin by Merck in 1987, other programs were established in endemic areas and focused on mass distribution of ivermectin (Ivermectin Distribution Program (IDP)) with assistance from nongovermental development organizations (NDGOs), in Latin America and in Africa.

However, the potential of onchocercosis eradication in some countries remains unclear. The conclusion of an expert panel at the Conference on Eradication of Onchocerciasis which convened in Atlanta Georgia in 2002 was that the disease was not eradicable in Africa with the current methods due to several major barriers: the unlikelihood that ivermectin as a single intervention will interrupt transmission; programmatic challenges in an extensive endemic area with mobile vectors and infected humans; poor health infrastructure and political instability in these countries; *Loa loa* co-infections in many onchocerciasis endemic areas; and inadequate funds and political support to expand the coverage to hypoendemic areas. Recrudescence has occurred when programs were disrupted from civil unrest. For the Americas and possibly Yemen, onchocerciasis transmission may be interrupted with current programs due to vector characteristics and/or geographic isolation of foci.

Suggested Reading

1. Bianco AE. Onchocerciasis-river blindness. In: Macpherson CNL, Craig PS, eds. Parasitic Helminths and Zoonoses in Africa. London: Unwin Hyman Press, 1991:128-203.
2. Brattig NW, Bazzocchi C, Kirschning CJ et al. The major surface protein of wolbachia endosymbionts in filarial nematodes elicits immune responses through TLR2 and TLR4. J Immunol 2004; 173:437-45.
3. Burnham G. Onchocerciasis. Lancet 1998; 351:1341-46.
4. Burnham G, Mebrahtu T. Review: The delivery of ivermectin (Mectizan). Trop Med Int Health 2004; 9:A26-A44 SUPPL.
5. Cooper PJ, Nutman TB. Onchocerciasis. Current Treatment Options in Infectious Diseases 2002; 4:327-35.
6. Dadzie Y, Neira M, Hopkins DR. Final Report of the Conference on the eradicability of Onchocerciasis 2002.
7. Duke BOL. Evidence for macrofilaricidal activity of ivermectin against female Onchocerca volvulus: further analysis of a clinical trial in the Republic of Cameroon indicating two distinct killing mechanisms. Parasitology 2005; 130:447-53.
8. Gilbert J, Nfon CK, Makepeace BL et al. Antiobiotic chemotherapy of onchocerciasis: in a bovine model, killing of adult parasites requires sustained depletion of endosymbiotic bacteria (Wolbachia species). J Infect Dis 2005; 192:1483-93.
9. Hise AG, Gillette-Ferguson I, Pearlman E. The role of emdosymbiotic Wolbachia bacteria in filarial disease. Cell Microbiol 2004; 6:97-104.
10. Hoerauf A, Brattig N. Resistance and susceptibility in human onchocerciasis—beyond Th1 vs Th2. Trends Parasitol 2002; 18:25-31.
11. Hougard JM, Alley ES, Yameogo L et al. Eliminating onchocerciasis after 14 years of vector control: a proved strategy. J Infect Dis 2001; 184:497-503.
12. Murdoch ME, Hay RJ, Mackenzie CD et al. A Clinical classification and grading system of cutaneous changes in onchocerciasis. Br J Dermatol 1993; 129:260-9.
13. Murdoch ME, Asuzu MC, Hagan M et al. Onchocerciasis: the clinical and epidemiological burden of skin disease in Africa. Ann Trop Med Parasitol 2002; 9(3):283-96.
14. Paisier AP, van Oortmarssen GJ, Remme J et al. The risk and dynamics of onchocerciasis recrudescence after cessation of vector control. Bull World Health Organ 1991; 69:169-78.
15. Rodger FC. The movement of microfilariae of Onchocerca volvulus in the human eye from lid to retina. Trans Roy Soc Trop Med Hyg 1959; 53:138-41.
16. World Health Organization. Onchocerciasis (river blindness). Report from the fourteenth inter American conference on onchocerciasis, Atlanta, Georgia, United States. [Congresses] Weekly Epidemiological Record 2005; 80:257-60.
17. World Health Organization. Onchocerciasis and its control. Report of a WHO Expert Committee on Onchocerciasis Control 1995; 852:1-103.
18. Zimmerman PA, Guderian RH, Araujo E et al. Polymerase Chain Reaction-based diagnosis of Onchocerca volvulus infection: improved detection of patients with onchocerciasis. J Infect Dis 1994; 169:686-9.

Chapter 8

Loiasis

Murliya Gowda

Introduction

Loa loa, a filarial nematode, is the causative organism of Loiasis. More commonly known as the African eye worm, infection with this organism is characterized by transient swelling of subcutaneous tissues, known as Calabar swellings, which occur as the adult worms migrate.

Epidemiology

Loa loa is endemic to the central and western regions of Africa. In these areas as many as 13 million people are affected. In endemic areas, up to 30% of long-term inhabitants can be infected. The vector, the *Chrysops* fly, breeds in mud near shaded water sources of the rain forest. Rubber tree plantations are especially prone to infestation with *Chrysops*. The flies are typically more common during the rainy season and are attracted to movement, smoke, dark skin and clothing.

There are no nonhuman reservoirs of *L. loa*. There is a form of *Loa* that is seen in nonhuman primates, but, in contrast to human *L. loa* infection, the organism tends to have a nocturnal periodicity and the flies that transmit this infection bite at night.

Life Cycle

The mango fly, or tabanid fly, which belong to the genus *Chrysops*, transmits *L. loa*. *Chrysops silacea* and *Chrysops dimidiata* are the primary vectors. The female fly, which feeds during daylight hours, acquires the organism when it ingests a microfilariae-containing blood meal from an infected individual. The microfilariae, which are sheathed and contain three or more nuclei in the caudad end (Fig. 8.1), maintain a diurnal rhythm such that they remain in the capillaries of the lung and other organs at night, but during the day, circulate in the peripheral blood and are thus available for ingestion by the day-feeding flies. After the microfilariae are ingested, they penetrate the wall of the stomach, and enter the fat body. Over the course of the next 8-12 days, the microfilariae increase in length, mature to the infective larval form and then migrate to the fly's mouth. When the *Chrysops* fly takes its next meal, the larvae are released into the host. Over the course of the next 1 to 4 years, the adult worms develop within the subcutaneous tissues of the host. The female adult worms are typically larger, measuring 0.5 mm in width and 50 to 70 mm in length whereas the male worms are 0.4 mm wide and 30 to 35 mm long. The adult worms reproduce and deposit microfilariae in

Medical Parasitology, edited by Abhay R. Satoskar, Gary L. Simon, Peter J. Hotez and Moriya Tsuji.

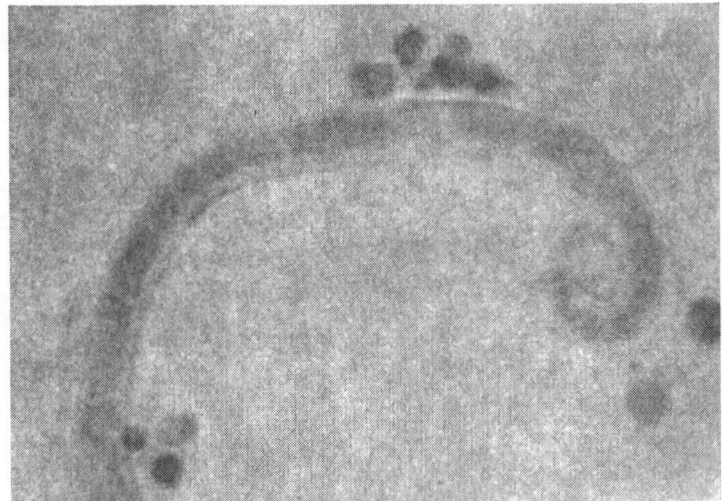

Figure 8.1. *Loa loa* microfilaria in blood.

8

the peripheral blood during the day thus completing the cycle. Adult worms can survive for up to 17 years in host tissues. The transparent adult worms can be seen migrating underneath the conjunctiva and through subcutaneous tissues causing migratory swelling and angioedema known as "Calabar swellings."

Immune Response

Immunological responses to parasite infection can be divided into three groups. Individuals who have high levels of microfilariae in the blood, but have weak responses to the parasite, characterize the first group. A second group includes individuals who develop the Calabar swellings and angioedema. Secretions from the adult worms and microfilarial antigens trigger host IgE responses and IL-5 production, resulting in host inflammatory responses and eosinophilia. The third subset of patients, typically live in hyperendemic areas and have neither microfilariae nor Calabar swellings. This group is thought to have protective immunity. Native individuals with *L. loa* infection tend to be asymptomatic whereas the hypersensitivity reactions are more common in expatriates and long-term visitors to the endemic areas.

Clinical Manifestations

Infection with *L. loa* can cause a broad range of symptoms from swelling in the face, extremities, and periorbital areas to more serious complications such as renal failure, central nervous system involvement. People infected with *L. loa* may not present with symptoms for years. Infected individuals may develop pain and pruritis along the path of the traveling adult worms. An area of nonpitting edema (Calabar of fugitive swelling) may soon develop along the migratory path of the worm and can last from two days to several weeks. The swellings are believed to be a hypersensitivity reaction secondary to antigenic substances released by the parasite. Repeated episodes of swelling or angioedema are common. Typically, the extremities, ankles, wrists and the periorbital areas are affected.

Eye involvement is the most well known clinical manifestation of *L. loa* infection. Movement of the adult worm under the conjunctiva causes conjunctivitis, as well as pruritis, pain and edema of the eyelid. Occasionally, the transparent adult worm can be seen migrating under the conjunctiva.

Rare ocular complications include macular deterioration and retinal artery occlusion. Other infrequent complications of loiasis infection include lymphadenitis, pulmonary infiltrates, hydroceles and joint involvement. Occasionally, intense swelling can cause nerve compression and subsequent neuropathies. The median nerve is the most common site of neural involvement and can lead to the development of carpal tunnel syndrome. With diethylcarbamazine therapy, the dying organism can activate host inflammatory responses, worsening swelling and nerve compression, thereby intensifying neurological symptoms. Another uncommon manifestation of loiasis is endomyocardial fibrosis, which has been epidemiologically linked to regions where *L. loa* is prevalent. One study describes a patient with loiasis and biopsy proven endomyocardial fibrosis who was treated with diethylcarbamazine with a subsequent decrease in eosinophilia and antifilarial antibodies.

More serious complications involving the kidneys may also occur in loiasis. Renal disease may worsen with diethylcarbamazine treatment, but are typically not permanent. Damage to the glomeruli as a result of immune complex deposition or direct injury from the renal filtration of microfilariae can lead to proteinuria and hematuria. Rarely, microfilariae are evident in the urine.

Central nervous system involvement, particularly meningoencephalitis, can occur in individuals with a high microfilarial burden; typically greater than 2,500 microfilaria/ml. Symptoms such as headaches can progress to meningoencephalitis or death. Seizures have also been reported. Microfilariae can be found in the cerebrospinal fluid in individuals with a high organism burden. Furthermore, treatment with diethylcarbamazine can exacerbate CNS symptoms. Neurological symptoms have also been described in patients who were treated for onchocerciasis with ivermectin, but were also co-infected with *L. loa*.

Diagnosis

Loiasis should be considered as a potential diagnosis in individuals from endemic regions of Africa who have swelling in the face, extremities, and periorbital areas. Eosinophilia should also raise the index of suspicion for this diagnosis. Elevated IgE levels are commonly observed in symptomatic patients and may be detected despite a paucity of microfilariae in peripheral blood samples. The diagnosis is established by detection of the microfilariae from a daytime sample of peripheral blood on Giemsa or Wright's stain. Characteristically, the microfilariae are sheathed and have three or more terminal nuclei (Fig. 8.1). In patients with low-level parasitemia, concentration methods such as microfiltration can be used to improve diagnostic sensitivity. Extraction of the adult worm from the conjunctiva or subcutaneous tissues is also diagnostic.

In the past, serological testing was infrequently used due to cross reactivity with other filarial antigens, especially in endemic populations in whom the incidence of antifilarial antibodies may be quite high. Indeed, in hyperendemic regions, up to 95% of population have developed antibodies to *L. loa* antigens by the age of two. However, in travelers to endemic areas who have unexplained eosinophilia, but no detectable microfilariae on peripheral smear examination, the presence of antifilarial antibodies may be sufficient evidence to warrant therapy for loiasis. More specific serological tests are available through research institutions. IgG4 anti-*Loa* antibodies have been used as a marker of infection, and a PCR testing of blood has also been developed.

Treatment, Prevention and Control

Diethylcarbamazine is the treatment of choice for loiasis. It is effective against both adult worms and microfilariae. Multiple courses diethylcarbamazine are frequently required to completely eradicate infection since recurrences have been documented several years after initial treatment.

Administration of diethylcarbamazine is associated with a variety of adverse effects which are believed to be secondary to antigens released from dying microfilariae. Severity of symptoms is related to organism burden. Serious renal and central nervous system complications tend to occur with higher organism burden. Milder adverse effects include fever, arthralgias and Calabar swellings, which tend to occur during the first few days of therapy. To reduce the incidence of treatment-induced complications, a test dose of 1 mg/kg is given, followed by escalating doses (1 mg/kg three times a day on day 2 and 1-2 mg/kg three times a day on day 3), until a maximum dose of 8-10 mg/kg/d is tolerated. A 21-day course of diethylcarbamazine at 8-10 mg/kg/d is standard treatment. Alternatively, some clinicians recommend 50 mg on day 1, 50 mg three times a day on day 2, 100 mg three times a day on day 3, and 8 mg/kg/d in three doses on days 4 through 14.

Antihistamine and corticosteroid therapy can also be administered to minimize symptoms. Prior to initiating diethylcarbamazine, ivermectin or albendazole can be used to reduce the microfilaria or adult worm burden respectively. In patients with a high organism levels, cytopheresis has been suggested as a means of reducing microfilaria burden prior to treatment with diethylcarbamazine. Adult worms in the eye must be removed surgically.

In those areas with extremely high prevalence, mass treatment with diethylcarbamazine may help to reduce transmission to uninfected individuals. Periodic chemotherapy of populations at risk has been utilized in sub-Saharan Africa, but the side effects of diethylcarbamazine may limit its usefulness in this setting. Albendazole or ivermectin may be more useful alternatives for such unmonitored therapy, although the latter is associated with encephalopathy in individuals with extremely high levels of microfilaremia.

Prevention methods include the use of insecticides and clearing forests in order to control the vector population. Insect repellants and protective clothing also provide added protection. In endemic areas, diethylcarbamazine prophylaxis is recommended for visitors. A weekly dose of 300 mg is sufficient and over a course of two years was not associated with serious side effects.

Selected Readings

1. Akue JP, Egwang TG, Devaney E. High levels of parasite-specific IgG4 in the absence of microfilaremia in Loa loa infection. Trop Med Parasitol 1994; 45:246.
2. Boussinesq M, Gardon J, Gardon-Wendel N et al. Clinical picture, epidemiology and outcome of Loa-associated serious adverse events related to mass ivermectin treatment of onchocerciasis in Cameroon. Filaria J 2003; 2:S4.
3. Carme B, Boulesteix J, Boutes H et al. Five cases of encephalitis during treatment of loiasis with diethylcarbamazine. Am J Trop Med Hyg 1991; 44:684-90.
4. Loiasis. In: Conner DH, Neafie RC, Meyers, WM, eds. Pathology of Tropical and Extraordinary Disease, Volume 2. Washington DC: Armed Forces Institute of Pathology, 1976:356-9.
5. The Filariae. In: Cupp EW, Jung RC, Beaver PC, eds. Clinical Parasitology, 9th Edition. Philadelphia: Lea & Ferbiger 1984:350-1,377-800.
6. Loa loa. In: Despommier DD, Gwadz RW, Hotez PJ, Knirsch CA, eds. Parasitic Diseases, 4th edition. New York: Apple Tree Productions, LLC 2000:143-6.
7. Horse flies, deer flies and snipe flies. In: James MT, Harwood RF, eds. Herm's Medical Entomology, 6th Edition. Toronto: Macmillan Company, 1969:228-9.
8. Klion AD, Nutman TB. Loiasis and mansonella infections. In: Guerrant RL. Walker DH, Weller PF, eds. Tropical Infectious Disease Principles, Pathogens and Practice. Vol 2. Philadelphia: Churchill Livingstone1999:861-72.
9. Klion AD, Otteson EA, Nutman TB. Effectiveness of diethylcarbamazine in treating loiasis acquired by expatriate visitors to endemic regions: long term follow-up. J Infect Dis 1994; 169: 604-10.
10. Mackenzie CD, Geary TG, Gerlach JA. Possible pathogenic pathways in the adverse clinical events seen following ivermectin administration to onchocerciasis patients. Filarial J 2003; 2:S5.
11. Nutman TB, Miller KD, Mulligan M et al. Diethylcarbamazine prophylaxis for human loiasis. Results of a double blind study. N Engl J Med 1988; 319:752-6.
12. Nutman TB, Miller KD, Mulligan M et al. Loa loa infection in temporary residents of endemic regions: recognition of a hyperresponsive syndrome with characteristic clinical manifestations. J Infect Dis 1986; 154:10-8.
13. Nutman TB, Zimmermann PA, Kubofcik J et al. Elisa-based detection of PCR products. A universally applicable approach to the diagnosis of filarial and other infections. Parisitol Today 1994; 10:239-43.
14. Ottenson EA. Filarial infections. Infect Dis Clin North Amer 1993; 7:619-33.
15. Loiasis. In: Strickland GT, ed. Hunter's Tropical Medicine and Emerging Infectious Diseases, 8th Edition. Philadelphia: W.B. Saunders, 2000:754-6.

CHAPTER 9

Dracunculiasis

David M. Parenti

Dracunculiasis (also know as dracontiasis) is caused by the "guinea worm" *Dracunculus medinensis*. It has been described in humans since antiquity with references to this infection being noted in the Bible and ancient Greek and Roman texts. A calcified adult worm has also been noted by X-ray in an Egyptian mummy and there are descriptions of the disease in ancient papyrus texts. It is transmitted by ingestion of a fresh water copepod (*Mesocyclops, Metacyclops, Thermocyclops*) containing infective larvae and is endemic only in areas where this intermediate host is found. Transmission occurs in underdeveloped regions with limited access to a safe water supply. Debilitation as a result of guinea worm infection is common, resulting in chronic pain, acute or chronic infection, impaired joint mobility and occasionally tetanus. Short term and long term disability from dracunculiasis leads to lost work days and decreased economic productivity.

Phylogenetically *Dracunculus* is related to the helminths in the Order Spirurida, which also contains the filariae. It is felt to be primarily a human parasite; but other *Dracunculus* species may infect humans and other animals. Recently ribosomal 18S rRNA sequencing has been able to distinguish at least some of these species and a worm identical to the human parasite has been obtained from a dog in Ghana. There is no clear evidence that animals are important reservoir hosts.

Epidemiology

Two important conditions need to be satisfied for transmission of *Dracunculus* to occur: emergent guinea worm lesions discharging larvae need to be in contact with drinking water and the intermediate copepod host needs to be present in the water supply. Contamination of step wells, cisterns and ponds is common in endemic areas. Transmission in ponds is highest just before the rainy season, when the density of copepods may be the highest.

Dracunculus infections were once distributed throughout equatorial Africa from the northwest (Senegal, Burkina Faso, Ghana, Nigeria) to Sudan and Uganda. It was also endemic in Iran, Afghanistan, Pakistan, India and the southernmost new Soviet republics. An estimated 3 million cases occurred worldwide in 1986. Transmission often occurs in remote areas with limited water supplies, or where larval contamination can easily occur. Although transmission does not occur in the United States, rare importations have occurred.

More recently, following the establishment of the WHO global eradication campaign, endemic regions have been limited to those listed in Table 9.1.

Medical Parasitology, edited by Abhay R. Satoskar, Gary L. Simon, Peter J. Hotez and Moriya Tsuji.

Table 9.1. Number of cases of Dracunculiaisis January-September 2005 (CDC)

Sudan	5,008
Ghana	2,936
Mali	475
Nigeria	116
Niger	66
Togo	58
Ethiopia	37
Burkina Faso	25
Cote d'Ivoire	9
Uganda	6
Benin	1

Life Cycle

The copepod is the intermediate host for *Dracunculus* and measures 1-3 mm in size. Following ingestion of an infected copepod the third stage infective larva is released. Whether the presence of reduced gastric pH is important in this process is unclear. The larva then penetrates through the intestinal mucosa and migrates to connective tissues, especially in the abdominal wall and thoracic wall and retroperitoneum, where it develops into an adult male or female worm.

The adult female worms are cylindroidal, 1-2 mm in width and up to 800 mm in length; the males much smaller at about 40 mm in length. The female is likely fertilized in this site before migration begins to the subcutaneous tissues. By 8-9 months after ingestion the uterus of the female worm becomes filled with eggs which then develop into first stage larvae. Full maturation takes up to 12 months, contributing to its seasonal endemicity. The gravid worm, containing 1-3 million first stage larvae, then embarks on a period of subcutaneous migration eventually taking it to the surface.

After migration the anterior end of the worm approaches the dermis, where first a papule and then a blister is formed. The lesions occur predominately on the distal lower extremities. On contact with fresh water the blister ruptures and first stage larvae (rhabditoid) are released as the uterus prolapses either through the mouth or body wall. The motile larvae are 15-25 µm × 500-750 µm in size and are subsequently ingested by the copepods. They penetrate through the intestinal wall, perhaps through use of a dorsal tooth that has been described and develop in the coelomic cavity. The larvae molt twice in the copepod over a period of 2-4 weeks and remain dormant as third stage (infective) larvae. Human infection again takes place after ingestion of infected copepods.

Clinical Manifestations

Initial ingestion of the infected copepod does not usually provoke any symptoms although urticarial reactions sometimes occur. Clinical disease is a direct result of the adult female worm which migrates in the subcutaneous tissue until stopping to discharge the larvae through the skin. The individual may notice a palpable or migrating worm in up to one-third of cases. Others may complain of allergic symptomatology prior to worm emergence: urticaria, infra-orbital edema, fever, dyspnea. The majority present with local signs of emergent worms.

Emergent lesions are primarily located in the lower extremities (> 90%), which are also the body areas most likely to come in contact with fresh water and thus allow completion of the life cycle. Worms may also emerge from the upper extremity, trunk, or head. Errant worms have appeared rarely in unusual locations such as the epidural space, testicle, orbit and eyelid.

Each individual usually has 1-3 worms emerge, but rarely multiple lesions may occur. The blister is usually painful and 1-3 cm in size and may also be pruritic. After emergence an ulcer forms and this remains an open wound until the worm

is expelled. The dying worm provokes a brisk inflammatory response which may lead to abscess formation. Abscesses may be seen in up to 10% of non-emergent worms. If the female emerges near or in a joint, usually the knee or ankle, a frank arthritis may ensue and larvae may be recovered from joint fluid. Nonemergent worms remain in the tissues to be contained by the host response and to become calcified.

Secondary infection is also a common complication of dracunculiasis (in up to 50% of cases) and a cause of significant morbidity. Infections may be caused by skin flora such as staphylococci or streptococci or enteric organisms such as *Escherichia coli*. Chronic skin ulcers of the extremities are an ideal portal of entry for *Clostridium tetani* and clinical tetanus is a not infrequent complication in endemic areas. In one study in Nigeria guinea worm lesions were the third most common portal of entry for the development of tetanus.

Diagnosis

After the appearance of the skin ulcer and worm, the diagnosis can be certain. Discharged material from the gravid uterus will reveal larvae characteristic of *Dracunculus*. Characteristic calcifications can also be identified in areas where the worms have died and calcified and represent a "sarcophagus around a long departed parasite". These are most common in the lower extremities where they lie along tissue planes in a linear fashion; but have also been noted in the upper extremities and in the trunk where they may have a coiled appearance.

In prepatent infection immunodiagnosis with ELISA or dot-ELISA may be useful, but there may also be cross reactivity with other helminths such as *Wuchereria* and *Onchocerca*. Eosinophilia may be present, particularly just prior to emergence. The differential diagnosis of dracunculiasis includes other migratory helminths such as *Ancylostoma braziliense* (cutaneous larval migrans) and *Stronglyoides* (larva curens). Nodular lesions reminiscent of unerupted worms may be seen with adult worms of *Onchocerca*, the tapeworm *Spirometra* (sparganosis) and cutaneous myiasis. Extracted adult worms of *Onchocerca* have been confused with *Dracunculus* in the Central African Republic, although they are smaller in diameter (0.3 mm vs 1-2mm).

Treatment

Treatment of dracunculiasis is generally focused on mechanical removal of the worm and treatment of secondary infection. The traditional method has been to wind the protruding worm on a stick and with gentle traction to remove several inches at a time. Extraction of the worm may occasionally be difficult. Incomplete removal or breakage of the worm with spillage of contents can lead to a more significant inflammatory response and appears to increase the risk of secondary infection. At least one study has suggested that removal of unerupted worms might be more easily accomplished with less disability. In this study in India 161 patients had surgical removal of worms prior to eruption, with an improvement in average time of disability from 3 weeks to 1 week.

Several antiparasitic agents have been studied in the treatment of dracunculiasis in humans and in animal models, including thiabendazole, mebendazole, niridazole, metronidazole and ivermectin, with little evidence for a cidal effect on the adult worms. A variety of treatment outcomes have been evaluated including time to worm expulsion, prevention of emergence of new lesions and ulcer healing. It has been suggested that metronidazole and some of the anthelminthics may modulate the inflammatory response and facilitate worm removal. Ivermectin has been tested in a single-blind placebo-controlled trial of 400 adults at risk for guinea worm infection. There was no effect on migration of the worms or prevention of emergent lesions. In a randomized, single blind, controlled trial topical antimicrobials have been shown to reduce the rate of secondary infection and improve wound healing.

Prevention

Because of the lack of effective anthelminthic agents, mass chemotherapy programs for treatment or prevention have not been feasible. Therefore the focus has been on elimination of the copepod intermediate host and establishment of safe supplies of potable water. Improved water sanitation can be accomplished by establishing a piped in water supply, construction of bore or tube wells, or other measures to prevent contamination. Implementation of individual or communal nylon or cloth filters (100-200 μm pore size) has also been effective in preventing transmission. Eradication of the intermediate host through water treatment with cyclopiscides such as the organophosphate temephos has also been used in many eradication programs.

Dracunculiasis is an ideal candidate for eradication: the disease is easy to diagnose, the adult worms have a limited lifespan, it has only a human host, it is prevalent in a limited geographic area, the intermediate host is not mobile and treatment of the intermediate host is relatively inexpensive. Eradication programs have included intensive cases finding, health education programs, utilization of nylon or cloth filters and application of cyclopiscides.

The campaign for global eradication of dracunculiasis began in 1980 as part of the International Drinking-Water Supply and Sanitation Decade (CDC) and was expanded by the WHO in 1986. Substantial progress has been made with a 98% decline in cases worldwide. Dracunculiasis eradication has been accomplished in Pakistan (1994) and India (1997). Those countries remaining with continued transmission are confined to Africa. Although many countries have continued to show a decline in cases during 2005, an increased number of cases has been noted in Mali and Ethiopia. Control measures in the Sudan have been particularly difficult because of sociopolitical unrest.

Suggested Reading

1. Beaver PC, Jung RC, Cupp EW. The Spirurida: Dracunculus and others. Clinical Parasitology, 9th Edition. Philadelphia: Lea & Febiger, 1984:335-40.
2. Belcher DW, Wunapa FK, Ward WB. Failure of thiabendazole and metronidazole in the treatment and suppression of guinea worm disease. Am J Trop Med Hyg 1975; 24:444-46.
3. Bimi L, Freeman AR, Eberhard ML et al. Differentiating Dracunculus medinensis from D. insignis, by the sequence analysis of the 18S rRNA gene. Ann Trop Med Parasitol 2005; 99:511-17.
4. Bloch P, Simonsen PE, Weiss N, Nutman TB. The significance of guinea worm infection in the immunological diagnosis of onchocerciasis and brancroftian filariasis. Trans R Soc Trop Med Hyg 1998; 92:518-21.
5. Cairncross S, Muller R, Zagaria N. Dracunculiasis (Guinea worm disease) and the eradication initiative. Clin Microbiol Rev 2002; 15:223-346.
6. CDC. Imported Dracunculiasis—United States, 1995 and 1997. MMWR 1998; 47:209-11.
7. CDC/WHO. Guinea Worm Wrap-Up #157, 10/26/05. http://www.cdc.gov/ncidod/dpd/parasites/dracunculiasis/wrapup/157.pdf
8. Cockshott P, Middlemiss H. Clinical Radiology in the Tropics. Edinburgh: Churchill Livingstone, 1979: 5-6.
9. Cox FEG. History of human parasitology. Clin Microbiol Rev 2002; 15:595-612.
10. Drugs for parasitic infections. The Medical Letter 2002:1-12.
11. Eberhard ML, Melemoko G, Zee AK et al. Misidentification of Onchocerca volvulus as guinea worm. Ann Trop Med Parasitol 2001; 95:821-26.
12. Hopkins DR, Azam M. Eradication of dracunculiasis from Pakistan. Lancet 1995; 346:621-24.
13. Hours M, Cairncross S. Long-term disability due to guinea worm disease. Trans R Soc Trop Med Hyg 1994; 8: 559-60.
14. Issaka-Tinorgah A, Magnussen P, Bloch P, Yakubu A. Lack of effect of ivermectin on prepatient guinea-worm: a single blind, placebo-controlled trial. Trans R Soc Trop Med Hyg 1994; 88:346-48.
15. Kaul SM, Sharma RS, Verghese T. Monitoring the efficacy of temephos application and use of fine mesh nylon strainers by examination of drinking water containers in guinea worm endemic villages. J Commun Dis 1992; 24:159-63.
16. Magnusssen P, Yakubu A, Bloch P. The effect of antibiotic- and hydrocortisone-containing ointments in prevention secondary infections in guinea worm disease. Am J Trop Med Hyg 1994; 51:797-79.
17. Muller R. Dracunculus and dracunculiasis. Adv Parasitol 1971; 9:73-151.
18. Neafie RC, Connor DH, Meyers WM. Dracunculiasis. In: Binford CH, Connor DH, eds. Pathology of Tropical and Extraordinary Diseases. Washington, DC: AFIP, 1976:397-401.
19. Rohde JE, Sharma BL, Patton H et al. Surgical extraction of guinea worm: disability reduction and contribution to disease control. Am J Trop Med Hyg 1993; 48:71-76.
20. Sullivan JJ, Long EG. Synthetic-fibre filters for preventing dracunculiasis: 100 versus 200 micrometres pore size. Trans R Soc Trop Med Hyg 1988; 82:465-66.
21. WHO. Weekly Epidemiological Record. 5/13/2005; 80: 165-76.

Chapter 10

Cutaneous Larva Migrans: "The Creeping Eruption"

Ann M. Labriola

Background

Cutaneous larva migrans (CLM), frequently termed "creeping eruption," is a parasitic skin infection that is caused by the filariform larvae of various animal hookworm nematodes. CLM has a worldwide distribution wherever humans have had skin contact with soil contaminated with infected animal feces. The disease most commonly occurs in subtropical and tropical regions, but may also occur in temperate climates particularly during the summer months and during rainy seasons. In the United States, CLM is seen in individuals living in the southeastern states, Florida and the Gulf Coast states, in travelers returning from sandy beaches and in military personnel returning from tropical postings. Children develop the disease by walking barefoot in sandy areas or playing in dirt or sandboxes that contain infected animal feces. Electricians, plumbers, utility workers and pest exterminators who have contact with soil under houses or at construction sites are at risk, as are fisherman, hunters, farmers and gardeners who handle contaminated soil.

Causative Agents

Ancylostoma braziliense, a hookworm of wild and domestic dogs and cats is the most commonly identified etiologic agent of CLM. *A. braziliense* is distributed throughout the tropics and subtropics, especially on the warm sandy beaches of the southeastern and Gulf Coast states, the Caribbean and southeast Asia where dogs and cats are permitted to defecate. *Ancylostoma caninum* (dog hookworm), *Ancylostoma ceylanicum* (dog and cat hookworm) and *Ancylostoma tubaeforme* (cat hookworm) may also produce cutaneous lesions. Whereas the skin lesions of *A. braziliense* may persist for months, the skin lesions that are seen with other *Ancylostoma* species resolve within several weeks.

A number of other nematodes have been associated with CLM-like lesions. *Strongyloides* species such as *S. papillosus, S. westeri, S. stercoralis, S. procyonis* and *S. myopotami,* nematodes found in the small intestine of mammals, migrate more quickly in human skin than hookworm larva to form skin eruptions called *larva currens* ("racing larva"). *Gnathostoma* species, a dog and cat roundworm found in Southeast Asia and *Dirofilaria repens,* a filarial nematode of dogs,

Medical Parasitology, edited by Abhay R. Satoskar, Gary L. Simon, Peter J. Hotez and Moriya Tsuji.

cats and wild carnivores, can travel through the skin and cause dermatitis, swelling and subcutaneous nodules. *Uncinaria stenocephala* (dog hookworm) and *Bunostomum phlebotomum* (cattle hookworm) may also cause cutaneous disease. *Ancylostoma duodenale* and *Necator americanus,* human hookworms, that, upon larval migration in the skin, produce a pruritic dermatitis ("ground itch") similar to CLM of animal hookworms. Free-living nematodes such as *Peloderma strongyloides* have also been described as a rare cause of cutaneous infection in humans.

Life Cycle

Unlike the classical hookworm life cycle, the cycle of the organisms that are typically associated with CLM is much abbreviated. *A. braziliense* third stage infective larvae (L3) enter the epidermis and migrate laterally in epidermis, generally unable to penetrate the basement membrane of the epidermal dermal junction. Humans are thus incidental dead-end hosts.

Disease Signs and Symptoms

The most common portals of entry by the larvae are the exposed areas of the body such as the dorsum of the feet, lower legs, arms and hands. The buttocks, thighs and abdomen also may be involved, probably due to lying directly on contaminated sand. CLM also occurs in the interdigital spaces of the toes, the anogenital region, knees and rarely on the face. Symptoms usually occur within the first hours of larval infection, although, occasionally, the larvae may lie dormant for several weeks. The first sign of infection is a stinging, tingling or burning sensation at the site of larval entry, followed by the development of an edematous, erythematous pruritic papule. The larvae migrate laterally from the papule and form raised 2-4 mm wide serpinginous tracks. These tracks mark the migratory route of the larvae and may advance from a few millimeters to several centimeters each day, hence the name "creeping eruption." As they burrow, the larvae produce hydrolytic enzymes that provoke the intense inflammatory reaction that characterizes CLM. In heavy infections, an individual may have hundreds of tracts. Severe inflammatory reactions may cause such an intense pruritis that individuals are unable to sleep, develop anorexia and even become psychotic. Secondary bacterial infections such as impetigo or cellulitis from scratching the skin are common complications.

CLM is a usually a self-limited disease. The larvae, unable to complete their life cycle, die in the epidermis within several weeks to months if left untreated. There are reports of larvae that have migrated for over one year. The skin lesions ulitmately resolve although there may be scarring. Rarely larva may migrate past the dermis to cause myositis, pneumonitis (Loffler's syndrome) or an eosinophilic enteritis (viz., *A. caninum*) of the small intestine. A pruritic folliculitis is another uncommon form of CLM that presents with serpiginous tracks interspersed with papules and pustules confined to a particular area of the body, usually the buttocks.

Diagnosis

The diagnosis of CLM is based on a history of exposure and the characteristic clinical appearance of the skin eruption. There may be a delay in diagnosis in chronic cases due to superimposed allergic dermatitis, secondary bacterial infection or lack of recognition of CLM by health care personnel. Peripheral eosinophilia and increased IgE levels are found in a minority of patients. Skin biopsies are usually not effective at establishing the diagnosis. Histologic examination usually reveals only an eosinophilic infiltrate or a nonspecific inflammatory response, although, occasionally, a skin biopsy taken at the leading edge of the track may contain a larva trapped in a follicular canal, stratum cornea or dermis.

The differential diagnosis of CLM should also include cercarial dermatitis ("swimmer's itch"), migratory myiasis, scabies, jelly fish sting, photoallergic dermatitis, epidermal dermatophytosis, erythema chronicum migrans of Lyme disease and photoallergic dermatitis. Vesicular lesions may be mistaken for viral infections or phytophotodermatitis.

Treatment

Oral albendazole (400 mg daily for 1-3 days) or ivermectin (200 µg/kg daily for 1-2 days) are the drugs of choice for the treatment of CLM. Oral thiabendazole is less effective than either albendazole or ivermectin. Whereas the cure rate with oral thiabendazole was 87% after 3-4 consecutive days of treatment, a single dose of ivermectin was effective in 81-100% of patients. Most studies of albendazole have shown cure rates 92-100%. A single randomized study of single dose ivermectin versus albendazole revealed cure rates of 100% for ivermectin with no relapses compared to a 90% cure rate with albendazole. However, half of the albendazole-treated patients relapsed. With treatment, symptoms of pruritus improve within 24-48 hours and the skin lesions usually resolve within one week. Thiabendazole is poorly tolerated compared to either albendazole or ivermectin. Side effects of thiabendazole include nausea, vomiting, headache and giddiness.

Treatment with topical thiabendazole is effective in treating patients with a few lesions. This approach avoids systemic side effects, but is less convenient and requires multiple applications. Cryotherapy or surgical excision is painful and not very effective because the larvae may be several centimeters beyond the end of the visible track. Its use is limited to pregnant women because of the uncertain teratogenic potential of ivermectin or albendazole.

Prevention and Prophylaxis

CLM can be prevented by avoiding skin contact with soil contaminated with infected animal feces. In geographic areas where the infection is endemic, one should avoid touching soil with bare hands, walking barefoot or lying directly on moist, warm, shady soil or warm, dry, sandy beaches protected from tidal movement. Beaches should also be kept free of dogs and cats. Pets should be examined and treated when infected with parasites and should receive periodic prophylaxis. Pet feces should be disposed of in a sanitary manner and sandboxes should be kept covered.

10

Suggested Reading

1. Albanese G, Venturi C. Albendazole: A new drug for human parasitoses. Dermatol Clin 2003; 21:283-90.
2. Blackwell V, Vega-Lopez F. Cutaneous larva migrans: clinical features and management of 44 cases presenting in the returning traveler. Br J Dermatol 2001; 145:434-7.
3. Bouchaud O, Houzé H, Schiemann R et al. Cutaneous larva migrans in travelers: A prospective study, with assessment of therapy with ivermectin. Clin Infect Dis 2000; 31:493-8.
4. Caumes E. Treatment of cutaneous larva migrans. Clin Infect Dis 2000; 30.5:811-4.
5. Caumes E, Ly F, Bricaire F. Cutaneous larva migrans with folliculitis: report of seven cases and review of the literature. Br J Dermatol 2002; 146:314-6.
6. Davies HD, Sakuls P, Keystone JS. Creeping eruption. A review of clinical presentation and management of 60 cases presenting to a tropical disease unit. Arch Dermatol 1993; 129:588-91.
7. Despommier DD, Gwadz RW, Hotez PJ et al. Parasitic Diseases. Clinical Parasitology; A Practical Approach, 4th ed. Philadelphia: Saunders, 1997.
8. Douglass MC. Cutaneous larva migrans. eMedicine.com. http://www.emedicine.com/derm/topic91.htm. Accessed 2005.
9. Edelglass JW, Douglass MC, Stiefler R et al. Cutaneous larva migrans in northern climates. A souvenir of your dream vacation. J Am Acad Dermatol 1982; 7.3:353-8.
10. Guill M, Odom R. Larva migrans complicated by Loffler's syndrome. Arch Dermatol 1978; 114:1525-6.
11. Herbener D, Borak J. Cutaneous larva migrans in northern climates. Am J Emerg Med 1988; 6.5:462-4[Medline].
12. Hotez PJ, Brooker S, Bethony JM et al. Current concepts: hookworm infection. N Engl J Med 2004; 351:799-807.
13. Jelinek T, Maiwald H, Nothdurft HD et al. Cutaneous larva migrans in travelers: synopsis of histories, symptoms and treatment of 98 patients. Clin Infect Dis 1994; 19.6:1062-6.
14. Jones CC, Rosen T, Greenberg C. Cutaneous larva migrans due to pelodera strongyloides. Cutis 1991; 48.2:123-6.
15. Kelsey DS. Enteric nematodes of lower animals transmitted to humans: zoonoses, (monogram online). In: Baron S, ed. Medical Microbiology, 4th ed. New York: Churchill Livingston; 1996. http://www.ncbi.nlm.nih.gov/books/bv.fcgi?call=bv.View..ShowTOC&rid=mmed.TOC&depth=10. Accessed 2005.
16. Kwon IH, Kim HS, Lee JH et al. A serologically diagnosed human case of cutaneous larva migrans caused by ancylostoma caninum. Korean J Parasitol 2003; 41:233-7.
17. Meyers WM, Neafie RC. Creeping eruption. In: Binford CH, Connor DH, eds. Pathology of Tropical and Extraordinary Diseases, 2nd Edition. Armed Forces Institute of Pathology, 1976:437-9.
18. Richey TK, Gentry RH, Fitzpatrick JE et al. Persistent cutaneous larva migrans due to ancylostoma species. South Med J 1996; 89.6:609-11.
19. Van den Enden E, Stevens A, Van Gompel A. Treatment of cutaneous larva migrans. N Engl J Med 1998; 339.17:1246-7.
20. Wang J. Cutaneous larva migrans. eMedicine.com. 16 March 2005. http://www.emedicine.com/ped/topic1278.htm. Accessed 2005.

CHAPTER 11

Baylisascariasis and Toxocariasis

Erin Elizabeth Dainty and Cynthia Livingstone Gibert

Baylisascariasis

Background

Causative Agent

Baylisascariasis is caused by the nematode parasite *Baylisascaris procyonis*. The North American raccoon (*Procyon lotor*) is the definitive host for *B. procyonis*. *B. procyonis* is also recognized as one of the most common causes of visceral larva migrans in animals, causing infection in over 100 species other than raccoons. Humans are infected as accidental intermediate hosts.

Geographical Distribution/Epidemiology

Raccoons are native to the Americas from Canada to Panama and have become increasingly concentrated in urban areas in recent years. *B. procyonis* is endemic in raccoons in most regions of the United States, with a higher prevalence on the West Coast, Midwest and Northeastern regions.

While the geographic distribution of the North American raccoon has been well established, the distribution of *B. procyonis* itself is not fully known. This uncertainty is due to the large abundance of eggs shed by raccoons that survive in the environment for extended periods of time and the large number of species that *B. procyonis* can potentially infect as accidental hosts.

Life Cycle and Mode of Transmission

Raccoons are infected by *B. procyonis* in one of two ways, depending on the age of the raccoon. Juvenile raccoons are typically infected by the ingestion of eggs in the environment. Younger raccoons have a higher prevalence of infection and have typically been shown to harbor a larger parasite burden. By contrast, adult raccoons are usually infected through ingesting larvae in the flesh of paratenic hosts, such as squirrels and rodents. In raccoons, infection with *B. procyonis* rarely causes clinical symptoms, as the adult worms are confined to the small intestine.

Once ingested, larvae migrate to the small intestine of the raccoon and remain in the lumen as they mature into adult worms. Maturation typically takes 1 to 2 months, depending upon the larval stage at the time of ingestion. Adult female worms have an extremely high fecundity rate, producing between 115,000 to 877,000 eggs/worm/d. An infected raccoon can shed as many as 45,000,000 eggs daily. An infectious dose of eggs is estimated to be less than 5,000. It takes 2 to 4 weeks for eggs to become infectious after being shed.

Medical Parasitology, edited by Abhay R. Satoskar, Gary L. Simon, Peter J. Hotez and Moriya Tsuji.

Communities of raccoons habitually defecate in discrete areas termed "latrines." This serves to harbor and accumulate increasingly large numbers of *B. procyonis* eggs and becomes the main means of transmission to intermediate hosts.

Humans are infected by ingesting *B. procyonis* eggs, typically at the sites of raccoon defecation. Pica and geophagia are risk factors for infection and are common in children under two years of age. After ingestion, larvae penetrate the intestinal mucosa and migrate rapidly to the liver and then to the lungs through the portal circulatory system. In the lungs, the larvae gain access to the pulmonary veins and are then distributed through the systemic circulation to the tissues. Larvae do not mature into adult worms in human hosts.

Clinical Features

In humans, baylisascariasis results in death or severe neurologic sequelae. To date, 13 confirmed cases of baylisascariasis in humans have been reported in the literature. In all cases, the victims were either young children or young adults with severe developmental disabilities. Disease results in neural larva migrans (NLM), ocular larva migrans (OLM) and visceral larva migrans (VLM). *B. procyonis* is distinct from other common parasites causing larva migrans by its propensity for continued larval growth in intermediate hosts, invasion of the central nervous system and its aggressive migration through host tissues.

B. procyonis is relatively unique among the zoonotic helminthes in its ability to cause neural larva migrans in addition to ocular and visceral larva migrans. It is estimated that 5 to 7% of ingested *B. procyonis* eggs travel to the central nervous system (CNS). When CNS involvement occurs, it most commonly presents as acute eosinophilic meningoencephalitis. Case reports in children have documented clinical signs including, sudden onset of lethargy, irritability, loss of motor coordination, weakness, generalized ataxia, stupor, coma, opisthotonus and death. Clinical symptoms may occur as early as 2 to 4 weeks post infection.

B. procyonis is the primary cause of the large nematode variant of diffuse unilateral subacute neuroretinitis (DUSN). This neuroretinitis is characterized by inflammatory and degenerative changes involving the retina and optic disk.

Based on the above data, the diagnosis of baylisascariasis should be considered in the setting of eosinophilia, both peripherally and in the CNS, DUSN, neurologic signs and a history of possible environmental exposure to raccoon feces.

Diagnosis

Laboratory Tests/Microscopy

Eosinophilic pleocytosis in the cerebral spinal fluid in combination with peripheral eosinophilia is highly suggestive of parasitic infection. Because of the prolonged, aberrant migration of *B. procyonis* larva, eosinophilic inflammation can be quite marked (28% or higher).

Definitive diagnosis is through morphologic identification of larvae in tissue sections, which is usually only accomplished at autopsy. Because *B. procyonis* does not establish an intestinal infection in humans, eggs will never be observed in feces.

Radiographic

Central nervous system infection with *B. procyonis* has been noted to cause diffuse, periventricular white matter changes on MRI, though this finding is not pathognomonic. Global atrophy as well as capsular and brain stem changes have been noted as late radiographic findings.

Molecular Diagnostic

Serologic testing for anti-bp antibodies in the CSF and serum by enzyme-linked immunosorbent assay (ELISA) or immunofluorescent antibody (IFA) has become the most important definitive diagnostic modality. This antibody assay has not been shown to have cross-reactivity with antibodies toward other ascarids, such as Toxocara species. In the United States, serologic testing is only available through the Department of Veterinary Pathobiology at Purdue University in West Lafayette, IN (phone 765-494-7558). *B. procyonis* infection also causes an elevation in isohemaglutinins due to a cross-reaction of larval proteins with human blood group antigens.

Treatment

Neural larva migrans has a universally poor prognosis. This is due in large part to failure of medical therapy when initiated at the onset of clinical signs and symptoms. Albendazole should be administered at a dose of 25 mg/kg for 20 days to a child with known exposure to raccoon stool. Ivermectin, mebendazole and thiabendazole can also be considered. Theoretically, the most promising anthelminthic agent is albendazole because of its ability to cross the blood-brain barrier. Whether albendazole would have any beneficial effect in treating CNS disease is unknown. In the past, corticosteroids have been administered to decrease the deleterious effects of the host inflammatory response to the migrating larvae. When a motile larva is found in the retina in a patient with ocular larva migrans, laser photocoagulation is curative.

Prevention and Prophylaxis

It is likely that *B. procyonis* will become a more significant pathogen for humans as the geographic distribution of the North American raccoon continues to expand and encroach upon urban human environments. Due to the relatively poor efficacy of treatment, preventative strategies are increasingly important. When possible, raccoons should be prevented from frequenting areas of human habitation. If an area is known to be contaminated by raccoon feces, the preferred method of disinfection is with direct flames.

Education of the public regarding the dangers of raccoon contact is paramount. Children should be closely monitored for ingestion of soil while playing outside. If ingestion of raccoon feces is suspected in a child, prophylactic treatment should consist of immediate administration of albendazole (25-50 mg/kg/d × 20 d; or 400 mg twice a day × 10 days).

Suggested Reading

1. Chronic and Recurrent Meningitis. Harrison's Online. McGraw-Hill Company, 2004-2005. www.accessmedicine.com. A comprehensive review of meningitis.
2. Gavin PJ, Kazacos KR, Shulman ST. Baylisascariasis. Clin Microbiol Rev 2005; 18:703-18. This is an excellent review.

11

3. Gavin PJ, Shulman ST. Raccoon roundworm (Baylisascaris procyonis). Pediatr Infect Dis 2003; 22:651-2.
4. Murray WJ, Kazacos KR. Raccoon roundworm encephalitis. Clin Infect Dis 2004; 39:1484-92.
5. Nash TE. Visceral larva migrans and other unusual helminth infections. In: Mandell GM, Bennett JE, Dolin R, eds. Principles and Practice of Infectious Diseases, 6th Edition. Philadelphia: Elsevier Churchill Livingstone 2005:3293-300. A comprehensive review of unusual helmith infections.
6. Page LK, Swihart RK, Kazacos KR. Implications of raccoon latrines in the epizootiology of baylisascaris. J Wildl Dis 1999; 35:474-80.
7. Park SY, Glaser C, Murray WJ et al. Raccoon roundworm (Baylisascaris procyonis) encephalitis: case report and field investigation. Pediatrics 2000; 106.http://www.pediatrics.org/cgi/content/full/106/4-e56.
8. Roussere GP, Murray WJ, Raudenbush CB et al. Raccoon roundworm eggs near homes and risk for larva migrans disease, California communities. Emerg Infect Dis 2003; 9:1516-22.
9. Rowley HA, Uht RM, Kazacos KR et al. Radiologic-pathologic findings in raccoon roundworm (Baylisascaris procyonis) encephalitis. Amv J Neuroradiol 2000; 21:415-20.
10. Sorvillo F, Ash LR, Berlin OG et al. Baylisascaris procyonis: An emerging helminthic zoonosis. Emerg Infect Dis 2002; 8:355-59.
11. Wise ME, Sorvillo FJ, Shafir SC et al. Severe and fatal central nervous system disease in humans caused by Baylisascaris procyonis, the common roundworm of raccoons: A review of current literature. Microbes Infect 2005; 7:317-23.

Toxocariasis

Many animal parasites are capable of infecting humans but rarely do so. Some helminths, however, infect humans more frequently and cause distinctive clinical syndromes. In the human host, especially children, the tissue migrating larvae of roundworms, *Toxocara canis* and *T. cati*, can cause serious complications including visceral larva migrans and ocular larva migrans. Originally toxocariasis was thought to be an uncommon pediatric infection. With improved serologic testing, toxocariasis is now known to be the most prevalent helminthic zoonosis in industrialized countries.

Background

Geographical Distribution and Epidemiology

Worldwide, toxocariasis is one of the most commonly reported zoonotic infections. *Toxocara* infect most domestic and many feral dogs and cats as well as foxes. In humans, infection is far more common in children than adults. Children usually acquire the infection through the ingestion of soil contaminated with embryonated *Toxocara* eggs.

Exposure is most likely to occur in playgrounds and sandboxes contaminated with cat and dog feces. There is evidence that the infection may also be acquired from direct contact with infected dogs, given the fact that the egg density on dog hair may be higher than that in soil.

Between 2% and 8% of healthy adults in Western countries in urban areas have serologic evidence of prior infection. The seroprevalence in children is much higher, ranging from 4 to 31% in developed countries to as high as 92.8% in rural areas in tropical countries. In a recent sero-epidemiological study of *T. canis* in schoolchildren in a mountainous region of Taiwan, 57.5% of children were seropositive. The two most significant risk factors for infection were living in household with dogs or playing in soil. In a similar study in Sorocaba, Brazil, 38.3% of children living in poorer outskirts of the city were infected compared to 11.1% in the central city. Overall, infection is most common in children with geographia or pica and exposure to puppies. Several studies have shown higher rates of infection in mentally retarded adults and children. In Israel, 8.5% of institutionalized mentally retarded adults had serologic evidence of infection. Several studies of mentally retarded children have found rates of infection from 10.6 to 20 percent. Although infection with *Toxocara* is common in cats, *Toxocara cati* is under recognized as a zoonotic infection.

Life Cycle and Mode of Transmission

In humans, *Toxocara* worms have a life cycle similar to that of *Ascaris lumbricoides* (Fig. 11.1). In the both animals and the aberrant host, ingestion of embryonated eggs initiates infection. Dogs and cats can also acquire the infection by eating earthworms or other paratenic hosts carrying embryonated eggs. Following ingestion, the eggs hatch in the small intestine to release larvae which penetrate the intestinal wall. The larvae then migrate via the bloodstream to the liver, lung and trachea. In the definitive host, particularly in dogs less than 6 months old, the larvae complete the life cycle after they are coughed up and swallowed, returning to the gastrointestinal tract. The larvae develop into the adult stage in the small intestine 60 to 90 days after hatching. Mating occurs in the intestine. Female worms may produce up to 200,000 eggs per day (Fig. 11.2). These nonembryonated eggs are then excreted into the soil

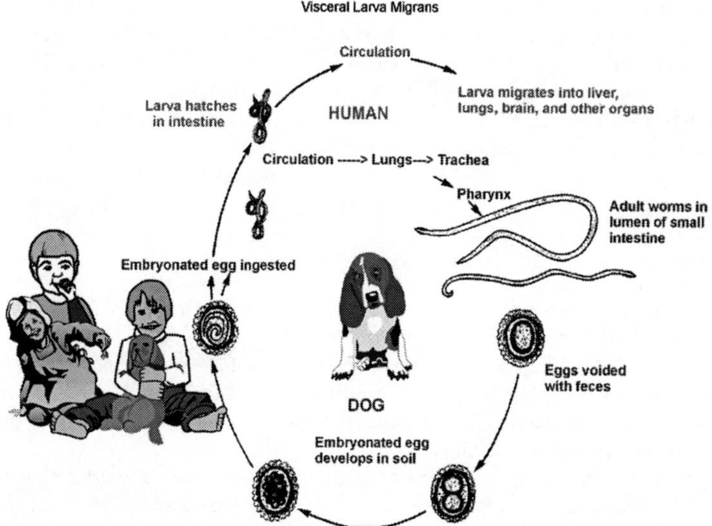

Figure 11.1. Life Cycle of *Toxocara canis*. Reproduced from: Nappi AJ, Vass E, eds. Parasites of Medical Importance. Austin: Landes Bioscience, 2002:86.

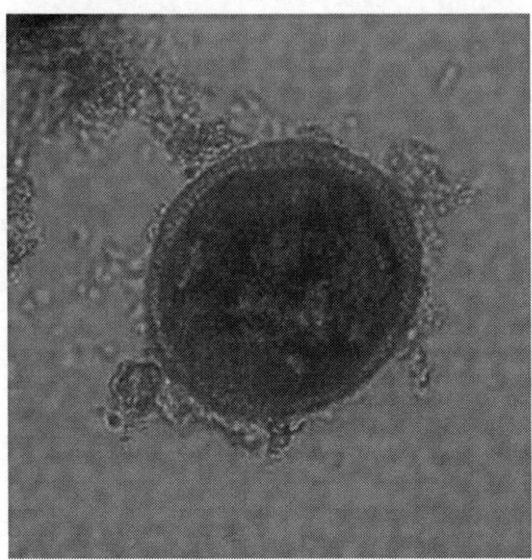

11

Figure 11.2. *Toxocara canis* egg.Reproduced from: Centers for Disease Control and Prevention (CDC) (http://www.dpd.cdc.gov/dpdx/).

where embryonation occurs within one to two weeks. Embryonated eggs can survive in the soil from days to months depending on the soil temperature. In pregnant female dogs and cats, dormant larvae, activated by hormonal stimuli, may develop and migrate transplacentally. Puppies infected in utero or transmammarily also shed eggs.

In humans, ingested eggs hatch in the small intestine and penetrate the intestinal wall in the same way as in the definitive host, but the larvae, unable to mature, migrate through the body for a prolonged period. Infective larvae, which do not mature into adult worms in humans, can persist for years after becoming encapsulated within granulomas. Hatched larvae have been isolated from the eye, liver, heart and brain. Besides ingestion of soil contaminated with embryonated eggs, infection can also occur after consumption of raw or undercooked meat infected with *Toxocara* larvae, as well as by ingestion of vegetables and salads grown in contaminated soil.

Clinical Features

The severity of disease and degree of host response vary dependent upon the tissue invaded by the migrating larvae, the number of larvae and the age of the host. The death of the juvenile larvae in tissue, particularly the lung, liver and brain, evokes an inflammatory response with marked eosinophilia, producing the symptoms of visceral larva migrans (VLM).

In the eye damage to the retina from the migrating larvae, ocular larva migrans (OLM), results in a granulomatous reaction and impaired vision. OLM is thought to occur mostly in children not previously sensitized to toxocariasis while VLM occurs following repeated exposure to migrating larvae.

Visceral Larva Migrans

VLM occurs primarily in young and preschool children. Most infections are asymptomatic, but fulminant disease and death do occur. Children come to medical attention with unexplained prolonged fever, cough, hepatosplenomegaly, wheeze and eosinophilia.

Other clinical signs and symptoms include lymphadenopathy, skin lesions, pruritus, anemia, failure to thrive, decreased appetite, nausea, vomiting, headache and pneumonia, as well as behavioral and sleep disturbances. In children with toxocariasis, pica or geographia are common. Myocarditis, respiratory failure or seizures may complicate overwhelming infection. Other less frequently reported complications included multiple ecchymoses with eosinophilia, pyogenic liver abscess, urticaria or prurigo, Henoch-Schönlein purpura, nephrotic syndrome, secondary thrombocytosis and eosinophilic arthritis. There is a report of VLM mimicking lymphoma with hilar and mediastinal lymphadenopathy and another of systemic vasculitis with lymphocytic temporal arteritis. Both of these two complications occurred in patients over 60 years of age.

Toxocariasis has been suggested to be an environmental risk factor for asthma among children living in urban areas, but supportive evidence for this hypothesis is lacking. Central nervous system involvement is a rare complication reported more frequently in adults. Toxocariasis should be considered as a causative agent in patients with eosinophilic meningoencephalitis or meningitis. *T. canis* has been reported to cause epileptic seizures, particularly late-onset partial epilepsy.

Ocular Larva Migrans

Ocular toxocariasis occurs primarily in young adults and older children who present with unilateral visual loss over days to weeks. OLM follows entrapment of a larva in the eye causing an intense eosinophilic inflammatory reaction. Posterior pole granuloma is the most common form of OLM in children between 6 and 14 years and causes decreased vision. Peripheral granuloma is usually seen at an older age in association with macular heterotropia, strabismus and decreased vision. Endophthalmitis occurs in younger children aged 2 to 9 years in whom there is marked visual impairment with evidence of vitritis and anterior uveitis. Retinal detachment may be seen on fundoscopic examination. OLM should be included in the differential diagnosis of any child with leukocoria.

Diagnosis

The presence of eosinophilia in a child with unexplained fever, abdominal pain, hepatosplenomegaly and multisystem illness raises the possibility of VLM, especially if there is a history of geographia or pica and contact with puppies. For a child with unilateral visual loss and strabismus, a diagnosis of OLM must be excluded. In OLM the blood eosinophil count is frequently not elevated. Leukocytosis, hypergammaglobulinemia, increased anti-*Toxocara* IgE serum concentration and elevated isohemagglutinin titers to A and B blood group antigens may be present in VLM.

The diagnosis of both VLM and OLM is usually based on serologic tests. However, serologic tests do not reliably distinguish between recent and past infection. The most frequently performed test is the enzyme-linked immunosorbent assay (ELISA) which uses *Toxocara* excretory-secretory antigens of the second-stage larvae.

This test is sufficiently specific to be the best indirect diagnostic assay. At a titer of greater than 1:32 the sensitivity of this test for diagnosing VLM is about 78%. The ELISA is less reliable for the diagnosis of OLM. The presence of elevated vitreous and aqueous fluid titers relative to serum titers supports the diagnosis of OLM.

Microscopy

A definitive diagnosis of VLM is based upon the visualization of the larvae in infected tissue such as lung, liver, or brain. In OLM the larvae may be seen in the enucleated eye.

Molecular Diagnostics and Radiographic

In OLM, ultrasound biomicroscopy has been used to detect the morphologic changes of peripheral vitreoretinal toxocariasis.

Ultrasonographic findings of hepatic toxocariasis complicating VLM may include ill-defined focal lesions, hepatosplenomegaly, the presence of biliary sludge and dilatation and periportal lymphadenopathy. If present, follow-up CT scan or MR imaging should be considered following treatment.

In patients with cerebral granulomatous toxocariasis, multiple subcortical, cortical, or white matter lesions that were hypoattenuating on CT scan, hyperintense on T2-weighted MR images and homogeneously enhancing have been reported. These findings are nonspecific. It has been suggested, nonetheless, that serial MR imaging may be used to monitor the course of disease treated with anthelminthic therapy.

Treatment

The decision to treat toxocariasis is made based on the type of infection and severity of clinical symptoms. Most patients do not require treatment.

Current Therapy and Alternative Therapies

Visceral Larva Migrans

Albendazole, in doses of 10 mg/kg or 400 mg, both given twice daily for 5 days is the treatment of choice for toxocariasis. Mebendazole is a second-line therapy since it is not absorbed outside of the gastrointestinal tract. These two agents have fewer side effects than thiabendazole and diethylcarbamazine, both of which require treatment for one to three weeks. Injury to the parasite may cause a more intense inflammatory response with worsening of symptoms for which antihistamines and corticosteroids may be of benefit.

Ocular Larva Migrans

The goal of therapy for the treatment of OLM is to decrease the severity of the inflammatory response. If given within the first four weeks, systemic or intraocular corticosteroids are the most effective intervention. There are reports of treatment with albendazole and corticosteroids although the ocular penetration of albendazole and other benzimidazoles has not been established. Diethlycarbamazine (DEC) may be the preferred agent for ocular disease, but the activity of DEC may be inhibited by corticosteroids. Accordingly, they should not be co-administered. If the migrating larva can be visualized, laser photocoagulation is effective and will destroy the organism. Treatment of chronic ocular infection, infection for more than 8 weeks, is problematic and should be managed by an experienced ophthalmologist.

Follow-Up after Treatment

Within a week of treatment there is usually a rise in eosinophilia accompanied by improvement in clinical parameters. By 4 weeks post-therapy, eosinophilia has usually resolved and the antitoxocara IgE has become negative.

Prevention and Control

Exposure to toxocariasis occurs primarily in overcrowded urban areas where children are in close contact with dogs and cats. Public education and control efforts should be directed at limiting exposure of children to soil contaminated with *Toxocara* eggs in public parks, playgrounds, sandboxes, home gardens and other areas where children congregate. Dogs and cats should be restricted from entering public areas where children play. Dog owners should clean up after their pets have defecated and have their pets wormed regularly. Children should wash their hands after playing in a park or coming in close contact with dogs, especially puppies and cats. Children with geographia or pica require medical evaluation.

Suggested Reading

1. Glickman LT, Schnatz PM. Epidemiology and pathogenesis of zoonotic toxocariasis. Epidemiol Rev 1981; 3:230-50.
2. Wolfe A, Wright IP. Human toxocariasis and direct contact with dogs. Vet Rec 2003; 152:419-22.
3. Fan Ck, Liao CW, Kao TC et al. Sero-epidemiology of Toxocara canis infection among aboriginal schoolchildren in the mountainous areas of northeastern Taiwan. Ann Trop Med Parasitol 2005; 99:593-600.
4. Coelho LM, Silva MV, Dini CY et al. Human toxocariasis: a seroepidemiological survey in schoolchildren in Sorocaba, Brazil. Mem Inst Oswaldo Cruz 2004; 99:553-7.
5. Kaplan M, Kalkan A, Hosoglu S et al. The frequency of Toxocara infection in mental retarded children. Mem Inst Oswaldo Cruz 2004; 99:121-5.
6. Huminer D, Symon K, Groskopf I et al. Seroepidemiologic study of toxocariasis and strongyloidiasis in institutionalized mentally retarded adults. Am J Trop Med Hyg 1992; 46:278-81.
7. Gillespie SH. Migrating worms. In: Cohen J, Powderly W, eds. Infectious Diseases, 2nd Edition. Oxford: Elsevier Inc., 2004:1633-5. A comprehensive review, including the pathogenesis.
8. Beaver PC. The nature of visceral larva migrans. J Parasitol 1969; 55:3-12.
9. Buijs J, Borsboom G, van Gemund JJ et al. Toxocara seroprevalence in 5-year-old elementary schoolchildren: relation with allergic asthma. Am J Epidemiol 1994; 140:839-47.
10. Sharghi N, Schantz PM, Caramico L et al. Environmental exposure to Toxocara as a possible risk factor for asthma: a clinic-based case-control study. Clin Infect Dis 2001; 32:111-6. The association of Toxocara infection with asthma is refuted in this study.
11. Nicoletti A, Bartoloni A, Reggio A et al. Epilepsy, cysticercosis and toxocariasis: a population-based case-control study in rural Bolivia. Neurology 2002; 58:1256-61.
12. Hemang P, Goldstein D. Pediatric uveitis. Pediatr Clin N Am 2003; 50:125-36. A comprehensive review of pediatric uveitis including toxocariasis and other infectious pathogens.
13. Good B, Holland CV, Taylor MRH et al. Ocular toxocariasis in schoolchildren. Clin Infect Dis 2004; 39:173-8.
14. Gillespie SH, Bidwell D, Voller A et al. Diagnosis of human toxocariasis by antigen capture enzyme linked immunoabsorbent assay. J Clin Pathol 1993; 46:551-4.
15. Xinou E, Lefkopoulos A, Gelagoti M et al. CT and MRI imaging findings in cerebral toxocaral disease. Am J Neuroadiol 2003; 24:714-8.
16. Strucher D, Schubarthi P, Gualzata M et al. Thiabendazole vs. albendazole in treatment of toxocariasis: a clinical trial. Ann Trop Med Parasitol 1989; 83:473-8.

CHAPTER 12

Lymphatic Filariasis

Subash Babu and Thomas B. Nutman

Background

The term "lymphatic filariasis" encompasses infection with three closely related nematode worms—*Wuchereria bancrofti, Brugia malayi* and *Brugia timori*. All three parasites are transmitted by the bites of infective mosquitoes and have quite similar life cycles in humans (Fig. 12.1) with the adult worms living in the afferent lymphatic vessels while their offspring, the microfilariae, circulate in the peripheral blood and are available to infect mosquito vectors when they feed. Though not fatal, the disease is responsible for considerable suffering, deformity and disability and is the second leading parasitic cause of disability with DALYs (disability-adjusted life years) estimated to be 5.549 million.

Lymphatic filariasis is a global health problem. At a recent estimate, it has been determined that over 2 billion people are at risk and at least 129 million people actually infected. *W. bancrofti* accounts for nearly 90% of these cases. *W. bancrofti* has the widest geographical distribution and is present in Africa, Asia, the Caribbean, Latin America and many islands of the Western and South Pacific Ocean. *B. malayi* is geographically more restricted, being found in Southwest India, China, Indonesia, Malaysia, Korea, the Philippines and Vietnam. *B. timori* is found in Timor, Flores, Alor, Roti and Southeast Indonesia.

Humans are the definitive host and mosquitoes are the intermediate hosts of *W. bancrofti* and Brugia spp. The life cycle of filarial parasites involves four larval stages and an adult stage. Infection begins with the deposition of infective stage larvae (L3) on the skin near the site of puncture during a mosquito bite. The larvae then pass through the puncture wound and reach the lymphatic system. Within the lymphatics and lymph nodes, the L3 larvae undergo molting and development to form L4 larvae. This takes about 7-10 days for both *W. bancrofti* and *B. malayi*. The L4 larvae undergo a subsequent molting/developmental step to form adult worms. This occurs about 4-6 weeks after L3 entry in the case of *B. malayi* and after several months in the case of *W. bancrofti*. The adult worms take permanent residence in afferent lymphatics or the cortical sinuses of lymph nodes and generate microscopic live progeny called "microfilariae". The female worms can give birth to as many as 50,000 microfilariae per day, which find their way into the blood circulation from the lymphatics. The adult worms are estimated to survive for a period of 5-10 years although longer durations have been recorded.

The major vectors of *W. bancrofti* are culicine mosquitoes in most urban and semi-urban areas, anophelines in the more rural areas of Africa and elsewhere and *Aedes* species in many of the endemic Pacific islands. For the Brugian parasites,

Medical Parasitology, edited by Abhay R. Satoskar, Gary L. Simon, Peter J. Hotez and Moriya Tsuji. ©2009 Landes Bioscience.

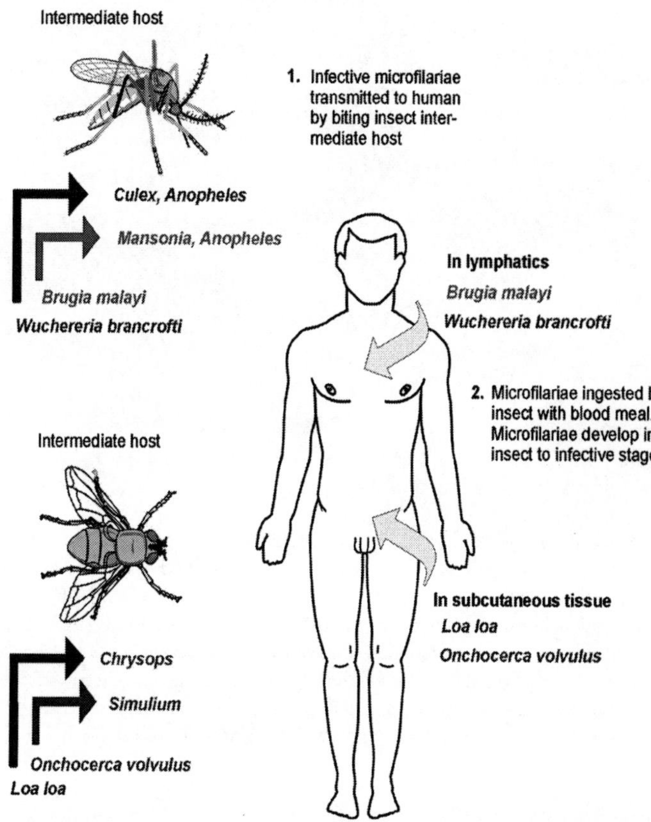

Figure 12.1. Generalized life cycle of filariids. Adapted from: Nappi AJ, Vass E, eds. Parasite of Medical Importance. Austin: Landes Bioscience, 2002:93.

Mansonia species serve as the main vector, but in some areas anopheline mosquitoes can transmit infection as well. *Culex quinquefasciatus* is the most important vector of *W. bancrofti* and is responsible for more than half of all lymphatic filarial infections. The microfilariae of *W. bancrofti* and *B. malayi*, for the large part, exhibit a phenomenon called nocturnal periodicity, i.e., they appear in larger numbers in the peripheral circulation at night and retreat during the day. Subperiodic or nonperiodic *W. bancrofti* and *B. malayi* are also found in certain parts of the world.

Filarial nematodes belong to the phylum Nematoda, class Secernentea, and superfamily Filarioidea. Adult *W. bancrofti* and *B. malayi* worms are long, slender, tapered and cylindrical worms. The males (4 cm × 0.1 mm for *W. bancrofti* and

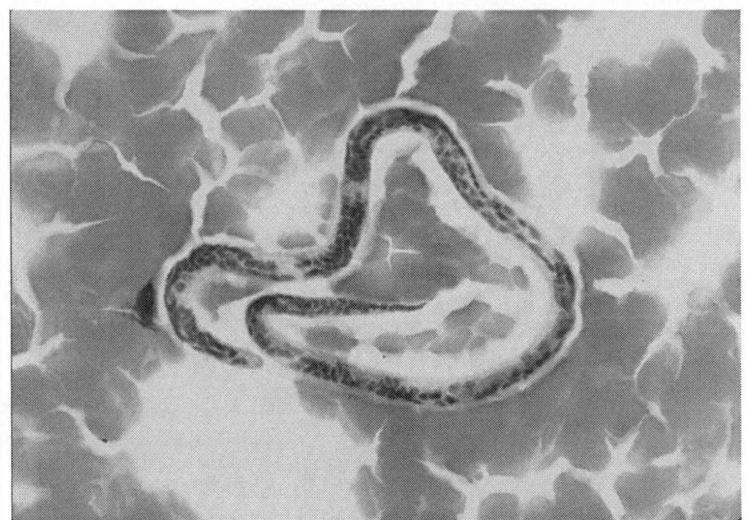

Figure 12.2. Microfilaria of *W. bancrofti* in a blood film stained with Giemsa. The stain reveals the nuclei and the sheath.

3.5 cm × 0.1 mm for *B. malayi*) are strikingly smaller than the females (6-10 cm × 0.2-0.3 mm for *W. bancrofti* and 5-6 cm × 0.1 mm for *B. malayi*). The microfilariae of *W. bancrofti* are ensheathed and measure about 245-300 µm by 7.5-10 µm (Fig. 12.2). The microfilariae of *B. malayi* are ensheathed and measure 175-230 µm by 5-6 µm. Most pathogenic human filarial parasites are infected with a bacterial endosymbiont called Wolbachia. It is an alpha-proteobacteria, related to Rickettsia, Erlichia and Anaplasma and is maternally inherited. It has been detected in all life cycle stages of the parasite and found to be essential for adult worm viability, normal fertility and larval development.

Clinical Manifestations

Lymphatic filariasis can manifest itself in a variety of clinical and subclinical conditions.

Subclinical (or asymptomatic) microfilaremia: In areas endemic for lymphatic filariasis, many individuals exhibit no symptoms of filarial infection and yet, on routine blood examinations, demonstrate the presence of significant numbers of parasites. These individuals are carriers of infection (and the reservoir for ongoing transmission) and have commonly been referred to as asymptomatic microfilaremics. The parasite burdens in these individuals can reach dramatically high numbers, exceeding 10,000 microfilariae in 1 ml of blood. With the advent of newer imaging techniques, it has become apparent that virtually all persons with microfilaremia have some degree of subclinical disease. These include profound changes such as marked dilatation and tortuosity of lymph vessels with collateral channeling and increased flow and abnormal patterns of lymph flow, a considerable degree of scrotal lymphangiectasia and microscopic hematuria and/

or proteinuria (indicative of low-grade renal damage). Thus, while apparently free of overt symptomatology, the asymptomatic microfilaremic individuals clearly are subject to subtle pathological changes.

Acute Clinical Disease

The acute manifestations of lymphatic filariasis are characterized by recurrent attacks of fever associated with the inflammation of lymph nodes (lymphadenitis) and lymphatics (lymphangitis). In brugian filariasis, episodes of fever, lymphadenitis and lymphangitis are common, while bancroftian filariasis present more insidiously with fewer overt acute episodes. The lymph nodes commonly involved are the inguinal, axillary and epitrochlear nodes and, in addition, the lymphatic system of the male genitals are frequently affected in *W. bancrofti* infection leading to funiculitis, epididymitis and/or orchitis.

It has been proposed that there are at least two distinct mechanisms involved in the pathogenesis of acute attacks. The more classical is acute filarial adenolymphangitis, which is felt to reflect an immune-mediated inflammatory response to dead or dying adult worms. The striking manifestation is a distinct well-circumscribed nodule or cord along with lymphadenitis and retrograde lymphangitis. Funiculo-epididymoorchitis is the usual presenting feature when the attacks involve the male genitalia. Fever is not usually present, but pain and tenderness at the affected site are common.

The other has been termed acute dermatotolymphangitis, a process characterized by development of a plaque-like lesion of cutaneous or subcutaneous inflammation and accompanied by ascending lymphangitis and regional lymphadenitis. There may or may not be edema of the affected limbs. These pathological features are accompanied by systemic signs of inflammation including fever and chills. This manifestation is thought to result primarily from bacterial and fungal superinfections of the affected limbs.

Manifestations of Chronic Infection

The chronic sequelae of filariasis are postulated to develop approximately 10 to 15 years after initial infection. In *Bancroftian filariasis,* the main clinical features are hydrocele, lymphedema, elephantiasis and chyluria. The manifestations in descending order of occurrence are hydrocele and swelling of the testis, followed by elephantiasis of the entire lower limb, the scrotum, the entire arm, the vulva and the breast. In *Brugian filariasis,* the leg below the knee and the arm below the elbow are commonly involved but rarely the genitals. Lymphedema can be classified or graded, a scheme proven very useful in clinical trials:

Grade 1

Pitting edema reversible on limb elevation.

Grade 2

Pitting/nonpitting edema not reversible on limb elevation and normal skin.

Grade 3

Nonpitting edema of the limb, not reversible on elevation with skin thickening.

Grade 4

Nonpitting edema with fibrotic and verrucous skin changes (elephantiasis) (Fig. 12.3A).

In men, scrotal hydrocele is the most common chronic clinical manifestation of bancroftian filariasis (Fig. 12.3B). Hydroceles are due to accumulation of edematous fluid in the cavity of the tunica vaginalis testis. Chronic epididymitis and funiculitis can also occur. Chyloceles can also occur. The prevalence of chyluria (excretion of chyle, a milky white fluid in the urine) is very low.

Tropical Pulmonary Eosinophilia

Tropical pulmonary eosinophilia (TPE) is a distinct syndrome that develops in some individuals infected with *W. bancrofti* and *B.malayi*. This syndrome affects males and females at a ratio of 4:1, often during the third decade of life. The majority of cases have been reported from India, Pakistan, Sri Lanka, Brazil, Guyana and Southeast Asia. The main clinical features include paroxysmal cough and wheezing that are usually nocturnal (and probably related to the nocturnal periodicity of microfilariae), weight loss, low-grade fever, adenopathy and pronounced blood eosinophilia (>3000 eosinophils/µL). Chest X-rays may be normal but generally show increased bronchovascular markings; diffuse miliary lesions or mottled opacities may be present in the middle and lower lung fields. Tests of pulmonary function show restrictive abnormalities in most cases and obstructive defects in half. Total serum IgE levels (10,000 to 100,000 ng/mL) and antifilarial antibody titers are characteristically elevated.

Other Manifestations

Lymphatic filariasis has been associated with a variety of renal abnormalities including hematuria, proteinuria, nephrotic syndrome and glomerulonephritis. Circulating immune complexes containing filarial antigens have been implicated in the renal damage. Lymphatic filariasis may also present as a mono-arthritis of the knee or ankle joint.

Uninfected, but exposed individuals (asymptomatic amicrofilaremia or endemic normals): In endemic areas, a proportion of the population remains uninfected despite exposure the parasite to the same degree as the rest of the population. This group has been termed endemic normal. The incidence of endemic normals in a population ranges from 0% to 50% in different endemic areas.

Diagnosis

The traditional method of diagnosing lymphatic filarial infections has been the detection of microfilariae in the peripheral blood collected during the night in areas of nocturnal periodicity and during the day in areas of subperiodic lymphatic filariae. The simplest method is a thick blood film of capillary blood stained with Giemsa stain, its disadvantage being poor sensitivity. The sensitivity of detection can be augmented by the use of concentration techniques such as the Knott's concentration method in which 1 ml of whole blood is added to 9 ml of a 2% formalin solution, centrifuged and the sediment examined for microfilariae. Another widely used concentration technique is the membrane filtration technique whereby 1-5 ml of blood is passed through a 3 or 5 µM polycarbonate membrane which retains the microfilariae.

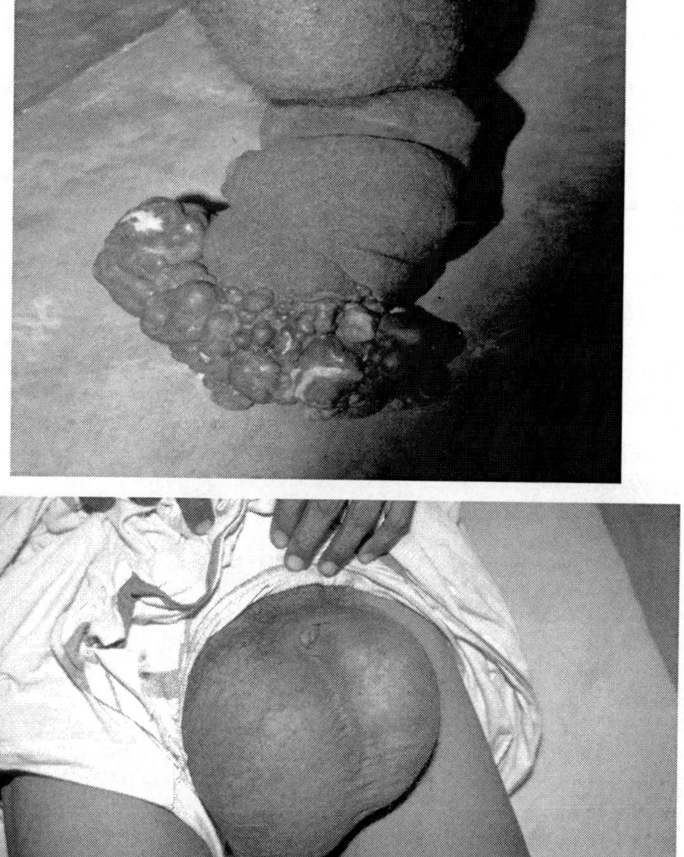

Figure 12.3. Chronic manifestations of lymphatic filariasis. A) Elephantiasis of the lower limb with verrucous and fibrotic skin changes. B) Hydrocele in a male patient.

For bancroftian filariasis, assays for the detection of circulating parasite antigens have been developed based on one of two well-characterized monoclonal antibodies, Og4C3 or AD12. The commercial Og4C3 ELISA (Trop Bio Og4C3 Antigen test, produced by Trop Bio Pty Ltd) has a sensitivity approaching 100% and specificity of 99-100%. The ICT filarial antigen test (Binax) is a rapid format card test with a sensitivity of 96-100% and specificity of 95-100%. It utilizes capillary or venous blood and is simple enough for field use.

Antibody-based assays for diagnosing filarial infection have typically used crude parasite extract and have suffered from poor specificity. Improvements have been made by the use of detection of antifilarial IgG4 antibodies in that they are produced in relative abundance during chronic infection. IgG4 antibodies correlate well with the intensity and duration of filarial exposure and the level of microfilaremia; these IgG4 antibodies also have very little cross-reactivity to nonfilarial helminths. In addition, IgG4 antibodies are also useful in the diagnosis of Brugian infections; indeed a diagnostic dipstick test has been used in areas endemic for Brugian filariasis based on a recombinant Brugian antigen.

PCR-based methods have been developed for the detection of *W. bancrofti* DNA in blood, plasma, paraffin embedded tissue sections, sputum, urine and in infected mosquitoes; and for *B. malayi* DNA in blood and in mosquitoes.

Finally, the examination of scrotum and breast using ultrasonography in conjunction with pulse wave Doppler techniques can identify motile adult worms within the lymphatics. The adult worms exhibit a characteristic pattern of movement known as the filarial dance sign and the location of these adult worm nests remains remarkably stable. On rare occasions, living adult worms reside in the lymphatics of inguinal crural, axillary and epitrochlear lymph nodes.

Treatment

Diethylcarbamazine (DEC, 6 mg/kg in three divided doses daily for 12 days), which has both macro- and microfilaricidal properties, remains the treatment of choice for the individual with active lymphatic filariasis (microfilaremia, antigen positivity, or adult worms on ultrasound), although albendazole (400 mg twice daily for 21 days) and ivermectin (400 μg/kg) also have activity against microfilariae.

In persons with chronic manifestations of lymphatic filariasis, treatment regimens that emphasize hygiene, prevention of secondary bacterial infections and physiotherapy have gained wide acceptance for morbidity control. Hydroceles can be drained repeatedly or managed surgically. In patients with chronic manifestations of lymphatic filariasis, drug treatment should be reserved for individuals with evidence of active infection as therapy has been associated with clinical improvement and, in some, reversal of lymphedema.

The recommended course of DEC treatment (12 days; total dose, 72 mg/kg) has remained standard for many years; however, data indicate that single-dose DEC treatment with 6 mg/kg may be equally efficacious. The 12-day course provides more rapid short-term microfilarial suppression. Regimens that utilize single-dose DEC or ivermectin or combinations of single doses of albendazole and either DEC or ivermectin have all been demonstrated to have a sustained microfilaricidal effect.

In the past few years, the use of antibiotics such as doxycycline, which have an effect on Wolbachia, to treat lymphatic filariasis has been investigated. Preliminary data suggest that 200 mg of doxycycline for 8 weeks is effective in eliminating microfilaremia.

Prevention and Control

DEC has the ability of killing developing forms of filarial parasites and has been shown to be useful as a prophylactic agent in humans. Like those integrated control programs used for onchocerciasis, long term microfilarial suppression using mass, annual distribution of single dose combinations of albendazole with either DEC or ivermectin is underway in many parts of the world. These strategies have as their basis the microfilarial suppression of >1 year using single dose combinations of albendazole/ivermectin or albendazole/DEC. An added benefit of these combinations is their secondary salutary effects on gastrointestinal helminth infections.

Vector control measures such as the use of insecticides, the use of polystyrene beads in infested pits, the use of *Bacillus sphaericus* as a larvicide, the use of larvivorous fish and the use of insecticide treated bednets have all been advocated as adjunct measure for control of filariasis.

Lymphatic filariasis is one of six potentially eradicable diseases (WHO International Task Force 1993) and the development of a global program to eliminate filariasis (GPELF) came about following a resolution by the WHO Assembly in 1997. The principal aims of the program are to interrupt transmission of infection and to alleviate and/or prevent disability. The recent advances in immunology, molecular biology and imaging technology have helped pave the way for the realization of the goal to eliminate of lymphatic filariasis.

Suggested Reading

1. Adebajo AO. Rheumatic manifestations of tropical diseases. Curr Opin Rheumatol 1996; 8:85-9.
2. Amaral F, Dreyer G, Figueredo-Silva J et al. Live adult worms detected by ultrasonography in human Bancroftian filariasis. Am J Trop Med Hyg 1994; 50:753-757.
3. Dreyer G, Amaral F, Noroes J et al. Ultrasonographic evidence for stability of adult worm location in bancroftian filariasis. Trans R Soc Trop Med Hyg 1994; 88:558.
4. Dreyer G, Medeiros Z, Netto MJ et al. Acute attacks in the extremities of persons living in an area endemic for bancroftian filariasis: differentiation of two syndromes. Trans R Soc Trop Med Hyg 1999a; 93:413-417.
5. Dreyer G, Santos A, Noroes J et al. Proposed panel of diagnostic criteria, including the use of ultrasound, to refine the concept of 'endemic normals' in lymphatic filariasis. Trop Med Int Health 1999b; 4:575-579.
6. Eberhard ML, Lammie PJ. Laboratory diagnosis of filariasis. Clin Lab Med 1991; 11:977-1010.
7. Horton J, Witt C, Ottesen EA et al. An analysis of the safety of the single dose, two drug regimens used in programmes to eliminate lymphatic filariasis. Parasitology 2000; 121 Suppl:S147-160.
8. Kazura J, Greenberg J, Perry R et al. Comparison of single-dose diethylcarbamazine and ivermectin for treatment of bancroftian filariasis in Papua New Guinea. Am J Trop Med Hyg 1993; 49:804-811.

9. Kimura E, Spears GF, Singh KI et al. Long-term efficacy of single-dose mass treatment with diethylcarbamazine citrate against diurnally subperiodic Wuchereria bancrofti: eight years' experience in Samoa. Bull World Health Organ 1992; 70:769-776.
10. Lagraulet J. Current status of filariasis in the Marquises and different epidemiological aspects. Bull Soc Pathol Exot Filiales 1973; 66:311-320.
11. Maxwell CA, Mohammed K, Kisumku U et al. Can vector control play a useful supplementary role against bancroftian filariasis? Bull World Health Organ 1999; 77:138-143.
12. Melrose WD. Lymphatic filariasis: new insights into an old disease. Int J Parasitol 2002; 32:947-960.
13. Michael E, Bundy DA, Grenfell BT. Re-assessing the global prevalence and distribution of lymphatic filariasis. Parasitology 1996; 112 (Pt 4):409-428.
14. Molyneux DH, Neira M, Liese B et al. Lymphatic filariasis: setting the scene for elimination. Trans R Soc Trop Med Hyg 2000; 94:589-591.
15. More SJ, Copeman DB. A highly specific and sensitive monoclonal antibody-based ELISA for the detection of circulating antigen in bancroftian filariasis. Trop Med Parasitol 1990; 41:403-406.
16. Ong RK, Doyle RL. Tropical pulmonary eosinophilia. Chest 1998; 113:1673-1679.
17. Ottesen EA, Nutman TB. Tropical pulmonary eosinophilia. Annu Rev Med 1992; 43:417-424.
18. Ottesen EA. The global programme to eliminate lymphatic filariasis. Trop Med Int Health 2000; 5:591-594.
19. Pani SP, Krishnamoorthy K, Rao AS et al. Clinical manifestations in malayan filariasis infection with special reference to lymphoedema grading. Indian J Med Res 1990; 91:200-207.
20. Pani SP, Srividya A. Clinical manifestations of bancroftian filariasis with special reference to lymphoedema grading. Indian J Med Res 1995; 102:114-118.
21. Partono F. The spectrum of disease in lymphatic filariasis. Ciba Found Symp 1987; 127:15-31.
22. Partono F, Dennis DT, Atmosoedjono S et al. The microfilaria of Brugia timori: morphologic description with comparison to Brugia malayi of Indonesia. J Parasitol 1977; 63:540-546.
23. Rahmah N, Taniawati S, Shenoy RK et al. Specificity and sensitivity of a rapid dipstick test (Brugia Rapid) in the detection of Brugia malayi infection. Trans R Soc Trop Med Hyg 2001; 95:601-604.
24. Taylor MJ, Hoerauf A. Wolbachia bacteria of filarial nematodes. Parasitol Today 1999; 11:437-442.
25. Taylor MJ, Makunde WH, McGarry HF et al. Macrofilaricidal activity after doxycycline treatment of Wuchereria bancrofti: a double-blind, randomised placebo-controlled trial. Lancet 2005; 365:2067-2068.
26. Vanamail P, Subramaniam S, Das PK et al. Estimation of age-specific rates of acquisition and loss of Wuchereria bancrofti infection. Trans R Soc Trop Med Hyg 1989; 83:689-693.
27. Weil GJ, Jain DC, Santhanam S et al. A monoclonal antibody-based enzyme immunoassay for detecting parasite antigenemia in bancroftian filariasis. J Infect Dis 1987; 156:350-355.
28. WHO. Lymphatic filariasis. World Health Organ Tech Rep Ser 1992; 821:1-71.
29. WHO. Lymphatic filariasis. Weekly Epidemiological Record 2003; 78:171-179.

Section II
Trematodes

CHAPTER 13

Clonorchiasis and Opisthorchiasis

John Cmar

Background

Causative Agents

The class Trematoda contains 13 species of parasitic flatworms that cause biliary tract disease in humans. Residing in the family Opisthorchiidae, the three most common are *Clonorchis* (formerly *Opisthorchis*) *sinensis*, *Opisthorchis felineus* and *Opisthorchis viverrini*. Adults of these species are dorso-ventrally flattened lancet-shaped hermaphroditic worms and typically reach 10-25 mm in length and 3-5 mm in width (Fig. 13.1). While similar in appearance, these species can be distinguished morphologically by the shape and appearance of the testes, as well as the arrangement of the vitelline glands. The eggs are 30 μm by 15 μm in size, ovoid and yellowish brown in color, with a well-developed operculum. In contrast to the adult forms, it is difficult to distinguish these species from each other on the basis of egg morphology.

Humans are among the piscivorous mammals that are the definitive hosts in whom the *C. sinensis* and *Opisthorchis* species undergo sexual reproduction. The flukes reproduce asexually in several species of snails, which are the first intermediate hosts. Various freshwater fish and crustaceans serve as the second intermediate hosts. The geographic distribution of human clonorchiasis is related to both the areas of population of the aforementioned snails and marine animals, as well as the local culinary and hygiene habits concerning said animals.

Geographical Distribution/Epidemiology

The 1994 report of the World Health Organization and the International Agency for Research on Cancer estimated the global number of *C. sinensis* infections to be 7 million; however, a series of more recent studies places the number at closer to 35 million, with over 15 million of those infected living in China. In addition, Japan, Taiwan, Hong Kong, Korea and Vietnam are also primary endemic countries for this infection. Prevalence rates within endemic regions can vary widely with local culinary customs and sanitation; various Chinese provinces have described prevalence ranging from <1 to 57 percent.

The two *Opisthorchis* species under consideration have different geographic distributions. *O. felineus* is endemic to Southeast Asia and Central and Eastern

Medical Parasitology, edited by Abhay R. Satoskar, Gary L. Simon, Peter J. Hotez and Moriya Tsuji.

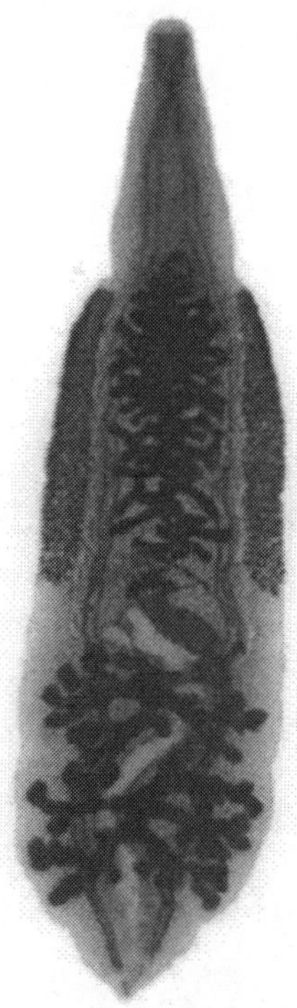

Figure 13.1. Adult *Clonorchis sinensis*, a hepatic fluke. Reproduced from: Nappi AJ, Vass E, eds. Parasites of Medical Importance. Austin: Landes Bioscience, 2002:44.

Europe, especially in Siberia and other former territories of the Soviet Union. Over 16 million people worldwide are thought to be infected, with prevalence rates of 40 to 95%. *O. viverrini* occurs primarily in Thailand, Laos and Kampuchea. Estimates of infection are 10 million people worldwide, with 24 to 90 percent prevalence in Thailand and 40 to 80 percent in Laos.

While these organisms may not have a major impact in non-endemic countries, clonorchiasis and opisthorchiasis can still occur there by one of several means. Travelers to an endemic area may return infected, acquiring the parasite from eating improperly prepared freshwater fish, or via the fecal-oral route in areas of poor sanitation. Immigrants from endemic countries to non-endemic countries can bring the disease with them; early studies of clinical clonorchiasis in Asian immigrants to North America described prevalence rates of up to 28%. Finally, freshwater fish and shrimp that are improperly pickled, dried, or salted for importation by non-endemic countries can harbor living encysted organisms and bring disease to those who have never visited an endemic region.

The adult flukes of *C. sinensis* and *Opisthorchis* species have a lifespan of up to 30 years in the bile ducts of humans, during which time immunity does not develop. As such, cumulative infections can occur, resulting in increased intensity of infection and greater worm burden. Thus, symptomatology is most common in older adults who may be far removed from their initial exposure.

Mode of Transmission

Clonorchiasis and opisthorchiasis are perpetuated in the following cycle (Fig. 13.2): Adult flukes, residing in the biliary tracts of a definitive host, produce eggs, which pass with bile into the feces and are ultimately expelled into a freshwater environment. The eggs hatch into miracidia and infect snails of the *Bithynia*

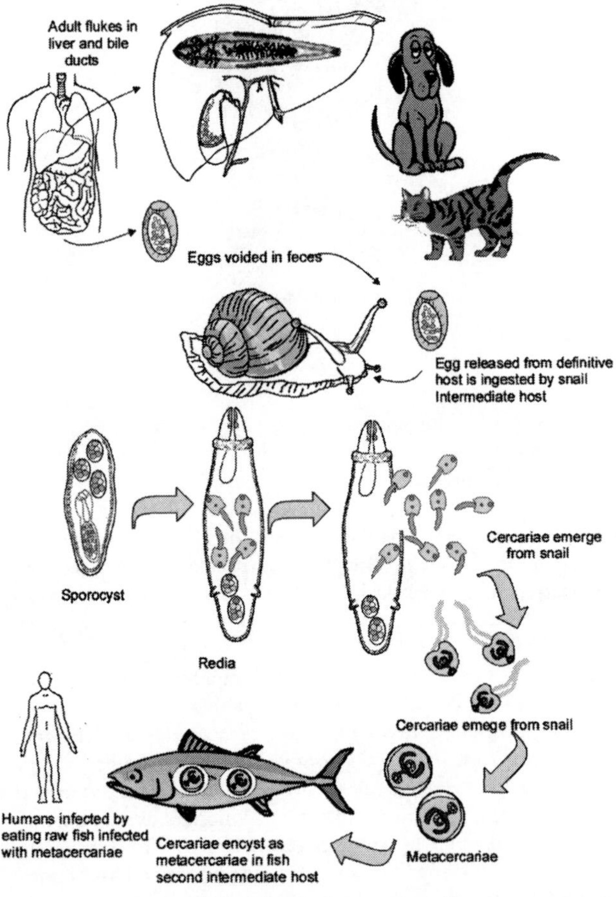

Figure 13.2. Life cycle of *Clonorchis sinensis*. Reproduced from: Nappi AJ, Vass E, eds. Parasites of Medical Importance. Austin: Landes Bioscience, 2002:45.

and *Parafossarulus* geni, which serve as the first intermediate hosts. The miracidia multiply in the snail and develop into cercariae.

After 4-6 weeks of gestation, the cercariae are expelled back into the freshwater environment, where they infect a second intermediate host. These are typically fish of the family Cyprinidae, but at least 113 species of freshwater fish from 13 families have been identified. The cercariae penetrate under the scales of the second intermediate host, where they then encyst in the muscle and form metacercariae. In this stage, the parasite lies dormant until ingested by the definitive host.

In addition to humans, naturally occurring definitive hosts include dogs, pigs, cats (both domestic and wild), martens, badgers, minks, weasels and rats. Once ingested, the metacercariae excyst in the duodenum or jejunum and migrate through the ampulla of Vater to adhere to the common bile duct. Following the epithelial lining of the biliary tree, they proceed into the intrahepatic ducts, typically the smaller branches in the left lobe. Other less common sites of residence include the gallbladder, pancreatic duct and, very rarely, the stomach. Once their final destination is reached, they mature into egg-producing adults worms within 4 weeks.

Disease Signs and Symptoms

Symptomatology of active clonorchiasis and opisthorchiasis, whether in an acute or chronic stage of infection, is dependent on the burden of adult worms in the biliary tree. So-called "light" infections, classified by either <10,000 eggs per gram of stool, or <100 adult worms, rarely cause symptoms. "Heavy" infections, which are more likely to cause symptoms, occur in only 10% of cases.

While *O. felineus* is the most likely of the three species under discussion to result in symptomatic acute infection, symptomatology in the acute phase is uncommon. The onset is usually 1-3 weeks after the ingestion of metacercariae and can persist for 2-4 weeks prior to resolution. Symptoms include fever, malaise, arthralgia/myalgia and anorexia, as well as abdominal pain and urticaria. Clinical signs are usually limited to tender hepatomegaly and lymphadenopathy. At this stage, peripheral eosinophilia is common and as symptoms resolve, eggs become detectible in the stool.

After infection has become established, further symptoms and complications can develop by a number of mechanisms. Mechanical obstruction of the biliary tract by adult worms results in intermittent but recurring symptoms, especially in those with "heavy" infection, including anorexia, weight loss, fatigue, abdominal pain, diarrhea and dyspepsia. Additionally, persistent irritation and damage to the biliary epithelium results in both intimal desquamation and proliferation, which eventually results in fibrotic, as well as hyperplastic and dysplastic, alterations in architecture. Immune-mediated tissue damage occurs locally, characterized by the periductal infiltration of lymphocytes and eosinophils. Ductal dilatation and stricture formation, pigment stone generation and hepatocellular fibrosis develop and the gallbladder becomes distended with stones. The ultimate serious sequelae of this include recurrent cholangitis, cholangiohepatitis and pancreatitis.

The most dire complication of clonorchiasis and opisthorchiasis is cholangiocarcinoma. While the specific carcinogenic mechanism is uncertain, postulated means include both intrinsic nitrosation and nitric oxide formation, as well as enzymatic activation. Weight loss and epigastric pain are the primary complaints, with ascities, jaundice and a palpable abdominal mass notable on exam. Overall survival is 6.5 months.

Diagnosis

Laboratory Tests, Microscopy

Classically, the diagnosis of clonorchiasis and opisthorchiasis is made by demonstrating eggs in the stool of infected hosts. They are typically not present until 4 weeks after establishment of infections, and in light infections may require specimen concentration to be detected. As the eggs of *C. sinensis* are difficult to distinguish from *Opisthorchis* spp., differentiation of species usually requires examination of expired adult flukes following therapy.

Biliary or duodenal fluid, sampled by endoscopic retrograde cholangiopancreatography (ERCP) or needle aspiration, can also be found to demonstrate eggs or adult worms.

Other Tests

Serum IgE levels may be elevated in liver fluke infection. Peripheral eosinophilia can be observed, but does not typically exceed 10-20% of the total WBC count. Alkaline phosphatase levels can be elevated in advanced liver disease, but other liver associated enzymes, such as the transaminases, are usually within normal limits.

Radiologic tests demonstrate the hepatic and biliary sequelae of chronic infection. Asymptomatic disease is associated with no radiologic abnormalities. Ultrasound can demonstrate nonshadowing echogenic foci within bile ducts that represent fluke aggregates. Other nonspecific changes by ultrasound can include hepatomegaly, ductal fibrosis and inflammation, as well as gallbladder irregularities and sludge. However, such changes are only seen in 50% of patients with active chronic disease. CT scanning is reportedly more sensitive, especially for detecting relapsing cholangitis.

Cholangiography can show multiple findings, depending on the stage of infection and worm burden. Multiple saccular or cystic dilatations of the intrahepatic bile ducts can appear as a "mulberry sign." The "arrow-head sign" is the rapid tapering of the intrahepatic bile ducts to the periphery. Portal and periportal fibrosis can result in a decrease in the number of intrahepatic radicles visualized. Individual adult worms can be seen as filamentous, wavy and elliptical filling defects. Duct wall irregularities, varying from small indentations—which, in series, can give rise to a "scalloped" appearance—to hemispherical filling defects.

Molecular Diagnosis

There are no widely available ELISA tests for liver flukes currently, although several are under investigation. Concerns surrounding one ELISA for *O. viverrini* include the inability of a positive result to distinguish between current and prior infection, as well as possible cross-reactivity with other parasitic infections. Another monoclonal antibody-based dot-ELISA has been demonstrated to have 100% sensitivity and specificity for *O. viverrini*, using purified antigen from that organism. An immunoblot assay for *C. sinensis*-specific excretory-secretory antigens has been shown to have 92% sensitivity in active infection.

PCR is also being investigated as a diagnostic modality for *O. viverrini* in stool. It has been described to have a sensitivity correlated with the amount of egg burden present, 100% for >1000 eggs per gram to 50% for <200 eggs per gram.

Treatment

The treatment of choice is praziquantel, 75 mg/kg divided into three doses given for 2 days. This has been demonstrated to have a nearly 100% cure rate, except for very heavy infections, for which 2 days of therapy can achieve similar success. Typically, eggs will disappear from the stool within 1 week after treatment, although clinical symptoms may take months to resolve due to residual damage to the biliary tract from the adult worms.

Side effects of treatment can be significant and include nausea, vomiting, headache, dizziness and insomnia. In an attempt to reduce the severity of adverse reactions, a lower dose regimen of 25 mg/kg/d for 3 days was investigated, but was found to have only an eradication rate of 29%. Similarly, single-dose praziquantel at 40 mg is often used in the setting of mass therapy for whole communities, both for convenience and for reduced side effects, but has only a 25% eradication rate.

Albendazole is an alternative to praziquantel therapy and is typically given 10 mg/kg/d for 7 days. It has been described as having eradication rates from 90-100%. While it does have a less severe side effect profile, its significantly longer therapy course and higher pill burden can make it less appealing.

With treatment, prognosis is good in light infections. Heavy infections, especially those that are long-standing, can occasionally end in death due to complications. Surgery is reserved for severe sequelae of clonorchiasis, such as emergent cholecystectomy with cholangitis, or palliative choledochojejunostomy in the setting of obstructive jaundice. Urgent biliary decompression may be required for acute cholangitis. When cholangiocarcinoma is suspected, endoscopic biopsies can be taken. Secondary bacterial infections can occur and should be managed with appropriate antibiotics.

Prevention and Prophylaxis

Proper preparation of freshwater fish, such as cooking or freezing, destroys the metacercariae and prevents infection. As fecal-oral inoculation can occur, transmission can also be reduced by strict hygiene measures. There is no role for chemoprophylaxis.

Suggested Reading

1. Kaewkes S. Taxonomy and biology of liver flukes. Acta Tropica 2003; 88:177-86.
2. Keiser J, Utzinger J. Emerging foodborne trematodiasis. Emerg Infect Dis 2005; 11:1507-14.
3. Leder K, Weller P. Liver flukes: Clonorchiasis and opisthorchiasis. UpToDate.com.
4. Lun ZR, Gasser RB, Lai DH et al. Clonorchiasis: A key foodborn zoonosis in China. Lancet Infect Dis 2005; 5:31-41.
5. Mairiang E, Mairiang P. Clinical manifestation of opisthorchiasis and treatment. Acta Tropica 2003; 88:221-7.
6. Reddy DN, Kumar YR. Endoscopic diagnosis and management of biliary parasitosis. UpToDate.com.
7. Rim HJ. Clonorchiasis: An update. J Helminthol 2005; 79:269-81.
8. Upatham ES, Viyanant V. Opisthoschis viverrini and opisthorchiasis: A historical review and future perspective. Acta Tropica 2003; 88:171-6.

CHAPTER 14

Liver Fluke: *Fasciola hepatica*

Michelle Paulson

Background

Geographical Distribution/Epidemiology

Fascioliasis, also known as sheep liver fluke infection, was once primarily thought of as a veterinary problem. Now it is emerging as a significant human parasitic disease throughout the world. It has been estimated that up to 17 million people are infected and that another 180 million are at risk. The pathogen is widespread, causing infections in Europe, Central and South America, Mexico, the Middle East and Asia. Areas with particularly high prevalence include the Altiplano region of Bolivia, the Mantaro Valley in Peru, the Abis area along the Nile River basis in Egypt and the Gilan province in Iran. Other rural areas may also have a high disease burden, but do not have the means to diagnose or report it. Contrary to expectation, human disease does not correlate to the areas where the most livestock are infected. Rather, it more closely correlates with the population of the intermediate host, the Lymnaeidae family of snails. In France, disease has primarily been associated with consumption of contaminated watercress. The United Kingdom has reported fascioliasis from vegetables. It has also been described within the United States, although the predominant threat is to livestock, rather than humans. Because of increasing commerce between countries, the risk of limited outbreaks within the United States exists.

In endemic areas children are most commonly affected. In one study of Peruvian children, disease was associated with consumption of alfalfa juice. It is also postulated that children are left to tend to animal herds, thereby increasing their length of exposure to infectious sources. Often a history of eating potentially contaminated vegetables cannot be elicited, suggesting that water contamination is likely.

Causative Agents

Fascioliasis is caused by a leaf-shaped trematode, *Fasciola hepatica*. The adult is brown and flat with average size of 2.5 × 1 cm. The eggs are yellow-brown, oval and measure 140 × 75 μm.

While termed the sheep liver fluke, it typically infects sheep, goats and cattle. In Bolivia, pigs and donkeys are also efficient reservoirs. In Corsica, the flukes have been isolated from black rats. Snails from the Lymnaeidae family are found in many different geographical areas, demonstrating an ability to adapt to a broad spectrum of environments, from the Andes Mountains to the French countryside. Humans obtain the disease by eating contaminated water plants, in particular

Medical Parasitology, edited by Abhay R. Satoskar, Gary L. Simon, Peter J. Hotez and Moriya Tsuji. ©2009 Landes Bioscience.

watercress (*Nasturtium officinale*), or by ingestion of contaminated water or foods prepared in that water.

Mode of Transmission

The human cycle begins when encysted metacercariae adhere to water plants, resulting in infection if the vegetation is not washed carefully prior to consumption. Once inside the small bowel, the metacercariae excyst and the larvae emerge. They travel through the lumen of the bowel, penetrate into the peritoneum, and invade the liver capsule. At this point, the larvae mature into adults, which gradually migrate into the biliary tract. The hermaphroditic flukes dwell in the biliary tree and produce eggs that may be passed in the stool. The eggs hatch in fresh water and the new miracidia infect snails, in particular those of the family *Lymnaeidae*, which are the intermediate host. The miracidia mature within the snails to form cercariae. They are released from snails in water, where they can attach to water-borne plants such as watercress. The cycle begins again when the contaminated vegetation or water are consumed.

Disease Signs and Symptoms

Fascioliasis occurs in two stages, which differ in symptoms based on the migration of *F. hepatica* through various organs.

The first stage, or prepatent or larval period, is marked clinically by abdominal pain, fever, weight loss and urticaria. Eosinophilia and elevations in liver transaminase enzymes may occur. A small portion of patients also experience cough and chest discomfort. This stage corresponds to the ingestion of the organism and penetration through the intestinal wall to the liver. The first stage can last for several months. Egg production during this time is minimal.

The second stage, referred to as the patent or biliary period, represents the maturation of larvae into adult flukes that pass into the biliary ducts. Symptoms during this phase are often subtle, vague and even asymptomatic. Patients may develop intermittent right upper quadrant pain, which can mimic cholecystitis. Eosinophilia may still be present. During this phase, ova are released and may be found on careful, repeated stool examination. See Figure 14.1.

Complications from chronic disease include anemia, cholangitis and biliary obstruction. Subcapsular liver hematomas and hemoperitoneum are also reported. Cases of invasion into inguinal lymph nodes have been described. Other ectopic sites include subcutaneous skin, brain and eyes. There is no known potential for malignancy of the biliary tract associated with chronic infection.

Diagnosis

Fascioliasis should be suspected in patients with abdominal pain and fever, in the setting of abnormal biliary tract findings on radiological studies. In the acute phase, eosinophilia provides a clue. Later in the disease, symptoms may only occur intermittently and are often vague, making the diagnosis more difficult. Abnormal liver function tests and elevated total bilirubin level may or may not be present. Coprologic analysis is limited, especially in the early stages of the disease, but is very specific when eggs are found. Even in later stages, multiple stool examinations may be necessary to increase sensitivity. Indirect hemagglutination tests were previously used to determine titers. The tests most

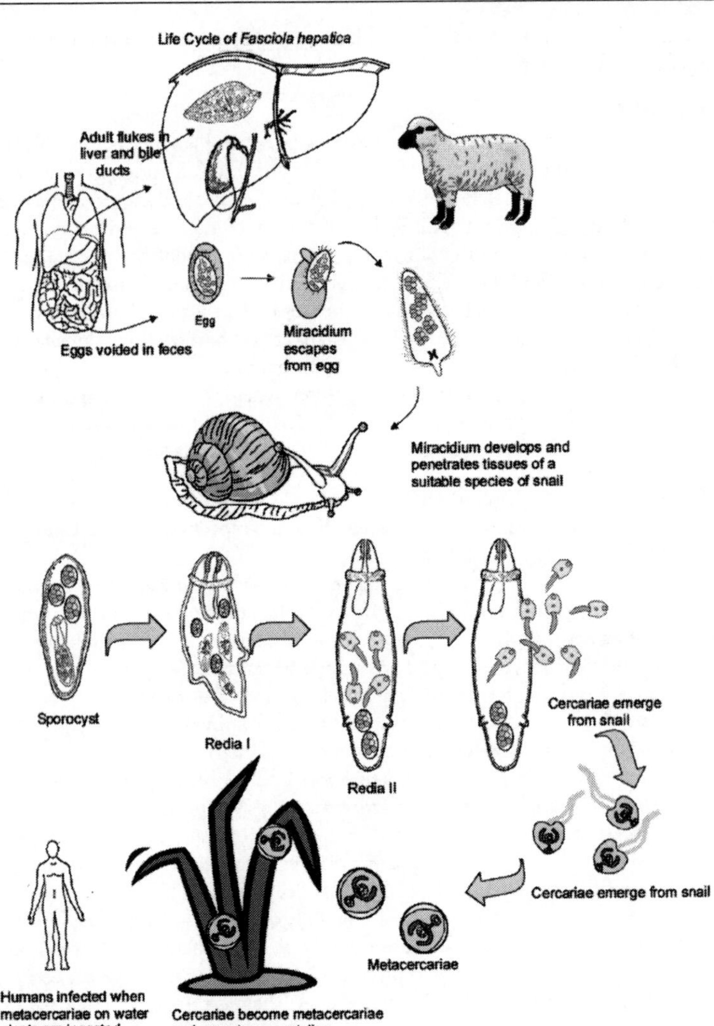

Figure 14.1. Life cycle of *Fasciola hepatica*. Reproduced from: Nappi AJ, Vass E, eds. Parasites of Medical Importance. Austin: Landes Bioscience, 2002:43.

commonly used to confirm the diagnosis are ELISAs, which detect various *F. hepatica* antigens. Earlier developed ELISAs utilized excretory-secretory antigens from adult organisms or coproantigens. One study described a sensitivity of 100% for detection of acute infection through use of excretory-secretory antigens. In chronic infections, however, the sensitivity was only 70%. The antigen tests reverted to normal generally within six months. The greatest drawback to these tests is cross-reactivity with other parasitic infections. More recently the cathepsin L1 (CL1) proteinase was found to react with serum IgG4 and was incorporated into an ELISA. This improved detection when compared to coprologic testing alone. Another ELISA under development uses the purified protein Fas2, which is an adult *F. hepatica* cysteine proteinase.

Radiological studies are useful in aiding diagnosis. Computerized tomography and ultrasound can assist in assessing biliary invasion. Common findings include thickening of the bile duct wall and biliary dilation. On ultrasound, mobile flukes may be demonstrated, but the more characteristic lesion is a crescent-shape in the biliary tract. A negative ultrasound examination does not exclude the possibility of infection. By CT, lesions are characterized as small, multiple hypodensities less than 10 mm in size. They can project in a tunneled, branching pattern. Larger abscesses have also been reported. Endoscopic retrograde cholangiography is not necessary for diagnosis in most cases but can be used to extract flukes from the bile ducts when they cause significant obstruction. Liver biopsy is not routinely indicated but may show eosinophils, histiocytes, granulomas and in some cases, even eggs.

Treatment

Triclabendazole, a benzimidazole, is the first line treatment for *F. hepatica*. It has an active sulfoxide metabolite with can inhibit fluke microtubules and protein synthesis. Therefore, it is a good drug to target both immature and mature forms of the trematode. Its primary use stems from treatment of fascioliasis in sheep and cattle beginning in 1983. It subsequently was used for human disease in 1989 during an epidemic in Iran. In 1997 it was officially listed on the World Health Organization's list of essential drugs. Currently, triclabendazole is the drug of choice selected by the Center for Disease Control and Prevention. The recommended dose is 10 mg/kg as a single dose. For more severe disease, an additional dose taken 12 hours later may be necessary. One study of Cuban patients who had previously been treated with anthelminthic medications showed that two doses of 10 mg/kg taken over 1 day eliminated egg excretion in 71 of 77 patients. The main side effect noted was abdominal discomfort, which was attributed to the expulsion of the organism through the bile ducts. This occurred 2-7 days after taking triclabendazole. Other reported side effects include dizziness, headache and fever. In another study, 134 Egyptian patients with disease, as established by egg count, were treated with triclabendazole. An initial dose provided cure in 86.6% of those infected and a subsequent dose in those who remained positive increased the number cured to 93.9%.

Nitazoxanide, an inhibitor of pyruvate:ferredoxin oxidoreductase enzyme, has also shown to be effective in small trials and case reports. One randomized, placebo controlled study demonstrated successful parasite elimination in 18 of 30 adults (60%) and 14 of 35 children (40%), whereas only 1 of 8 adults and 0 of 8 children were cured with placebo alone. The adult dose used in this trial was 500 mg taken twice a day with meals for 7 days. Most side effects reported were mild and included abdominal pain, diarrhea and headaches.

Bithionol, at a dose of 30-50 mg/kg on alternate days for 10-15 doses is an alternative choice, but carries significant side effects such as nausea, vomiting, abdominal pain and urticaria.

Praziquantel, which is used in other fluke infections, is not effective in fascioliasis. Other medications used to treat parasitic infections such as metronidazole and albendazole are also not efficacious.

Larger therapy trials need to be undertaken to better establish drug efficacy of current regimens, while development of new antifasciola medications continues. Treatment success can be based on several endpoints: lack of further detection of fecal eggs, antigen reversion to negative, absence of radiological evidence of persistent infection and resolution of symptoms. ELISA testing may take up to a year or more to normalize.

Prevention and Prophylaxis

Early detection is one strategy to prevent chronic fascioliasis. Despite its worldwide prevalence, it is often goes unsuspected in nonendemic areas. It is therefore important to develop a simple, cost effective method for rapid diagnosis. In endemic regions public health measures have been instituted to raise awareness of symptoms that can aid in earlier detection. Health care workers themselves need to be better trained to recognize disease. Other measures have attempted to better define the population at risk and not limit disease potential to only persons who report consumption of watercress; frequently this history cannot be elicited.

Another approach to disease prevention focuses on improving sanitation. Improvement of inspection and transport of vegetation can further help to reduce risk. Educating the public on the proper ways of cooking and cleaning vegetation also decreases disease. Better sanitation, in particular decreasing outdoor defecation and thus shedding of viable eggs, will also be key in limiting fascioliasis.

Governments need to establish guidelines with the assistance of medical specialists for routine treatment of livestock. Safe means of controlling snail populations also need to be explored. The Egyptian Ministry of Health and Population launched a program targeting school children from high prevalence areas, screening 36,000 children and subsequently treating 1280. The prevalence of the disease then fell from 5.6 to 1.2%. Similar programs need to be explored further to evaluate their feasibility in other endemic regions.

Suggested Reading

1. Arjona R, Riancho JA, Aguado JM et al. Fascioliasis in developed countries: a review of classic and aberrant forms of the disease. Medicine 1995; 74:13-23.
2. Curtale F, Hassanein Y, Savioli L. Control of human fascioliasis by selective chemotherapy: design, cost and effect of the first public health, school-based intervention implemented in endemic areas of the Nile Delta, Egypt. Trans Royal Soc of Trop Med Hyg 2005; 99:599-609.
3. El-Morshedy H, Farghaly A, Sharaf A et al. Triclabendazole in the treatment of human fascioliasis: a community-based study. East Mediterr Health J 1999; 5:888-94.
4. Espino AM, Diaz A, Perez A et al. Dynamics of antigenemia and coproantigens during a human Fasciola hepatica outbreak. J Clin Mic 1998; 36:2723-6.
5. Espinoza JR, Timoteo O, Herrera-Velit P. Fas-2 ELISA in the detection of human infection by Fasciola hepatica. J Helminth 2005; 79:235-40.
6. Farag HF. Human fascioliasis in some countries of the Eastern Mediterranean Region. East Mediterr Health J 1998; 4:156-60.

7. Favennec L, Jave Ortiz J, Gargala G et al. Double-blind, randomized, placebo-controlled study of nitazoxanide in the treatment of fascioliasis in adults and children from northern Peru. Aliment Pharmacol Ther 2003; 17:265-70.
8. Graham CS, Brodie SB, Weller PF. Imported Fasciola hepatica infection in the United States and treatment with triclabendazole. Clin Inf Dis 2001; 33:1-6.
9. Harris NL, McNeely WF, Shepard JO et al. Case 12-2002, Weekly clinicopathological exercises. NEJM 2002; 346:1232-9.
10. Keiser J, Utzinger J. Emerging foodborne trematodiasis. EID 2005; 10:1507-14.
11. Mas-Coma MS, Esteban JG, Bargues MD. Epidemiology of human fascioliasis: a review and proposed new classification. Bull WHO 1999; 77:340-6.
12. Millán JC, Mull R, Freise S et al. The efficacy and tolerability of triclabendazole in Cuban patients with latent and chronic Fasciola hepatica infection. Am J Trop Med Hyg 2000; 63:264-9.
13. O'Neill SM, Parkinson M, Dowd AJ et al. Short report: immunodiagnosis of human fascioliasis using recombinant Fasciola hepatica cathepsin L1 cysteine proteinase. Am J Trop Med Hyg 1999; 60:749-51.
14. Rossignol JF, Abaza H, Friedman H. Successful treatment of human fascioliasis with nitazoxanide. Trans Royal Soc of Trop Med Hyg 1998; 92:103-4.
15. Saba R, Korkmaz M, Inan D et al. Human fascioliasis. Clin Microbiol Infect 2004; 10:385-7.
16. Sezgin O, Altintas E, Disibeyaz S et al. Hepatobiliary fascioliasis: clinical and radiologic features and endoscopic management. J Clin Gastroenterol 2004; 38:285-91.
17. Shehab AY, Hassan EM, Abou Basha LM et al. Detection of circulating E/S antigens in the sera of patients with fascioliasis by IELISA: a tool of serodiagnosis and assessment of cure. Trop Med Int Health 1999; 4:686-90.
18. Talaie H, Emami H, Yadegarinia D et al. Randomized trial of a single, double and triple dose of 10 mg/kg of a human formulation of triclabendazole in patients with fascioliasis. Clin Exp Pharm Phys 2004; 31:777-82.
19. Trueba G, Guerrero T, Fornasini M et al. Detection of Fasciola hepatica infection in a community located in the Ecuadorian Andes. Am J Trop Med Hyg 2000; 62:518.

Chapter 15

Paragonimiasis

Angelike Liappis

Introduction

Paragonimiasis is a parasitic disease acquired when trematodes of the genus *Paragonimus* are accidentally consumed by a human host. The adult *Paragonimus* (or lung fluke) commonly localizes to the bronchioles after ingestion. While pulmonary involvement is the most common clinical complication of infection, the disease is often asymptomatic. When manifested by pleuritic chest pain and hemoptysis, pulmonary paragonimiasis may mimic the clinical presentation of tuberculosis in countries where both diseases are endemic. A careful epidemiologic history and a high clinical suspicion for diagnosis are necessary when considering paragonimiasis in a patient with suspected pulmonary disease.

Geographic Distribution and Epidemiology

Paragonimus is a parasite of mammalian carnivores and omnivores whose diet includes fresh-water crustaceans (crabs, crayfish). First described in tigers more than a century ago, a variety of wild and domesticated carnivores have been found to harbor Paragonimus species worldwide.

With a geographic range which spans most of Asia and parts of Africa, *Paragonimus westermani* is the most commonly identified lung fluke. Globally, a quarter of the more than 50 known species of this parasite have been associated with human disease. Regional endemic species causing clinical infection include *P. mexicanus* (Central America), *P. skrjabini* (China), *P. miyazakii* (Japan), *P. uterobilateralis* (West Africa) and *P. kellicotti* (North America).

Humans acquire the parasite from the ingestion of uncooked or insufficiently cooked fresh-water crustaceans or, rarely, through the flesh of mammalian reservoirs which harbor immature flukes. In experimental models, feeding *Paragonimus* metacercariae to rats results in the development of mature flukes in the lung, but immature, underdeveloped parasites in the muscles. Clinical human disease has been associated with the consumption of undercooked wild boar in Asia.

In North America, *P. kellicotti* is a rare cause of human infection. The majority of paragonimiasis diagnosed in the United States is attributable to exposures to imported, high risk food items or to those patients traveling from countries where the local, endemic *Paragonimus* species are associated with human disease and have had the proper epidemiologic exposure history.

In countries where the parasite is endemic, social customs and regional preparation techniques have been implicated as factors which increase the risk of human infection. Raw crab and/or crayfish preparations eaten locally or frozen for export are potential sources of exposure. In some cultural settings, the consumption of

Medical Parasitology, edited by Abhay R. Satoskar, Gary L. Simon, Peter J. Hotez and Moriya Tsuji. ©2009 Landes Bioscience.

pickled or raw seafood is a traditional food preparation technique. Infections among families and other groups who routinely share meals have been described. Clinical paragonimiasis has been linked to several traditional dishes, including a raw seafood salad served in the Americas known as *seviche* and *drunken crab* (raw crabmeat marinated in wine or alcohol) common to several Asian countries.

The infection rate of freshwater crabs, collected from regional waterways or sold in local markets, are as high as 6 to 12% in Africa and Asia, respectively. Paragonimiasis is estimated to infect several million people worldwide, with projections of those at risk of acquiring disease reaching 200 million.

Pathogenesis

Adult worms (approximately 0.8-1.6 cm in length × 0.4-0.8 cm in width) develop in the pulmonary bronchioles, where they can encapsulate (Fig. 15.1). After approximately 6 weeks, dark brown, operculate eggs (80-120 μm × 50-60 μm) are produced. The parasite eggs pass from stool and respiratory secretions into freshwater and develop into miracidia, passing through two intermediate invertebrate hosts. Over several weeks to months, the parasite matures in freshwater snails to become cercaria which are either eaten or penetrate the secondary, crustacean hosts. The cercaria mature within the muscles and viscera of a crayfish or crab into metacercariae over 6-8 weeks. The cycle is completed when these infective metacercariae are consumed by the mammalian host and penetrate the intestinal wall, migrating through the peritoneum into the lungs and other organs.

Migration and maturation in the lungs of the definitive host may take several weeks. The parasite localizes to the bronchioles where it may cause an acute pneumonitis, associated with hemorrhage and an eosinophilic infiltrate. Adult flukes mature within the lumen of the bronchioles or may encyst in the lung parenchyma. Infection can persist for several years after a host leaves an endemic area.

Paragonimus has evolved a number of mechanisms to evade host defense. The parasites elaborate several excretory-secretory products (ESP) including a number of cysteine proteinases and, in a recent report, copper/zinc superoxide dismutase. Superoxide dismutase is constitutively expressed during maturation and protects the parasite from toxic cellular enzymes elaborated by the host. *Paragonimus* ESPs also modulate other aspects of the host inflammatory response. ESPs have been found to regulate the survival of eosinophils, to play a role in the degradation of host antiparasitic immunoglobulin, to regulate migration and tissue penetration, to facilitate encystment within the lung and other tissues and to modulate the production of chemokines such as interleukin-8.

Clinical Presentation and Diagnostic Evaluation

Paragonimiasis may present with a range of clinical findings. The majority of patients infected will have few clinical symptoms. The most common finding may be a nonspecific elevation of the peripheral eosinophil count. In one report, 80% of patients with paragonimiasis had eosinophilia detected on peripheral blood smear.

More severe disease is associated with higher organism burden. Symptomatic infection is a manifestation of inflammatory, often hemorrhagic lesions produced as the organisms migrate from the bowel, traverse the peritoneal cavity and lodge in the lungs and/or other organs. The local inflammatory reaction includes eosinophilic

15

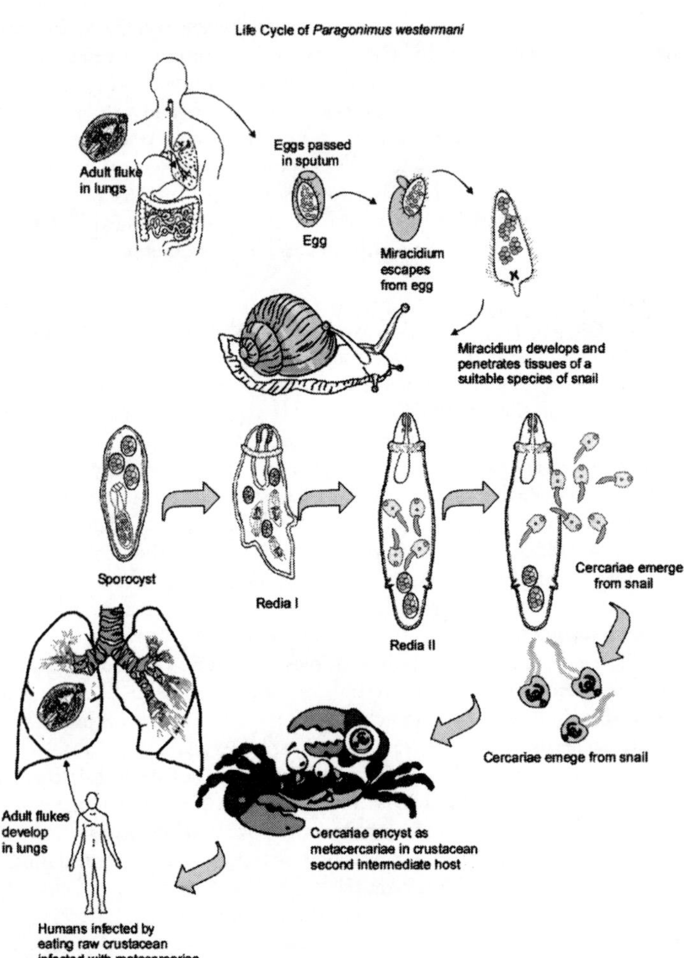

Figure 15.1. Life Cycle of *Paragonimus westermani*. Reproduced from: Nappi AJ, Vass E, eds. Parasites of Medical Importance. Austin: Landes Bioscience, 2002:47.

exudates and, in some cases, a granulomatous reaction to the eggs. Examination of expectorated sputum under light microscopy should be obtained for the presence of the brown, operculate eggs as well as for eosinophils and Charcot-Leyden crystals. When secretions are swallowed, eggs may also be found in the stool of infected patients. Both sputum and stool should be sent for microscopic evaluation when pulmonary paragonimiasis is suspected.

Hemoptysis and cough are the most common clinical symptoms of pulmonary paragonimiasis. In countries where both paragonimiasis and tuberculosis are endemic, it is not uncommon for patients to be treated empirically for tuberculosis based on their similar clinical presentations. The parasitic diagnosis is often obtained after a more thorough inspection of the sputum and/or the stools are sent for examination. In addition to mimicking tuberculosis, pulmonary paragonimiasis can be mistaken for bronchiectasis or other respiratory diseases that may present with increased sputum production and/or pleuritic chest pain. A careful epidemiologic history and high clinical suspicion for diagnosis are necessary when considering this disease in a patient with pulmonary disease.

Examination of pleural fluid may also demonstrate eosinophilia and the eggs on cytologic examination. Other pleural fluid indicators mimic bacterial infection with low pH, low glucose (<10 mg/dl) and elevated protein and lactate dehydrogenase, suggesting an exudative process. The parasite and the eggs have been identified on microscopic examination of bronchoalveolar lavage and fine needle aspiration of pulmonary lesions. In rare cases, the adult worms may be seen in the sputum or expectorated with coughing after the initiation of therapy.

The organism's migration through the peritoneal cavity may lead to intra-abdominal seeding. Clinical cases of hepatic, omental and retroperitoneal collections have been described. Eggs reaching the circulation or fluke migrations outside the pulmonary cavity may direct disease to distal sites, leading to ectopic foci of infection. The immature flukes of *P. skrjabini,* native to China, have been associated with cutaneous involvement known as *trematode larval migrans*. Unlike the other species of *Paragonimus*, this particular species is associated with migratory subcutaneous nodules and, only rarely, pulmonary disease.

Worldwide, other focal skin infections have been reported, as well as intraocular, pericardial or brain involvement with *P. westermani* and a variety of other *Paragonimus* species. Cerebral involvement is one of the most common extrapulmonary sites. When affecting the brain, paragonimiasis may present as a mass lesion with or without seizure, chronic meningeal irritation or acute meningitis; it is associated with a high mortality due to the higher incidence of intracranial hemorrhage.

The diagnosis of paragonimiasis may be aided by radiologic imaging; however the findings are generally nonspecific. In a retrospective review of over 70 patients with pleuropulmonary paragonimiasis, chest radiographs demonstrated pleural lesions in over half of the cases reviewed, with less than 20% showing evidence of pleural effusion. Nonspecific findings of the parenchymal lung disease on radiographs include air space consolidation, cystic changes and peripheral linear densities. The radiographically opaque linear densities have been attributed to the parasite's migration tracks through the tissues.

Computed tomography (CT) has been used to aid in the diagnosis of pulmonary and extrapulmonary paragonimiasis. CT is more sensitive in demonstrating small pleural effusions and better depicts the mass lesions seen on radiographs. Masses usually consist of one or more peripheral, irregularly shaped densities which calcify with time. Characteristically, the mass lesions may display either a low attenuation signal at their center or show evidence of an air-fluid level, associated with ring-like enhancement at the edges. In setting of pulmonary paragonimiasis, these lesions have been described as "worm cysts." Recently, Kuroki et al evaluated eight patients with high-resolution CT and were better able to demonstrate bronchial thickening and ground-glass opacities in association with the usual CT findings of pulmonary paragonimiasis.

Radiologic imaging studies serve as an important adjunct to the diagnosis of paragonimiasis, but definitive diagnostic evaluation today requires microbiologic and, increasingly, serologic methods. It may take up to 3 weeks for flukes to mature and produce eggs visible for microscopic diagnosis. To augment microscopic examination in cases where infection may be early or those samples difficult to access (extrapulmonary disease), immunologic tests have been developed.

The diagnosis of patients with early pulmonary disease or those with extrapulmonary disease has been improved with the development of serologic testing. While a simple intradermal test has been developed, the cutaneous reaction cannot accurately distinguish between active or past infection despite good sensitivity. Enzyme linked immunosorbent assays (ELISA) with greater specificity and sensitivity for infection have been developed, including an ELISA using the parasite's excretory-secretory antigens. PCR assays are under development and are not yet commercially available; however these assays hold the promise of being able to differentiate among the various *Paragonimus* species in addition to diagnosing infection.

15

Treatment and Prevention

The current treatment of choice for paragonimiasis is praziquantel. It is given orally at 25 mg/kg, three times a day for two days for those patients with pulmonary disease. Longer treatment courses may be necessary for extrapulmonary infection, particularly lesions involving the brain. Surgical resection, combined with medical therapy, may be necessary in severe plueropulmonary disease.

Praziquantel is associated with headache and drowsiness, but is generally well tolerated. It is highly effective against all species of Paragonimus, with a cure rate exceeding 95%. Bithionol, which was used in the past, required up to 25 days of therapy, but has been largely replaced by praziquantel due to compliance and toxicity issues. Approved only for veterinary use, triclabendazole as a single 10 mg/kg dose is also effective and is used outside the US in the treatment of pulmonary disease.

The in vitro effect of praziquantel on *Paragonimus* has been examined by electron microscopy. Parasites under treatment exhibit severe vacuolization in a variety of organs, including the tegument and reproductive system. The worms are both immobilized and unable to produce eggs after praziquantel therapy. The partially disintegrated or intact, immobilized adults or the eggs may persist in tissues for weeks after treatment. Resolution by radiologic imaging may be delayed despite clinical response. The dead and dying parasites and/or eggs create a persistent antigenic challenge, decreasing the utility of following posttreatment serologies.

In endemic areas the main risk for human acquisition of disease is food preparation technique. Prevention can best be achieved by avoiding undercooked freshwater crabs and crayfish and in improving the monitoring of imported of pickled or frozen raw products imported from endemic countries. Transmission to humans can be minimized by targeting treatment of municipal water supplies in endemic areas. By reducing fecal contamination, transmission in both animal and human hosts can be reduced.

Suggested Reading

1. Blair D, Xu ZB, Agatsuma T. Paragonimiasis and the genus Paragonimus. Adv Parasitol 1999; 42:113-222.
2. Bunnag D, Cross JH, Bunnag T. Lung Fluke Infections: Paragonimiasis, In: Strickland, GT ed. Hunter's Tropical Medicine and Emerging Infectious Diseases, 8th Edition. Philadelphia: WB Saunders, 2000:847-51.
3. Cho SY, Hong ST, Rho YH et al. Application of micro-ELISA in serodiagnosis of Human paragonimiasis. Kisaengchunghak Chapchi 1981; 19:151-6.
4. Harinasuta T, Pungpak S, Keystone JS. Trematode infections. Opisthorchiasis, clonorchiasis, fascioliasis and paragonimiasis. Infect Dis Clin North Am 1993; 7:699.
5. Im JG, Whang HY, Kim WS et al. Pleuropulmonary paragonimiasis: radiologic findings in 71 patients. Am J Roentgenol 1992; 159:39-43.
6. Keiser J, Utzinger J. Chemotherapy for major food-borne trematodes: a review. Expert Opin Pharmacother 2004; 5:1711-26.
7. Kuroki M, Hatabu H, Nakata H et al. High-resolution computed tomography findings of P. westermani. J Thorac Imaging 2005; 20:210-3.
8. Lee SH, Park HJ, Hong SJ et al. In vitro effect of praziquantel on Paragonimus westermani by light and scanning electron microscopic observation. Kisaengchunghak Chapchi 1987; 25:24-36.
9. Li AH, Na BK, Kong Y et al. Molecular cloning and characterization of copper/zinc-superoxide dismutase of Paragonimus westermani. J Parasitol 2005; 91:293-9.
10. Maleewong W. Recent advances in diagnosis of paragonimiasis. Southeast Asian J Trop Med Public Health 1997; 28:134-8.
11. Min DY, Lee YA, Ryu JS et al. Caspase-3-mediated apoptosis of human eosinophils by the tissue-invading helminth Paragonimus westermani. Int Arch Allergy Immunol 2004; 133:357-64.
12. Moyou-Somo R, Kefie-Arrey C, Dreyfuss G et al. An epidemiological study of pleuropulmonary paragonimiasis among pupils in the peri-urban zone of Kumba town, Meme Division, Cameroon. BMC Public Health 2003; 3:40.
13. Narain K, Devi KR, Mahanta J. Development of enzyme-linked immunosorbent assay for serodiagnosis of human paragonimiasis: Indian J Med Res 2005; 121:739-46.
14. Narain K, Rekha Devi K, Mahanta J. A rodent model for pulmonary paragonimiasis. Parasitol Res 2003; 91:517-9.
15. Park H, Kim SI, Hong KM et al. Characterization and classification of five cysteine proteinases expressed by Paragonimus westermani adult worm. Exp Parasitol 2002; 102:143-9.
16. Pezzella AT, Yu HS, Kim JE. Surgical aspects of pulmonary paragonimiasis. Cardiovasc Dis 1981; 8:187-94.
17. Shin MH, Lee SY. Proteolytic activity of cysteine protease in excretory-secretory product of Paragonimus westermani newly excysted metacercariae pivotally regulates IL-8 production of human eosinophils. Parasite Immunol 2000; 22:529-33.
18. Shin MH, Seoh JY, Park HY et al. Excretory-secretory products secreted by Paragonimus westermani delay the spontaneous cell death of human eosinophils through autocrine production of GM-CSF. Int Arch Allergy Immunol 2003; 132:48-57.
19. Singh TS, Mutum SS, Razaque MA. Pulmonary paragonimiasis: clinical features, diagnosis and treatment of 39 cases in Manipur. Trans R Soc Trop Med Hyg 1986; 80:967-71.
20. Spitalny KC, Senft AW, Meglio FD et al. Treatment of pulmonary paragonimiasis with a new broad-spectrum anthelminthic, praziquantel. J Pediatr 1982; 101:144-6.
21. The Trematodes. In: Markell EK, John DT, Krotoski WA, eds. Medical Parisitology, 8th Edition. Philadelphia: WB Saunders, 1999:225-31.
22. Velez ID, Ortega JE, Velasquez LE. Paragonimiasis: a view from Columbia. Clin Chest Med 2002; 23:421-31, ix-x.

CHAPTER 16

Intestinal Trematode Infections

Sharon H. Wu, Peter J. Hotez and Thaddeus K. Graczyk

Introduction and General Epidemiology

The intestinal trematodes are estimated to account for almost 1.3 million of the 40-50 million food-borne trematode infections world-wide, particularly within endemic foci in Southeast Asia and the Western Pacific region where they are significant causes of pediatric malnutrition. Rural poverty and inadequate sanitation represent major risk factors for this food and waterborne infection, in addition to unique cultural traditions of raw food consumption, agricultural use of "night soil" (the use of human feces as fertilizer), promiscuous defecation and an absence of health education. Transmission occurs focally in endemic areas, which are characterized by remote, rural and semi-urban communities that practice small-scale farming. Infections rarely cause mortality; however, they inflict considerable morbidity, particularly seen in school-aged children with chronic infections resulting in malnutrition and stunted growth.

Human intestinal trematodiases are caused by some 70 different species spanning 13 families within the Digenea subclass, parasitising all major vertebrate groups as definitive hosts, gastropods and other mollusk groups as first intermediate hosts and humans as second intermediate hosts. This discussion focuses on major intestinal flukes belonging to the families Fasciolidae (e.g., *Fascioloposis buski*), Echinostomatidae (*Echinostoma spp.*) and Heterophyidae (*Heterophyes heterophyes* and *Metagonimus yokogawai*), which are responsible for the bulk of intestinal trematodiases in humans.

Fasciolidae

Etiology and Definitive Hosts

Fasciolopsis buski is the only species within the Fasciolidae family that regularly infects humans and it is one of the largest, typically ranging from 8-10 cm in length and 1-3 cm wide. Pigs are the most important reservoir though they carry only 3 to 10 adult flukes which produce fewer eggs/adult than in humans, whereas up to 800 flukes/child has been reported with a mean egg production of 16,000 eggs/fluke per day.

Geographical Distribution and Epidemiology

Fasciolopsis buski is endemic in the Far Eastern and Southeast Asian countries of mainland China (including Hong Kong), Taiwan, Thailand, Vietnam, Lao DR, Cambodia, Philippines, Singapore, Indonesia and Malaysia, as well as in India (Assam, Maharashtra, Tamil Nadu and Uttar Pradesh) and Bangladesh.

Medical Parasitology, edited by Abhay R. Satoskar, Gary L. Simon, Peter J. Hotez and Moriya Tsuji.

Fasciolopsiasis is predominantly a neglected tropical disease of school-aged children in rural and remote areas with freshwater habitats and bodies of stagnant water. In such areas, endemic foci range in prevalence from 57% of children in China reported from as many as 10 different provinces, 25% in Taiwan, 50% in Bangladesh, 60% in India and 10% in Thailand. Two studies have reported a slightly greater predilection in females than males in Thailand, the opposite trend reported in Taiwan. The infection is underreported and the exact global prevalence is unknown. However, an estimated 200,000 and 10,000 infections occur in China and Thailand, respectively. Endemic areas are characterized by standing or slow-moving freshwater habitats of food plants such as the water caltrop, water chestnut, watercress, water bamboo, water lily, gankola, or morning glory, feeding both livestock and humans. Infection occurs through ingestion of metacercarial cysts on raw or undercooked water plants, drinking contaminated water and the handling or processing of water plants through the use of teeth to shuck the outer layers.

Life Cycle and Transmission

Indiscriminate defecation and inadequate sanitation play a large role in continuing the life cycle and transmission of this intestinal fluke. Humans shed parasite eggs through the feces and upon contact with water, eggs hatch after a period of 3-7 weeks, releasing miracidium. These then swim to find their first intermediate hosts: snails of the genera *Segmentina, Hippeutis and Gyraulus*, among others. Miracidia attach and penetrate the snails' mantle, tentacles or feet, with sporocyst development occurring within 2 days. Mother radiae develop rapidly inside the sporocyst, emerging within 9-10 days after miracidial penetration. They then migrate to the ovotestis of the snail, where daughter radiae develop and mature, each harboring up to 45 cercariae. These then emerge from the daughter radiae (average patency period 21 days postinfection) and the snails begin to shed cercariae in irregular patterns, all snails dying after their first heavy shedding of 50 cercariae or more. *Fasciolopsis buski* parasitism of the snail causes physiological and histological damage to the ovotestis resulting in castration and death. Cercariae released into water then encyst on fresh water plants or on the water surface, surviving from 64-72 days. Upon ingestion, typically while peeling the plants using one's teeth, the encysted metacercariae exyst in the duodenum and attach to the intestinal wall where, after about 3 months, they develop into adult worms. Although pigs are a potential source of eggs in the environment, the overall contribution of zoonotic transmission to humans is unknown.

Human Disease

The site of infection occurs mainly in the duodenum and jejunum, though heavy infections can affect the entire intestinal tract including the stomach. Clinical symptoms are related to parasite load, with light infections generally being asymptomatic or associated with mild bouts of diarrhea, abdominal pain, anemia, anorexia, eosinophilia, headache and dizziness. Moderate to heavy infections can result in a severe enteritis with extensive intestinal and duodenal erosions, ulcerations, hemorrhaging and abscesses. Fasciolopsis enteritis can result in severe epigastric and abdominal pain, acute ileus, bowel obstruction, nausea, fever, vomiting and marked eosinophilia and leukocytosis. Ascites, general edema and facial edema can result, possibly from elevated immunologic response to absorption of toxic and allergic worm metabolites. In very heavy infections, intestinal obstruction can result in death. Fasciolopsiasis has been shown to decrease intestinal absorption of vitamin B12, though having no effect on carbohydrate, fat and protein absorption. Feces are characterized by a yellow-brown or green color with pieces of undigested food.

Diagnosis and Treatment

Diagnosis of fasciolopsiasis is done by fecal examination to identify the characteristic large, ellipsoidal, operculate eggs, or expelled adult worms. In some cases, patients vomit adult worms, which are subsequently identified. The treatment of choice for fasciolopsiasis is praziquantel, which induces rapid vacuolisation and disintegration of the tegument of the parasite. For children and adults the recommended dose is 75 mg/kg/d in 3 doses given over a period of one day. However, some investigators recommend a single praziquantel dose of 15 mg/kg for children at bedtime, that being the lowest dose with no side effects, and 25 mg/kg for adults. In the past, niclosamide has been recommended as an alternative treatment.

Economic Impact

Fasciolopsiasis is considered to be a major cause of malnutrition in children in endemic areas of underdeveloped countries. During this particularly crucial time of growth and development, improving children's health could have positive economic impact through increasing school attendance, learning and healthy behaviors leading to less-parasitized adults. Measures of disease mortality and morbidity specific for *F. buski* are not available yet, but like other food-borne zoonotic trematodiases, fasciolopsiasis is aggravated by socio-economic factors, particularly practices of free-food markets which lack regulatory inspection and sanitation. It has also important to consider the effect of polyparasitism on child growth. In China, soil-transmitted helminthiases and foodborne trematodiases are prevalent among children in schistosome-endemic areas.

Control and Prevention

The main goal of current control strategies for other helminthiases (e.g., schistosomiasis, soil-transmitted helminthiases) is to prevent disease through large-scale treatment programs rather than reducing or eradicating transmission. Control of intestinal trematode infections requires a slightly different approach. Since morbidity is directly associated with intensity of infection, prevalence is not the best indicator of disease morbidity within a population. A reduction in prevalence may not reflect a reduction of infection intensity after anthelminthic treatment alone, but rather the effects of several interventions acting in concert. A decreasing prevalence pattern of intestinal trematodiases observed currently in Southeast Asian countries proves to be the result of industrialization of animal production, food processing, health education, environmental alteration and changes in solid waste management. WHO and FAO have developed a conceptual framework for integrated food-borne trematode control comprised of (1) diagnosis through routine stool examination, community participation and self-reliance; (2) treatment with anthelminthics to reduce morbidity and human host reservoirs; and

(3) prevention though health education and reduction of risk behavior to break the cycle of transmission.

Prevention of food-borne trematode infections involves economic development, industrialization, agricultural guidelines, changes in eating behavior, changes in defecation behavior and sanitation along with sustained praziquantel treatment of existing infections and reinfections. The simplest method of prevention is health education to thoroughly cook raw vegetables before ingestion and the containment of fecal matter separate from food vegetation. Because the disease is most prevalent in school-aged children, aggressive school-based education programs have been implemented to target those who may not be as entrenched in their eating habits and hygienic customs.

Echinostomatidae

Etiology and Definitive Hosts

Human echinostomiasis is caused by up to 20 different species across eight genera. It differs from other food-borne trematodiases in that its range of definitive and intermediate hosts is quite broad, making parasite transmission especially difficult to interrupt. The reservoir for echinostomes include domesticated and wild waterfowl, other birds, mammals and any aquatic organisms that maintain the intermediate hosts. Intermediate hosts include snails, bivalves, crustaceans, fish, amphibians, rats, dogs, humans and a variety of other mammals. The adult flukes are typically 3-10 mm in length, 1-3 mm wide with a large ventral sucker and numerous collar spines.

Geographical Distribution and Epidemiology

Echinostomiasis is endemic in China, Taiwan, India, Korea, Malaysia, the Philippines and Indonesia, with some foci occurring in Thailand, Hungary, Italy, Romania and Egypt. Prevalence of infection in certain endemic foci is estimated to be 5% in China, 65% in Taiwan, 22% in Korea, 44% in the Philippines and 50% in northern Thailand, with 150,000 people infected in China and 56,000 in Korea. Three species of echinostomes have been reported in Korea: *E. hortense* with over 100 proven cases and 50,000 estimated cases, *E. cinetorchis* with four proven cases and 1,000 estimated cases and *E. japonicus* with four proven cases and 5,000 estimated cases. Infection is associated with consumption of the intermediate host, usually raw or inadequately cooked fish, snails, shrimp, frogs, tadpoles, clams and mussels. Endemic foci are characterized by fresh or brackish water habitats containing intermediate hosts that are consumed raw or undercooked and socio-economic factors that bring populations into contact with aquatic environments for food sources.

Life Cycle and Transmission

Humans ingest metacercariae in raw or undercooked intermediate hosts which then exyst in the intestinal tract, maturing into adults. The adults lay eggs which hatch in the environment, releasing miracidia which then infect an intermediate host, usually a snail or fish. Cercariae develop and are released, encysting in the environment for the next host. Humans are accidental intermediate hosts and echinostomes demonstrate a sylvatic cycle as well as a human cycle.

Human Disease

Clinical symptoms associated with light to moderate infections include anemia, headache, dizziness, stomachache, gastric pain and loose stools while heavy infections can cause eosinophilia, abdominal pain, profuse watery diarrhea, anemia, edema and anorexia. The worms cause pathological changes in the intestinal lumen. Erosions, destruction of villi, loss of mucosal integrity and catarrhal inflammation have all been reported.

Diagnosis and Treatment

Diagnosis of echinostomiasis is done by identifying eggs upon fecal examination. Praziquantel is probably the drug of choice for the treatment of human echinostomiasis, although it has been reported that infections can also be treated with mebendazole, albendazole, or niclosamide.

Economic Impact

Echinostomiasis is an underreported infection in endemic areas, which tend to be remote places where the burden of infection is carried by the rural poor and women of child-bearing age. Like fasciolopsiasis, echinostomiasis is a disease of poverty and poor sanitation. Education, industrialization, changes in ecology and agricultural practices have reduced prevalence of some trematodiases in endemic areas. In Lindu Valley, Indonesia, predation of cercaria-infected *Corbicula* clams by *Tilapia mossambica* fish drastically reduced echinostomiasis infection in the region.

Control and Prevention

The control of echinostomiasis, like that of fasciolopsiasis, involves interrupting the life cycle through diagnosis, treatment and prevention of reinfection. Within its endemic boundaries, however, echinostomiasis is difficult to control or eradicate due to the operation of three independent transmission cycles, i.e., human (anthroponotic), zoonotic (domesticated livestock) and sylvatic (wildlife). Thus, any successful control program involving anthelminthic treatment would have to be sustained to cure re-infections in humans as well as animal populations. Treatment with broad-spectrum anthelminthics would also cure concurrent helminthiases in polyparasitized individuals. Prevention of echinostomiasis follows a similar strategy for fasciolopsiasis, relying on health education targeted to school-aged children to change people's eating habits. Industrialization and economic development would have the greatest impact on reducing transmission since vector and intermediate host control is not possible.

Heterophyidae

Etiology and Definitive Hosts

While there are many genera of heterophyids containing species that cause human trematodiasis, this discussion focuses on *Heterophyes heterophyes* and *Metagonimus yokogawai*, the two most prevalent species infecting humans. Definitive host range for *H. heterophyes* and *M. yokogawai* include various fish-eating mammals such as dogs, cats, foxes, jackals, birds and humans; first intermediate hosts are littorine snails; and second intermediate hosts are shore-fish and brackish water fishes such as cunners, gudgeon, charr, perch, shad, mullet and goby. These worms are small (<0.5 mm in length), egg-shaped and possess a unique morphological feature of a *gonotyl* or genital sucker.

Geographical Distribution and Epidemiology

H. heterophyes have been reported in Egypt, Sudan, Iran, Turkey, Tunisia, China, Taiwan, the Philippines, Indonesia, Africa and India. In one Egyptian village, the highest prevalence of infection, 37%, was found among individuals aged 15-45 years, greater for females than males, followed by 28% prevalence in children under 5 years of age. Over 10,000 people are estimated to be infected in Egypt. *H. heterophyes* in China occurs in approximately 230,000 infected people.

M. yokogawai is mainly distributed in China, Japan, Korea, Taiwan, Indonesia, Russia, Israel, the Balkans and Spain. An estimated 500,000 people are infected in South Korea, 150,000 in Japan and 12,000 in Russia. Epidemiologic studies have compared prevalence of food-borne trematode infections in villages close to water bodies and found a relative risk of 5.01 to 7.44 in Republic of Korea for *M. yokogawai* infection associated with proximity to fresh water. Twenty-five percent of food fishes like perch and mullet studied from Jinju Bay, Korea, were infected with heterophyid metacercaria.

Life Cycle and Transmission

Transmission of heterophyids occurs through consumption of raw, undercooked or improperly processed metacercaria-infected fishes.

Human Disease

H. heterophyes and *M. yokogawai* invade intestinal mucosa of the small intestine, causing inflammation, ulceration and superficial necrosis. Heavy infections are associated with diarrhea, mucus-rich feces, pain, dyspepsia, anorexia, nausea and vomiting. Worm eggs may also enter the lymphatic vessels via the crypts of Lieberkühn, enter the circulation and may occlude cardiac vessels leading to heart damage and fibrosis. Rare cases of egg emboli in the spinal cord and brain have been reported.

Diagnosis and Treatment

Diagnosis is done by fecal examination and treatment is praziquantel administered in three doses of 25 mg/kg in a single day (75 mg/kg/d in 3 doses × 1 d) or as a single dose.

Control and Prevention

Heterophyiases and metagonimiases overlap geographically with other food-borne trematodiases, particularly the liver flukes that cause clonorchiases and opisthorchiases. Control and prevention are similar to those strategies employed for fasciolopsiasis and echinostomiasis. In addition, the major soil-transmitted helminthiases and schistosomiases are coendemic and while the number of these infections has decreased due to large-scale deworming programs in many areas, overall the number of cases of food-borne trematode infections have not decreased significantly. This could be due to a number of factors, such as the rapidly growing aquaculture industry in Asian countries playing a key role in transmission. Prevention would then require strict regulation of this expanding sector of food production along with praziquantel-based treatment programs.

Suggested Reading

1. Control of foodborne trematode infections. Report of a WHO Study Group. World Health Organ Tech Rep Ser 1995; 849:1-157.
2. Abdussalam M, Kaferstein FK, Mott KE. Food safety measures for the control of foodborne trematode infections. Food Control 1995; 6:71-79.
3. Keiser J, Utzinger J. Emerging foodborne trematodiasis. Emerg Infect Di 2005; 11:1507-14.
4. Mas-Coma S, Valero B. Fascioliasis and other plant-borne trematode zoonoses. Int J Parasitol 2005; 35:1255-78.
5. Manning GS, Ratanarat C. Fasciolopsis buski (Lankester, 1957) in Thailand. Amer J Trop Med Hyg 1970; 19:613-9.
6. Sadun EH, Maiphoom C. Studies on the epidemiology of the human intestinal fluke, Fasciolopsis buski (Lankester) in Central Thailand. Am J Trop Med Hyg 1953; 2:1070-84.
7. Joint WHO/FAO workshop on food-borne trematode infections in Asia. WHO, Regional Office for the Western Pacific 2002.
8. Graczyk TK, Alam K, Gilman RH et al. Development of Fasciolopsis buski (Trematoda: Fasciolidae) in Hippeutis umbilicalis and Segmentina trochoideus (Gastropoda: Pulmonata). Parasitol Res 2000; 86:324-32.
9. Lo CT. Life history of the snail, Segmentina hemisphaerula (Benson) and its experimental infection with Fasciolopsis buski (Lankester). J Parasitol 1967; 53:735-8.
10. Jaroonvesama N, Charoenlarp K, Areekul S. Intestinal absorption studies in Fasciolopsis buski infection. Southeast Asian J Trop Med Public Health 1986; 17:582-5.
11. Le TH, Nguyen VD, Phan BU et al. Case report: unusual presentation of Fasciolopsis buski in a Viet Namese child. Trans R Soc Trop Med Hyg 2004; 98:193-4.
12. Gupta A, Xess A, Sharma HP et al. Fasciolopsis buski (giant intestinal fluke)—a case report. Indian J Pathol Microbiol 1999; 42:359-60.
13. Keiser J, Utzinger J. Chemotherapy for major food-borne trematodes: a review. Exp Opin Pharmacother 2004; 5:1711-26.
14. Zhou H, Ohtsuka R, He Y et al. Impact of parasitic infections and dietary intake on child growth in the Schistosomiasis-endemic Dongting Lake region, China. Am J Trop Med Hyg 2005; 72:534-9.
15. Graczyk TK, Fried B. Echinostomiasis: a common but forgotten food-borne disease. Am J Trop Med Hyg 1998; 58:501-4.
16. Chai JY, Lee SH. Intestinal trematodes of humans in Korea: Metagonimus, Heterophyids and Echinostomes. Kisaengchunghak Chapchi 1990; 28:103-22.
17. Kim DG, Kim TS, Cho SH et al. Heterphyid metacercarial infections in brackish water fishes from Jinju-man (Bay), Kyongsangnam-do, Korea. Korean J Parasitol 2006; 44:7-13.
18. Abou-Basha LM, Abdel-Fattah M, Orecchia P et al. Epidemiological study of heterophyiasis among humans in an area of Egypt. Eastern Med Health J 2000; 6:932-8.
19. Goldsmid JM. Ecological and cultural aspects of human trematodiasis (excluding schistosomiasis) in Africa. Cent Afr J Med 1975; 21:49-52.
20. Olson PD, Cribb TH, Tkach VV et al. Phylogeny and classification of the Digenea (Platyhelminthes: Trematoda). Int J Parasitol 2003; 33:733-55.
21. Bhatti HS, Malla N, Mahajan RC et al. Fasciolopsiasis—a re-emerging infection in Azamgarh (Uttar Pradesh). Indian J Pathol Microbiol 2000; 43:73-6.
22. The Medical Letter on Drugs and Therapeutics. Drugs for Parasitic Infections 2004:4, www.medicalletter.org
23. Cross JH. Changing patterns of some trematode infections in Asia. Arzneimittelforschung 1984; 34:1224-6.
24. Brooker S, Whawell S, Kabatereine NB et al. Evaluating the epidemiological impact of national control programmes for helminths. Trends Parasitol 2004; 20:537-45.

16

CHAPTER 17

Schistosomiasis: *Schistosoma japonicum*

Edsel Maurice T. Salvana and Charles H. King

Background

Schistosoma japonicum, also known as the oriental blood fluke, is one of the causative agents of chronic intestinal schistosomiasis in humans. This condition, which frequently progresses to liver fibrosis, portal hypertension and splenomegaly, is endemic to southeastern and eastern Asia, particularly parts of China, Sulawesi, Indonesia, Thailand, Laos, Cambodia and the Philippines. At one time, it was endemic in southern Japan, where Katsurada first described the adult form of the *S. japonicum* parasite. However, transmission of this infection was completely eliminated in Japan as of 1977.

In the late 1950s, an estimated 11.6 million people in China were infected with *S. japonicum*. A national control program was launched, with emphasis on intermediate host snail control through environmental management, and this decreased the numbers of infected individuals dramatically. Recently between 2000 and 2003, there has been an upsurge of cases in China, which suggests a reemergence of the disease. More than 1.3 million people in the world are currently infected with *S. japonicum*; the ~843,000 in China and over 500,000 in the Philippines make up the greater part of present *S. japonicum* cases.

S. japonicum is a trematode parasite or fluke that requires a snail of the genus *Oncomelania* as an intermediate host. Compared to *S. mansoni* and *S. haematobium*, *S. japonicum* is associated with more severe disease manifestations. Ecologically, it is perhaps the most difficult schistosome species to control, in part due to its higher egg production and its ability to infect a wider range of definitive mammalian hosts, including common domestic animals such as dogs, cats, pigs and cows, as well as feral animals such as mice, rabbits and monkeys. Nevertheless, nations such as Japan, China and the Philippines have made significant progress in *S. japonicum* control, mostly linked to economic development and focused efforts at improved sanitation. Geographically distinct strains of *S. japonicum* have different preferred hosts and appear to have varying degrees of pathogenicity. Moreover, different geographic strains elicit distinct immune responses, which are not necessarily cross-protective. This phenomenon means that a universal vaccine against *S. japonicum* will be difficult to develop.

Several *S. japonicum*-like flukes have been identified in Thailand and Malaysia that are of uncertain clinical significance for humans. In addition, in 1978, a new species of *Schistosoma* was recognized within the Mekong River basin in Laos to Cambodia. This new species, *S. mekongi,* utilizes the snail *Neotricula aperta* as its

Medical Parasitology, edited by Abhay R. Satoskar, Gary L. Simon, Peter J. Hotez and Moriya Tsuji.

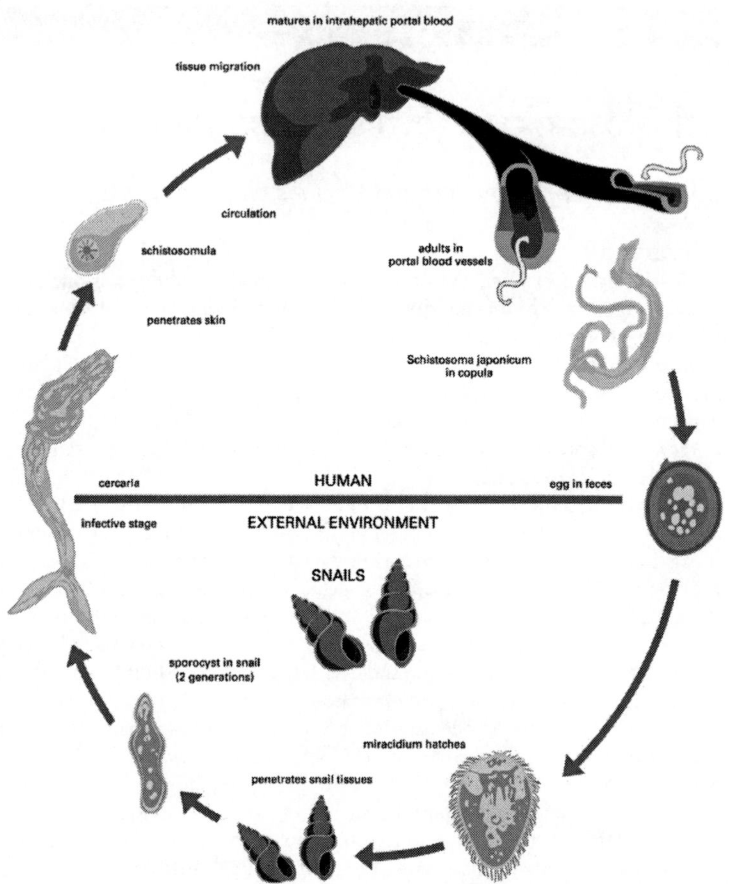

Figure 17.1. Life cycle of *Schistosoma japonicum*. Modified with permission from: Belizario VY, de Leon WU, eds. Philippine Textbook of Parasitology, 2nd Ed. Manila: University of the Philippines, 2004:196.

17

intermediate host. In human infection, its clinical manifestations and pathogenesis are similar to those of *S. japonicum*.

The *S. japonicum* life cycle (Fig. 17.1) includes both parasitic and free-living stages. The infective stage for humans is the cercaria (pl. cercariae), which is free-living and free-swimming, but short-lived (24-72 h). The *S. japonicum* cercariae gain entry to the host through penetration of skin immersed in water. The cercariae then transform into larval schistosomula, which penetrate the circulation through subcutaneous vessels and reach the pulmonary circulation. In the lungs, the schistosomules elongate, break into the pulmonary veins and then travel through the heart to the systemic capillary bed. If the schistosomulum reaches the splanchnic vessels, it moves across the capillary bed to the portal circulation (otherwise, it returns to the heart to circulate again). From the mesenteric capillaries, the schistosomulum travels to the liver, where it goes into the intrahepatic branches of the portal vein and matures into an adult schistosome. Adult *S. japonicum* are dioecious, i.e., either male or female and these migrate back to the mesenteric vessels to pair, mate and begin oviposition in the wall of the bowel. Eggs leave the human body in feces and when exposed to fresh water, these hatch to release miracidia, which in turn infect intermediate host snails, which complete the parasite life cycle by releasing cercariae 4-12 weeks after infection.

Areas with high rates of *S. japonicum* infection have habitats that are rich in water and provide a favorable breeding ground for snails, as well as opportunities for the cercariae to infect definitive hosts. Rural areas with rice fields and seasonal flooding are ideal for the parasite to thrive. Control of infection has been attempted in the past through eradication of the intermediate snail hosts, and this was done successfully in Japan. Lately, the focus has been on mass treatment of affected human populations, especially in areas with high disease prevalence.

Disease Signs and Symptoms

Acute Schistosomiasis

The penetration of cercariae into the skin causes localized inflammation and pruritus known as cercarial dermatitis or "swimmer's itch." This is not unique to *S. japonicum* and characterizes the trematode cercariae across the parasite genus and family, including those bird schistosomes such as *Trichobilharzia* that cause zoonotic swimmer's itch. *S. mansoni* and *S. haematobium* are more likely to cause dermatitis than *S. japonicum*.

More commonly, successful cercarial entry to the body and development of larvae in the bloodstream brings about a self-limiting febrile illness known as "snail fever" or Katayama fever. This can be accompanied by arthralgias, myalgias and abdominal pain, as well as findings of lymphadenopathy, hepatosplenomegaly and eosinophilia.

As the maturing worms migrate through the circulatory system, they may rarely end up in aberrant areas such as the brain or spinal cord, where direct occlusion of blood vessels may cause transient ischemic attacks, strokes, paraparesis, seizures, or hydrocephalus.

Repeated infections can stimulate partial immunity in the host and, due to a compensatory immunomodulation of antiparasite immune responses, the dermatitis and systemic manifestations of acute infection may actually be decreased when infected individuals are exposed to subsequent cercarial penetration. A proportion of immune patients may spontaneously clear their infections without treatment, but the remainder go on to develop chronic schistosomiasis with variable degrees of morbidity.

Chronic Schistosomiasis

Schistosoma japonicum causes chronic pathology when the adult worms find their way into the portal circulation. A mated pair of worms produces 300 to 3,000 eggs per day, which are released into the capillaries and portal veins. In order to be expelled through the intestinal wall, an egg must be inserted into a terminal vein within the bowel wall and then be ulcerated (via immune-mediated inflammation) through the mucosa into the bowel lumen. Bleeding and polyp formation in the bowel wall are thus common complications of egg transfer from the venule to the bowel lumen. The process of local inflammation can lead to protein loss, iron loss, anemia of chronic disease, diarrhea and in some cases, intestinal obstruction.

More than half of the worm eggs become permanently trapped in host tissues, where they each evoke an immune-mediated inflammatory granuloma. The significant numbers of eggs ultimately deposited in tissues cause widespread fibrosis. This is especially apparent in the liver, where a classic pattern of periportal "pipestem" fibrosis is manifested, which leads to portal hypertension and its sequelae of varix formation, ascites and splenomegaly.

As portal hypertension increases, eggs are shunted into the pulmonary circulation, where local fibrosis causes pulmonary hypertension, which may lead to cor pulmonale. Eggs reaching the central nervous system lead to local granuloma formation, which can cause seizures, hydrocephalus and space occupying lesions.

Pediatric Infection

Infection with *S. japonicum* has detrimental effects on the growth of children in endemic areas. A recent study in China noted that height, weight and mid-upper arm circumference were all significantly reduced in *S. japonicum*-infected children compared to controls and this effect was most severe among girls. Comparable studies in the Philippines have shown *S. japonicum*-infected children to have an increased prevalence of nutritional morbidity, as well as poorer performance on standardized tests of learning. Because of their chronic, long-term persistence over many years, these 'subtle' morbidities constitute a significant lifetime burden of disease for endemic populations.

Diagnosis

Acute Infection

Acute infection with schistosomiasis is difficult to diagnose definitively. Clinical symptoms are nonspecific. A history of exposure of skin to water in endemic areas followed by a suitable clinical syndrome should increase the index of suspicion for acute schistosomiasis. Antischistosome serologic tests may be performed at this juncture, although a positive result does not distinguish a recent infection from a remote one. However, among returning travelers who have no history of prior exposure, a negative serology is quite helpful in ruling out the possibility of schistosome infection.

Chronic Infection

Direct stool examination using the Kato-Katz technique is the method of choice for determining the presence of infection and the associated egg density in affected humans. *S. japonicum* eggs have a distinct appearance, which is ovoidal with a small knob near one of the poles (Fig. 17.2). The typical egg measures around 100 μm on the long axis and 60 μm on the short axis.

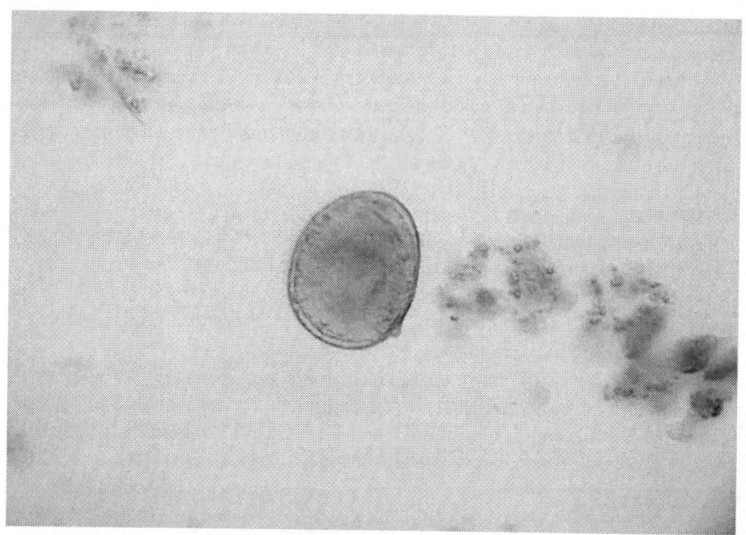

Figure 17.2. *Schistosoma japonicum* egg.

Concentration techniques are helpful for processing large stool volumes but are not completely sensitive for light infections. Common concentration techniques include formaldehyde-ether concentration technique, merthiolate-formaldehyde concentration technique and merthiolate-iodine-formaldehyde concentration technique.

In consideration of cases of very light infection, a rectal biopsy can be useful if repeated stool examinations remain negative in the face of a high clinical suspicion of infection. An estimation of the viability of any recovered eggs is appropriate, as large numbers of eggs can concentrate in the rectal mucosa and remain there even if active infection has ceased. Liver biopsy can also demonstrate worm eggs in the parenchyma, but this procedure is more invasive, risky and requires specialized equipment.

Immunologic diagnosis is available using both *S. japonicum* egg and adult worm extracts, although the World Health Organization recommends that testing be performed with crude egg antigens for greater specificity. Currently available tests include the circumoval precipitin test (COPT), indirect hemagglutination test and ELISA against soluble schistosome antigens.

The COPT tests patient's serum for reactivity with *S. japonicum* eggs. Upon serum exposure, a precipitate forms on the egg surface that can be visualized through the microscope. Like the ELISA for anti-*S. japonicum* antibodies, the COPT does not distinguish an active from a remote infection.

Antigen detection through ELISA is emerging as the most specific test for detecting active infection. Monoclonal antibodies against specific *S. japonicum* antigens are used to detect circulating antigens, including gut-associated antigens such as circulating cathodic antigen (CAA). The CAA assay is particularly useful because the antigen can be detected in urine, which should facilitate its use in large-scale field studies.

Liver ultrasound and computed tomography can identify a distinctive pattern of periportal fibrosis and a "network" pattern in *S. japonicum*. This can be useful in distinguishing liver disease secondary to the parasite from other etiologies. However, more imaging studies are required to validate these findings.

Treatment

Praziquantel remains the mainstay of therapy for active *S. japonicum* disease. While individuals with *acute* infection can initially be observed and treated symptomatically, evidence of chronic infection, as manifested by recovery of eggs from stool or tissue requires anthelminthic therapy to prevent significant morbidity, as well as to prevent further transmission.

For *S. japonicum* treatment, praziquantel should be given at an oral dose of 60 mg/kg in two to three divided doses over the period of one day. This regimen results in an 80-90 % cure rate. Common side-effects are mild and include abdominal discomfort, low-grade fevers, nausea, dizziness and diaphoresis. There is evidence that treatment of chronic schistosome infection, if given early in the course of the disease, can reverse infection-associated morbidity such as hepatosplenomegaly. Among patients with extensive fibrosis and obstruction, however, treatment may have little effect, with only minimal regression of clinical and ultrasound findings.

Alternative drug regimens include niridazole and antimony-based therapy, but these drugs are not FDA-approved and are associated with much lower cure rates and with significant toxicity. Artemisinin derivatives such as artemether have been shown to have antiparasitic activity, particularly against juvenile forms of the worm, but they are less active against the mature, adult forms of the parasite. Combination treatment with artemether and praziquantel has been more highly effective in reducing worm burdens in laboratory animals and is believed to have considerable potential for interrupting transmission in high-transmission communities.

Recently, calcium-channel blockers have been shown to affect the eggshell formation in *S. mansoni* and these agents may hold some promise in reducing or interrupting *S. japonicum* transmission. Several unique schistosomal proteins involved in eggshell synthesis have been identified and are currently being studied as targets for treatment.

Prevention and Prophylaxis

The mainstay of epidemiologic control involves interruption of transmission. In areas of high prevalence, mass chemotherapy is the main control strategy. Although mass treatment may reduce transmission, it does not assure significant interruption of transmission. Environmental control of the *Oncomelania* host snail is a strategy that has been used successfully in Japan and parts of China. Elimination of snails involves elimination of breeding sites as well as the use of chemical molluscicides to kill snails at transmission sites. Improved sanitation (to prevent *S. japonicum* eggs in feces from reaching fresh water snail habitat) is likewise an important control measure.

No effective chemoprophylaxis currently exists for *S. japonicum*. Some field studies involving artemether for prophylaxis have been promising, but results from large-scale intervention trials are still pending. There have been studies involving chemical repellants to prevent penetration of cercariae but these have been largely unsatisfactory. Therefore, travelers have no protective options and should be strongly cautioned against participating in wading or swimming activities in any unchlorinated fresh water found in *S. japonicum*-endemic areas.

Vaccine Research

An effective vaccine against the different geographic strains of *S. japonicum* will be a major breakthrough in the control of schistosomiasis in Asia. This is because current chemotherapy does little to interrupt transmission and does nothing to prevent reinfection. A number of vaccines based on novel target antigens are in development. A transmission-blocking veterinary vaccine that targets infection in water buffaloes has completed field trails and may have a major role in integrated schistosomiasis control programs in the future. Recent elucidation of the *S. japonicum* genome is expected to further boost research into finding targets for drug treatment and vaccine development.

Suggested Reading

1. Blas B, Rosales M, Lipayon I et al. The schistosomiasis problem in the Philippines: a review. Parasitol Int 2004; 53:127-34.
2. Bonn D. Schistosomiasis: a new target for calcium channel blockers. Lancet Infect Dis 2004; 4:190.
3. Coutinho HM, McGarvey ST, Acosta LP et al. Nutritional status and serum cytokine profiles in children, adolescents and young adults with Schistosoma japonicum-associated hepatic fibrosis, in Leyte, Philippines. J Infect Dis 2005; 192:528-36.
4. Ezeamama AE, Friedman JF, Acosta LP et al. Helminth infection and cognitive impairment among Filipino children. Am J Trop Med Hyg 2005; 72:540-8.
5. Garcia EG, Belizario VY Jr. Trematode infections: blood flukes. In: Belizario VY, de Leon WU, eds. Philippine Textbook of Medical Parasitology, 2nd Edition. Manila: The Publications Program University of the Philippines, 2004:195-211.
6. Hirayama K. Immunogenetic analysis of postschistosomal liver fibrosis. Parasitol Int 2004; 53:193-6.
7. King CH. Acute and chronic schistosomiasis. Hosp Pract 1991; 26:117-30.
8. Li Y, Herter U, Ruppel A. Acute, chronic and late-stage infections with Schistosoma japonicum: reactivity of patient sera in indirect immunofluorescence tests. Ann Trop Med Parasitol 2004; 98:49-57.
9. McManus DP. Prospects for development of a transmission blocking vaccine against Schistosoma japonicum. Parasite Immunol 2005; 27:297-308.
10. Ohmae H, Sy OS, Chigusa Y et al. Imaging diagnosis of schistosomiasis japonica—the use in Japan and application for field study in the present endemic area. Parasitol Int 2003; 52:385-93.
11. Olveda RM. Disease in schistosomiasis japonica. In: Mahmoud AAF, ed. Schistosomiasis. Tropical Medicine: Science and Practice. London: Imperial College Press, 2001:361-89.
12. Ross AGP, Bartley PB, Sleigh AC et al. Current Concepts: Schistosomiasis. N Engl J Med 2002; 346:1212-20.
13. Vennervald BJ, Dunne DW. Morbidity in schistosomiasis: an update. Curr Opin Infect Dis 2004; 17:439-47.
14. Zhou H, Ohtsuka R, He Y et al. Impact of parasitic infections and dietary intake on child growth in the schistosomiasis-endemic Dongting Lake Region, China. Am J Trop Med Hyg 2005; 72:534-9.
15. Zhou XN, Wang LY, Chen MG et al. The public health significance and control of schistosomiasis in China—then and now. Acta Trop 2005; 96:97-105.

CHAPTER 18

Schistosomiasis: *Schistosoma mansoni*

Wafa Alnassir and Charles H. King

Background

Archeological evidence in Africa and China indicates that the parasitic trematode infection schistosomiasis has been part of human life for at least four millennia. In 1852, Dr. Theodore Bilharz, a German physician working in Egypt, first described the presence of adult worms in postmortem examinations of affected patients and since then, the disease is often referred to generically as 'Bilharziasis'. It was not until a half-century later (in 1907) that Dr. Patrick Manson defined the existence of different *Schistosoma* species and later in 1908, Sambon established the distinct species we now call *Schistosoma mansoni*.

Epidemiology

Internationally, *Schistosoma mansoni* is the most prevalent of the schistosome species that affect the intestines and liver. An estimated 62 million persons are infected worldwide. *S. mansoni* is known to occur in 52 countries, including sub-Saharan Africa (where around 85% of the global burden is concentrated), North African and Eastern Mediterranean countries, South American countries (Brazil, Venezuela, Surinam) as well as several Caribbean countries (Saint Lucia, Montserrat, Martnique, Guadeloupe, Dominican Republic and Puerto Rico) (Fig. 18.1).

Life Cycle

Among all species of human schistosome parasites (*S. mansoni, S. haematobium, S. japonicum, S. intercalatum, S. mekongi*), the schistosome life cycle is very similar, with the exception that different species differ in the final location where the adult worms prefer to reside within the human body. Adults of *S. mansoni* and *S. japonicum* favor the intestines, whereas *S. haematobium* prefer the urinary tract. The adult form of *S. mansoni* most frequently lives in the mesenteric veins of the small and large intestines and its eggs typically pass out of the body in the feces. However, in heavy infection *S. mansoni* will find space in other pelvic abdominal organs and *S. mansoni* eggs can be passed in the urine as well.

The life cycle of the schistosome worms involves an adult dioecious (i.e., either male or female) sexual stage within the definitive human host and an asexual reproductive stage within an intermediate host snail. When parasite eggs (ova) reach fresh water, they hatch and free-swimming miracidial forms are released. These penetrate the bodies of suitable aquatic snails (for *S. mansoni,* these are *Biomphalaria* species snails (Fig. 18.3)) and for the next 3-5 weeks they multiply asexually to

Medical Parasitology, edited by Abhay R. Satoskar, Gary L. Simon, Peter J. Hotez and Moriya Tsuji.

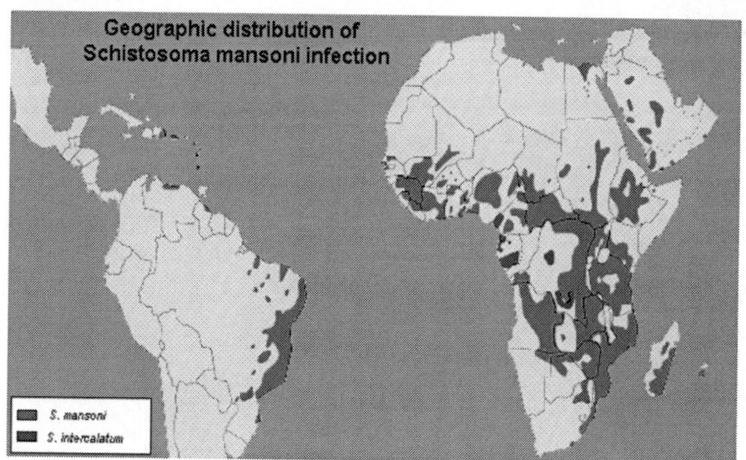

Figure 18.1. Geographic distribution of *Schistosoma mansoni* infection. Adapted with permission from the *Atlas of Medical Parasitology*, Carlo Denegri Foundation.

form hundreds of fork-tailed cercariae. The cercariae leave the snail and swim to a human or nonhuman animal, where they penetrate the skin (Fig. 18.2).

Inside the human body, the larval forms, or schistosomula, are transported through blood or lymphatics to the right side of the heart, then to the lungs, where they mature for a period of 7-10 days. They then break into the pulmonary veins and travel through the systemic circulation to reach the portal circulation and the liver, where they develop into adult worms (Fig. 18.3). Ultimately, the mature adult worms migrate from the liver through the portal veins to the mesentery and the wall of the bowel. These adult worms do little damage to the host. They feed on circulating erythrocytes and plasma glucose but usually don't cause symptoms.

The time between the first skin penetration by a cercaria to the first ova production by an adult worm is around 4-6 weeks. The female *S. mansoni* adult worm mates with the male worm and lays around 100-300 ovoid, laterally spined eggs each day (Fig. 18.3). This morphological appearance of the eggs allows very clear diagnostic identification of the infecting species. The female worm lives approximately 3-8 years and, when mated, lays eggs throughout her adult life. About half of all eggs reach the lumen of the bowel by a process of inflammatory ulceration through the bowel wall. The remainder of schistosome eggs remains trapped within the human host's body and result in acute and chronic granulomatous inflammation (Fig. 18.3).

The host immune response to schistosome eggs is the hallmark of disease due to schistosomiasis, in that immunological reaction and granuloma formation in the tissues is responsible for the morbidity and mortality of chronic schistosomiasis.

In terms of disease risk, no clear-cut racial predilection exists, although different populations have been found to have varying risk for fibrotic liver disease. Men often have a higher incidence of infection and disease than women, most likely because of their increased exposure to infested water via bathing, swimming and agricultural

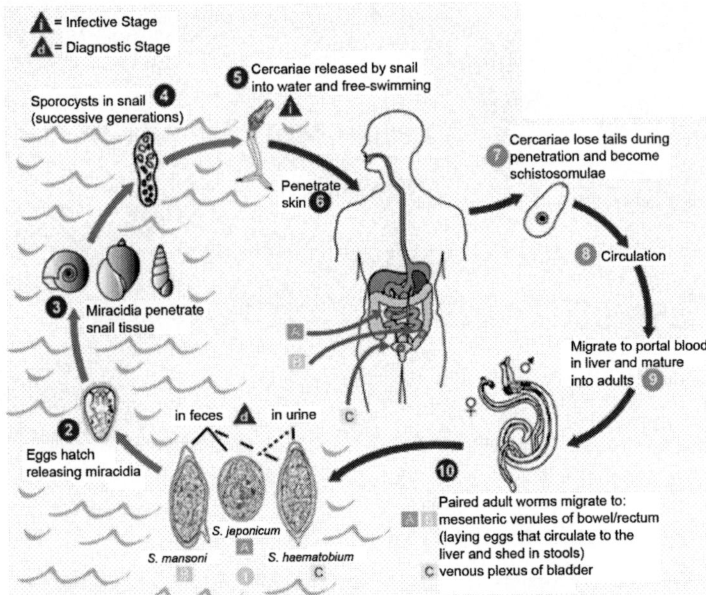

Figure 18.2. Life cycle of schistosomiasis. Reproduced from: Centers for Disease Control and Prevention (CDC) (http://www.dpd.cdc.gov/dpdx).

activities. In terms of age-related risk, schistosome exposure can start shortly after birth, but tends to increase during late childhood, with maximum exposure occurring among people aged 10-14 years. There is typically a lower incidence of infection and disease among adults, probably due to a combination of acquired partial immunity and an age-related decrease in exposure to infected water.

Typical Risk Factors

People who live or travel in areas where schistosomiasis is endemic and who are exposed to or swim in standing or running freshwater wherever the appropriate type of *Biomphalaria* snails are present, are at risk for infection. Transmission is intermittent, but under the right circumstances only a brief (<2 min) exposure can result in infection.

Disease Signs and Symptoms

Acute Manifestations

After cercarial penetration, there can be localized pruritus lasting for few hours ('swimmers itch') followed by urticarial rash or papuloerythematous exanthem. Most people will be initially asymptomatic after this first stage of infection. However, for some infected individuals (especially in non-immune individuals), symptoms of 'snail fever' or Katayama syndrome may develop after an incubation

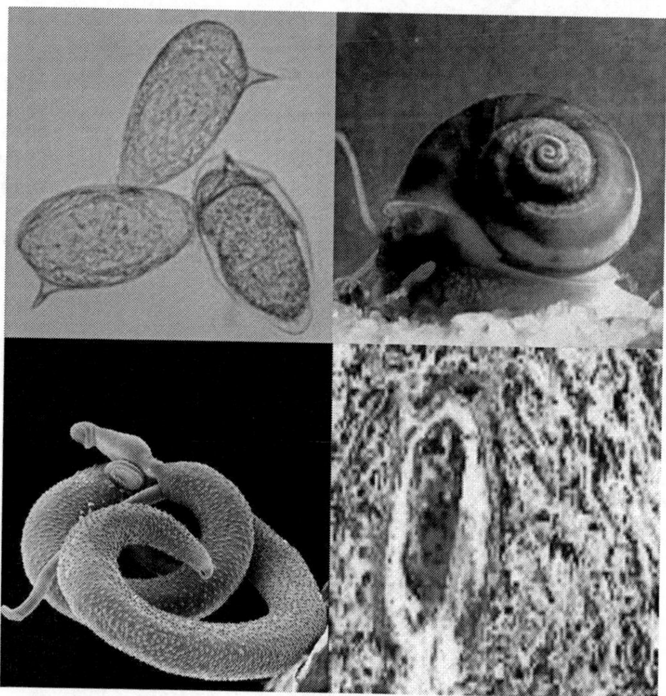

Figure 18.3. Upper left) *S. mansoni* eggs (reproduced from: Centers for Disease Control and Prevention (CDC) (http://www.dpd.cdc.gov/dpdx)). Upper right) *Biomphalaria* snail (reproduced from: *WHO/TDR*). Lower left) Female *S. mansoni* worm mated and resident in the groove of an adult male, (reproduced with permission from Davies Laboratory Uniformed Services University, Bethesda, MD). Lower right) *S. mansoni* pathology in liver tissue (reproduced with permission from the *Atlas of Medical Parasitology*, Carlo Denegri Foundation).

period of 1-2 months. This syndrome is associated with abrupt onset of fever, often accompanied by headache, shivering, anorexia, myalgia and right upper quadrant pain and less commonly with nausea, vomiting, diarrhea, cough and mild bronchospasm. Some may also develop hypersensitivity reaction to initial egg deposition, in the form of urticaria, generalized pruritus, facial edema, erythematus plaques or purpuric lesions.

Signs of chronic systemic inflammatory illness may persist and weight loss is the rule. Clinical signs include hepatosplenomegaly, which may disappear in few months. Laboratory findings include leukocytosis up to 50,000 per μL, with pronounced eosinophilia (typically, 20-30% up to 70%), increased serum gamma globulins, an increase in the sedimentation rate and possibly interstitial infiltrates on the chest X-ray.

Chronic Manifestations

Intestinal schistosomiasis: Due to chronic egg-mediated inflammation of the bowel wall, patients with intestinal schistosomiasis may present with fatigue, vague abdominal pain, diarrhea alternating with constipation and sometimes dysentery-like illness with bloody bowel movements. Eggs in the gut wall induce inflammation, hyperplasia, ulceration, microabscess formation and polyposis. Detection of occult blood in the stools is common. Intestinal obstruction secondary to marked polyposis is a rare complication.

Hepatosplenic schistosomiasis: This complication occurs in 4-8% of patients with chronic infection and is associated with pronounced host inflammatory response and/or sustained heavy infection. With development of portal hypertension secondary to egg-mediated periportal fibrosis (Symmers' pipe-stem fibrosis), secondary phenomena such as ascites, pedal edema, hepatosplenomegaly and variceal shunting will develop. The liver is usually hard with a nodular surface and frequently a prominent left lobe. Hematemesis secondary to esophageal varices is frequent and may prove fatal.

Other Manifestations and Complications

1. Cardiopulmonary schistosomiasis: As presinusoidal portal hypertension develops (see above) it fosters the development of portosystemtic collateral vessels that allow schistosome eggs to embolize into the pulmonary circulation, where the eggs can set up an inflammatory reaction with granuloma formation, which may block the small pulmonary arterioles. Intractable pulmonary hypertension and cor pulmonale may result.
2. CNS schistosomiasis may develop due to aberrant worm migration or egg deposition. A presentation with focal or generalized seizures can occur with *S. mansoni,* but transverse myelitis is the most common neurological manifestation of infection by this species. Myeloradiculopathy may also occur.
3. In high-risk populations, anemia, growth retardation in children and malnutrition are associated with *S. mansoni* infection, probably as a consequence of the chronic inflammation that is associated with infection.
4. Proteinuria can be found in 20-25% of *S. mansoni* cases with hepatosplenic disease. In addition, a distinct form of glomerulopathy is associated with chronic *S. mansoni* infection.

Interaction with Other Infections

Human Immunodeficiency Virus

HIV infection may exert a deleterious effect on the natural course of schistosomiasis in different ways: time until reinfection with schistosomiasis is shorter in HIV-positive than in HIV-negative individuals. On the other hand, treatment and response to praziquantel, a drug requiring a functioning immune system to be effective, is not impaired in patients with HIV co-infection. Evidence also suggests that *Schistosoma* infection may render the host more susceptible to HIV infection by interfering with specific immune responses.

Salmonella

Prolonged or refractory septicemic salmonellosis is described in *S. mansoni*-endemic areas. Treatment of schistosomiasis is often required to cure the *Salmonella* infection.

HBV Infection

Significant interactions between schistosomiasis and hepatitis B have been reported. Those people with coinfection are thought to have more severe disease and a worse prognosis.

Malaria

Malaria and schistosomiasis are co-endemic in many areas and studies have shown that co-infection with malaria may increase the level of morbidity in hepatosplenic schistosomiasis and alter the host immune response towards *Schistosoma* antigens.

Diagnosis

Routine Laboratory Testing

Peripheral eosinophilia is usually present in acute schistosomiasis, but in the chronic form, the peripheral eosinophilic counts may be minimal, even though pronounced tissue eosinophilia persists.

The detection of *S. mansoni* eggs in standardized fecal smear preparations (Kato-Katz method) is the diagnostic method of choice. The extent of egg production fluctuates over time, and as many as three separate stool specimens may be required for diagnosis in some patients. The use of formalin-based techniques for sedimentation and concentration may increase the diagnostic yield.

Other Laboratory Testing

Egg viability testing may help to assess the presence or absence of active infection, particularly after treatment. Testing involves mixing stool samples with room temperature distilled water and observing the excreted *S. mansoni* eggs for hatching of live miracidia.

Stool occult blood can be noted in intestinal schistosomiasis, although it is an inconstant finding and may be due to other disease processes.

Liver function testing is usually within normal limits until the end-stage of the disease. If LFTs are abnormal, then one should consider co-infection, e.g., viral hepatitis.

Anti-*S. mansoni* serology—Antibody testing is a useful epidemiological tool, but cannot differentiate active from past infection and, unlike egg counts, it does not allow quantification of infectious burden. Serology may be used to help in the diagnosis of patients from nonendemic areas, because negative antibody titers are expected and negative testing may help to exclude diagnosis.

Other investigational blood testing involves detection of circulating parasite antigens such as CAA and CCA (circulating anodic and cathodic proteoglycan antigens); these are thought to indicate active infection and also to quantify the intensity of infection. Studies are underway to evaluate the sensitivity and specificity of these tests under field conditions.

Imaging

Ultrasound of the liver and spleen is an early and accurate diagnostic method for detecting *S. mansoni*-associated periportal fibrosis and hepatosplenomegaly and also serves to grade or stage the hepatosplenic disease. Portable ultrasound machines are becoming relatively inexpensive and scanning may serve as an effective diagnostic method for *S. mansoni*-related disease in endemic areas.

Invasive procedures—Rectal biopsies and mucosal biopsies of the bowel are effective in visualizing eggs and helpful when the stool sample testing is negative, typically in the case of light infection. It is recommended to obtain multiple biopsy samples and to crush them between two slides to increase volume of sampling and to increase the likelihood of finding the eggs in the tissues. In advanced disease, upper gastrointestinal endoscopy may be used to assess and treat esophageal varices.

Treatment

Acute Schistosomiasis

For acute schistosomiasis mansoni, the treatment is usually supportive and focused on controlling symptoms. In the case of severe and persistent symptoms, specific antiparasitic treatment combined with anti-inflammatory therapy may play a role. Prednisone 1 mg/kg a day before antiparasite treatment, for 1 wk, starting one day before the antiparasitic treatment, followed by 0.5 mg for the second week and 0.25 mg for the third week, may increase the chances of cure and improve symptoms.

Chronic Schistosomiasis

The aim of chemotherapy in chronic schistosomiasis is two-fold. The first goal is to cure the disease, or at least minimize its associated morbidity. The second goal is to limit transmission of the parasite within the endemic area. Praziquantel and oxamniquine are commonly used for *S. mansoni* treatment, but because praziquantel is active against all *Schistosoma* species, it is, overall, the preferred treatment for schistosomiasis. Clinical studies have shown that artemether, which is an effective antimalarial treatment, is also active against immature forms of all three major schistosome parasites. Trials that involve the combination of artemether and praziquantel treatment indicate that combination therapy might have a more beneficial effect than monotherapy in terms of cure rate and morbidity reduction.

Praziquantel (Biltricide and generics), is a pyrazinoisoquinolone derivative. It is currently the mainstay of treatment of schistosomiasis mansoni, as it is the most effective and most readily available agent. It plays a critical part of community-based schistosomiasis control programs. Treatment is usually well tolerated, although side effects can include dizziness, drowsiness, headache, nausea, vomiting, abdominal pain, pruritus, hives and diarrhea. Such side effects are not long lasting and usually resolve in less than 2 hours. Praziquantel may provoke significant inflammatory symptoms when given to patients with neurocysticercosis, so that the drug should be used with caution in areas where *S. mansoni* and *T. solium* are both endemic.

Praziquantel's mechanism of action is complex. It is thought to damage the worm's outer tegument membrane (the natural covering of the worm body) and expose the worm to the body's immune response, which ultimately results in worm death. In treating *S. mansoni* infection, the praziquantel-mediated cure rate is equal to or greater than 85-90%. Among those persons not cured, the worm load and associated egg burden are markedly decreased. The recommended dose for praziquantel treatment is a single treatment of 40 mg/kg as a single dose or divided into two doses over 1 day. When possible, rescreening for persistent infection 6-8 weeks after initial therapy, followed by repeat dosing as needed, will provide optimal cure rates and morbidity control.

Oxamniquine (Vansil), is a tetrahydroquinoline antischistosomal agent that is solely effective against *S. mansoni*. In animal models, the drug is seen to cause paralysis of the worm with a consequent shift of worms into the liver. This appears to be associated with damage the tegumental surface of male schistosome worms, so that the host immune system is able to kill them. It also stops female worms from producing eggs, with a subsequently marked reduction in host morbidity. Dosing is 15 mg/kg once (in S. America) or twice daily for 2 days (in Africa). Overall cure rate is 60-100%. This drug is not available in the United States.

Artemisinin derivatives, artemether and artesunate, are sesquiterpene lactones derived from the active components of the plant *Artemisia annua*, and these drugs are best known for their antimalarial properties. The antischistosomal activities of artemisinin derivatives were discovered in the 1980s, with initial interest focusing on its effects on *S. japonicum*. Animal studies have since suggested that these drugs are also effective against *S. mansoni* and clinical testing of artmether for prevention of schistosomiasis mansoni suggests that the artemesinins are safe and effective drugs for treatment of early stages of infection.

18

Combination chemotherapy with praziquantel and oxamniquine has been tested—initial studies suggest some synergy, but direct comparison against mono-therapy has not been done and most of the populations studied have had concurrent infections with *S. haematobium*, which is not susceptible to oxamniquine. Further studies are warranted.

Initial laboratory and nonrandomized studies of combined praziquantel and artemesinin derivatives suggested greater effectiveness for the combination therapy. A recent randomized, double blind, placebo-controlled trial in Gabon indicated that the overall cure rate with combination therapy was not substantially higher compared to praziquantel alone, but in terms of the egg count reduction, the praziquantel-artemsinin combination had greater beneficial effect.

Emergence of drug resistance or tolerance is a threat to all chemotherapy-based control programs. Risk of schistosome resistance to praziquantel has been recently reviewed by Cioli who concluded that resistance is likely to exist, but it is currently of low-level and its clinical importance is, so far, minimal. The conclusion from recent studies has been that there is no conclusive evidence of emerging resistance of *S. mansoni* to praziquantel, and the observational studies that previously suggested drug failure may be explained by the specific transmission charcteristics. Unfortunately, there is no reliable testing available to determine praziquantel resistance in vitro.

Resistance to oxamniquine is undisputedly documented in vitro and in vivo, but its impact has been limited and it is not yet considered to be a problem for control program operations.

Advanced S. Mansoni*-Associated Disease*

For patients with late, severe complications of *S. mansoni* infection such as GI bleeding, intestinal obstruction, cardiopulmonary syndromes, or CNS disease, in-patient treatment and supportive measures (beyond antiparasite therapy) are needed. Endoscopy and variceal sclerosis or surgical treatment (portosystemic shunting) are indicated for severe bleeding secondary to portal hypertension.

Prognosis after Treatment

Response to treatment is evaluated by assessing the amount of decrease in egg excretion. In the first two weeks after treatment, excreted egg counts may not decrease because eggs that were laid before the treatment can require up to 2 weeks to be shed. Even with effective therapy, some viable eggs can be excreted for up to 6-8 weeks after treatment. Newer tests that measure circulating parasite antigens (CCA, CAA), when measured 5-10 days after treatment, may better assess acute antiworm therapeutic response. Persistent circulating antigen, or the persistent excretion of eggs at 6-8 weeks, indicates residual infection. Such patients should be retreated as needed to effect a cure.

Significant Points about Therapy

- Early intestinal, hepatic and portal vein disease usually improves with treatment.
- In the absence of other forms of liver disease, hepatosplenic schistosomiasis can have a relatively good prognosis because hepatocyte function is preserved. Late disease (in which variceal bleeding occurs) may be irreversible and result in fatality due to severe hemorrhage and its acute complications. In community treatment, approximately 40% of hepatosplenic disease is reversed with therapy.
- *S. mansoni*-associated cor pulmonale is a form of advanced schistosomiasis and usually does not improve significantly with antiparasitic treatment. Adjunctive treatment with selective vasodilators (sildenafil) might prove beneficial.
- Spinal cord schistosomiasis carries a guarded prognosis. Pharmacologic or surgical decompression as appropriate, combined with specific antiparasite praziquantel treatment, should be administered as soon as possible.

Prevention and Prophylaxis

Over the last two-to-three decades, large-scale national and regional schistosomiasis control programs have been implemented in a number of areas with varying levels of success. The most important examples of successful schistosomiasis control have been reported from Brazil, Venezuela, Cambodia, China, Egypt and the Philippines. Given the strong link between environmental factors, poverty and schistosomiasis transmission, it is imperative to take full advantage of the renewed contemporary international emphasis on the provision of of clean water and sanitation, as this represents a fundamental basis for schistosomiasis transmission control. A combined transmission reduction strategy would mutually reinforce chemotherapy-based morbidity control and provide the optimal preventive strategy for resource-poor developing areas.

Figure 18.4. Left panel) Spray mollusciciding for snail control. Right panel) Water contact activities associated with risk for *S. mansoni* infection. Reproduced with permission from the *Atlas of Medical Parasitology*, Carlo Denegri Foundation.

Programs for controlling *Schistosomiasis mansoni* within an endemic area should consider (and optimally include) all of the following:

- Population-based chemotherapy.
- Provision of a safe water supply and washing/swimming facilities to reduce exposure.
- Health education that promotes improved sanitation, including means for reduction in local water contamination by schistosome egg-containing urine or stool.
- Control of intermediate host snails (*Biomphalaria* spp.).
- Instruction to travelers and immigrants to avoid contact with fresh water in endemic areas.
- As local transmission falls, anticipate increased incidence of acute schistosomiasis in the setting of any recent freshwater contact. Make early treatment for *S. mansoni* available if diagnostic test results are positive or clinical suspicion is high.
- Consider clinical trials (involving human volunteers) towards development of an effective vaccine against *Schistosomiasis mansoni*.

Suggested Reading

1. Case records of the Massachusetts General Hospital. Weekly clinicopathological exercises. Case 21-2001. A 31-year-old man with an apparent seizure and a mass in the right parietal lobe. N Engl J Med 2001; 345:126-31.
2. Carod Artal FJ, Vargas AP, Horan TA et al. Schistosoma mansoni myelopathy: clinical and pathologic findings. Neurology 2004; 63:388-91.
3. Cioli D. Praziquantel: is there real resistance and are there alternatives? Curr Opin Infect Dis 2000; 13:659-63.
4. Corachan M. Schistosomiasis and international travel. Clin Infect Dis 2002; 35:446-50.
5. Gellido CL, Onesti S, Llena J et al. Spinal schistosomiasis. Neurology 2000; 54:527.
6. Ferrari ML, Coelho PM, Antunes CM et al. Efficacy of oxamniquine and praziquantel in the treatment of Schistosoma mansoni infection: a controlled trial. Bull World Health Organ 2003; 81:190-6.

7. King CH. Acute and chronic schistosomiasis. Hosp Pract 1991; 26:117-30.
8. Mahmoud AA, ed. Schistosomiasis. London: Imperial College Press, 2001.
9. Richter J, Hatz C, Haussinger D. Ultrasound in tropical and parasitic diseases. Lancet 2003; 362:900-2.
10. Ross AG, Bartley PB, Sleigh AC et al. Schistosomiasis. N Engl J Med 2002; 346:1212-20.
11. Utzinger J, Keiser J, Shuhua X et al. Combination chemotherapy of schistosomiasis in laboratory studies and clinical trials. Antimicrob Agents Chemother 2003; 47:1487-95.
12. Utzinger J, N'Goran EK, N'Dri A et al. Oral artemether for prevention of Schistosoma mansoni infection: randomised controlled trial. Lancet 2000; 355:1320-5.
13. van Lieshout L, Polderman AM, Deelder AM. Immunodiagnosis of by determination of the circulating antigens CAA and CCA, in particular in individuals with recent or light infections. Acta Trop 2000; 77:69-80.
14. Vennervald BJ, Dunne DW. Morbidity in schistosomiasis: an update. Curr Opin Infect Dis 2004; 17:439-47.
15. Wynn TA, Thompson RW, Cheever AW et al. Immunopathogenesis of schistosomiasis. Immunol Rev 2004; 201:156-67.

CHAPTER 19

Schistosomiasis: *Schistosoma haematobium*

Vijay Khiani and Charles H. King

Background

Schistosoma haematobium is a parasitic trematode that infects over 111 million people, mostly in Africa and the Middle East. It is the most common cause of urinary schistosomiasis, which is caused by antiparasite inflammation in the wall of the human host's bladder, ureters, or kidneys. Because the syndrome is unfamiliar to most physicians in the United States, they may often overlook the diagnosis of *S. haematobium* infection and its associated disease when presented with 'occult' cases of hematuria, hematospermia, pelvic inflammation, or infertility. While the typical returning traveler or immigrant is more likely to be infected with hepatitis, malaria, or intestinal helminths, it should be noted that schistosome infection is not uncommon after either short- or long-term travel to endemic areas.

S. haematobium infection is prevalent in areas that have a reservoir of human infection, the presence of an intermediate *Bulinus* species snail host and the poor socioeconomic conditions or poor sanitation that allow urinary contamination of local freshwater. These factors are all essential components of the parasitic trematode life cycle, which requires transmission from a definitive human host (in which adult worms undergo sexual reproduction) to an intermediate snail host (in which asexual multiplication of larvae occurs) and then the reverse process of snail-to-human transmission.

The mode of transmission of *S. haematobium* to humans begins with exposure to fresh water that contains infected snail intermediate hosts (Fig. 19.1). Of note, the transmission of schistosomiasis is NOT person-to-person. Infected snails release free-swimming cercariae, which seek and penetrate human skin. The cercariae are typically 2 mm in length and have acetabular and head glands to aid in attachment and penetration of the skin. Upon penetrating the epidermal layer, the cercariae reach the dermis and transform into immature larval forms called schistosomula. Over a period of two months, these schistosomula migrate first to the lungs (through venous circulation) and then to the left heart and into the systemic circulation. The schistosomula proceed to the liver, where they mature into either male or female adult worms. The *S. haematobium* worms, now 1 to 2 cm in length, migrate to the veins of the bladder (or, less frequently, the bowel), where the worms lie together in pairs and the female worms lay their eggs in the wall of the nearby urinary collecting system. The mature female worms deposit several hundred eggs each day. Approximately 50% of these eggs will become inadvertently trapped in host

Medical Parasitology, edited by Abhay R. Satoskar, Gary L. Simon, Peter J. Hotez and Moriya Tsuji.

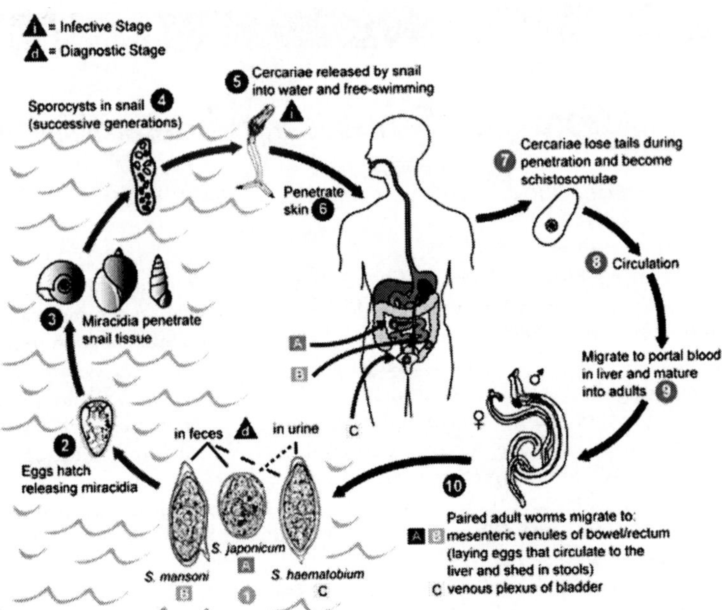

Figure 19.1. Transmission and life cycle of human schistosomes. Reproduced from: Centers for Disease Control and Prevention (CDC), Atlanta, GA (http://www.dpd.cdc.gov/dpdx).

tissues, while the other half manage (via local inflammation and ulceration) to penetrate through to the lumen of the ureter or bladder. From there, the eggs pass in the urine into the environment to reach bodies of fresh water. These eggs hatch and release motile ciliated forms called *miracidia* into the water, which seek out and infect the intermediate *Bulinus* species snail hosts by direct penetration. In the snail, the miracidia develop into sporocysts, which complete the life-cycle by a process of asexual reproduction that yields multiple infectious cercariae after a period of 1-2 months.

Most infected humans (50 to 75%) carry light infection with relatively low worm burdens, while a minority (1 to 5%) carry very heavy infection (defined as excreting > 400 eggs/10 mL urine), which results in the production of thousands of eggs daily. As a consequence, heavily infected hosts are most prone to the progressive disease sequelae of *S. haematobium,* including severe hematuria, fibrosis, ureteral stricture and calcification of the urinary tract. However, any level of infection can cause disease, as a function of the relative intensity of the affected host's anti-egg immune response. In some endemic areas, there is a definite link between *S. haematobium* and squamous cell carcinoma of the bladder and S. *haematobium* has been formally listed as a carcinogen by the WHO. Genetic variation of the parasite and host are also likely factors in determining the pathogenesis and clinical outcomes of infection and disease.

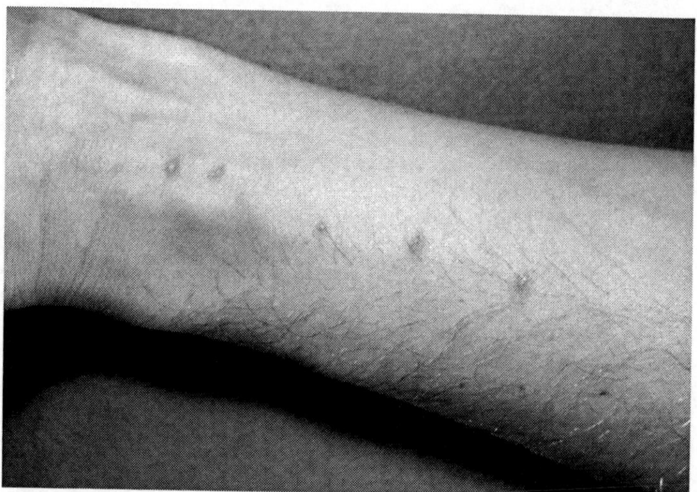

Figure 19.2. Cercarial dermatitis cause by exposure to schistosome-infested water. Reproduced from: Centers for Disease Control and Prevention (CDC) (http://phil.cdc.gov/phil/details.asp).

Disease Signs and Symptoms

Immediate Manifestations

The earliest form of infection-associated disease for patients affected by *S. haematobium* is cercarial dermatitis (Fig. 19.2), which is, in essence, a maculopapular, blistering eruption in the area of skin exposed to parasite penetration. This presentation, which can be due either to an immediate or delayed hypersensitivity, usually begins 1-2 days after exposure, lasts for a few days and is self-limited.

Acute Schistosomiasis

Acute schistosomiasis, also known as Katayama fever or snail fever, is most often seen in travelers who visit endemic areas for the first time, contract a schistosome infection and then present with signs of fever, lymphadenopathy, hepatosplenomegaly and blood eosinophilia some 4-6 weeks later. This process is usually triggered during the early egg deposition period of the parasite's life cycle and tends to affect the adolescent or young adult patient most frequently.

This acute process is related to an initial allergic hypersensitivity to the parasite, as well as subsequent formation of soluble immune complexes that can cause a serum-sickness type of illness. Acute schistosomiasis is usually a self-limited process, but can become quite severe and debilitating. In some cases, because the process of inflammation is antigen driven, acute schistosomiasis may even worsen with antiparasitic therapy. In such cases, it is recommended to start with symptomatic therapy to effect control of the process of inflammation, a step which usually requires the use of corticosteroids. This is then followed by specific antiparasitic therapy (praziquantel) to eliminate the causative infecting worms.

Chronic Schistosomiasis

Chronic schistosomiasis in humans is a consequence of prolonged (multi-year) and repeated infection with schistosome parasites. In endemic areas, chronic schistosomiasis is often established by age five, but as a consequence of repeated exposure, the intensity of worm burden increases over time, with a peak, maximum burden of infection experienced at ~13 years of age. The chronic form of disease develops as many *S. haematobium* eggs remain trapped in the host tissues and become surrounded by delayed-type hypersensitivity granulomatous inflammation (Fig. 19.3). This inflammation is formed by host lymphocytes, eosinophils and macrophages and ultimately kills and destroys the parasite eggs. However, the resolution of the inflammation is associated with collagen deposition and scar formation. Gradual accumulation of this scar tissue within bladder and ureters can lead to deformity, contractile incoordination and obstruction, resulting in hydroureter, hydronephrosis and ascending bacterial infection due to urinary reflux. Inflammation can result in local ulceration and significant blood loss in the urine and/or feces. Additionally, the chronic presence of inflammation may be associated with dyserythropoesis (anemia of chronic disease) and signs of iron and/or protein-calorie malnutrition.

The severity of the manifestations associated with chronic schistosomiasis is related both to the extent of exposure to the infection and to the intensity of host immune response. For those with less intense exposure, the early signs and symptoms include dysuria, proteinuria and bladder polyps. Later signs and

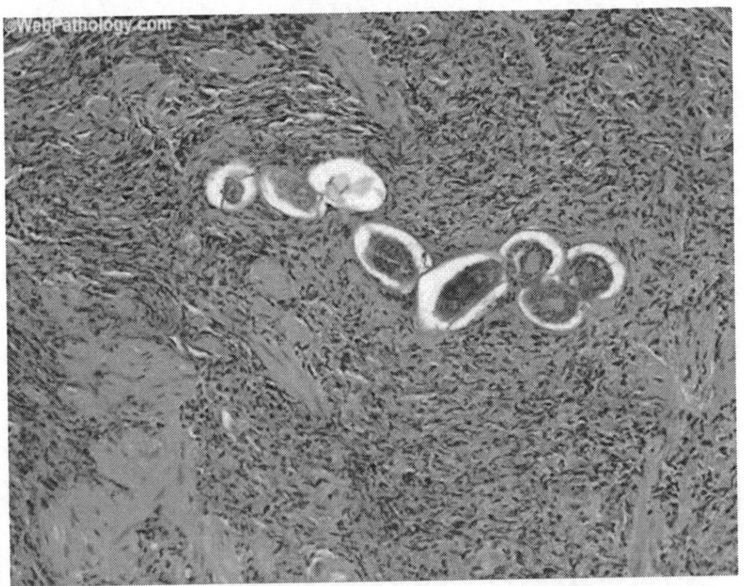

Figure 19.3. *S. haematobium* eggs surrounded by host inflammation in bladder wall. Reproduced from: http://webpathology.com/image.cfm?case=51&n=6, with kind permission of Dharam M. Ramnani, MD.

symptoms include hydronephrosis, hydroureter, bladder calcification, increased risk of urinary tract infections and ultimately squamous cell carcinoma of the bladder. For hosts with eggs in the bladder or lower ureters, over 50% of the patients have symptoms of dysuria, frequency and terminal hematuria. Cor pulmonale can result via direct passage of eggs through the ureteral veins to the systemic veins, resulting in egg transport to the pulmonary circulation and trapping in the lung tissues. Central nervous system disorders can result from eggs passing into the brain, spinal cord, meninges, or ventricles, or from aberrant migration of worms into CNS tissues. The resulting inflammation can lead to seizures, spinal cord compression, or hydrocephalus.

Diagnosis

The most important aspect of making the diagnosis of *S. haematobium* is to appreciate risk of exposure. Possible urinary schistosomiasis can be established by means of a thorough history and physical examination. Specifically, the history should focus on specifics of travel history, including the geographic areas visited and known or possible exposures to freshwater bodies while abroad. Upon completion of the history and examination, one may begin to pursue tests and studies to arrive at the final diagnosis.

The majority of cases of *S. haematobium* are diagnosed by the finding of parasite eggs in urine or feces. Concentration methods are helpful in increasing the sensitivity of urine testing. Sedimentation or membrane filtration techniques are often used when analyzing the urine for *S. haematobium*. Cystoscopy is rarely required, but may be helpful in recovering parasite eggs in cases of light infection with significant pathology. *S. haematobium* eggs are spindle shaped, usually around 140-150 by 60 micrometers (Fig. 19.4). The eggs have a terminal spine, which is similar to the appearance of another, less common human parasite, *S. intercalatum*. The two species can be differentiated by the Ziehl-Nielsen test, which is negative for eggs of *S. haematobium*.

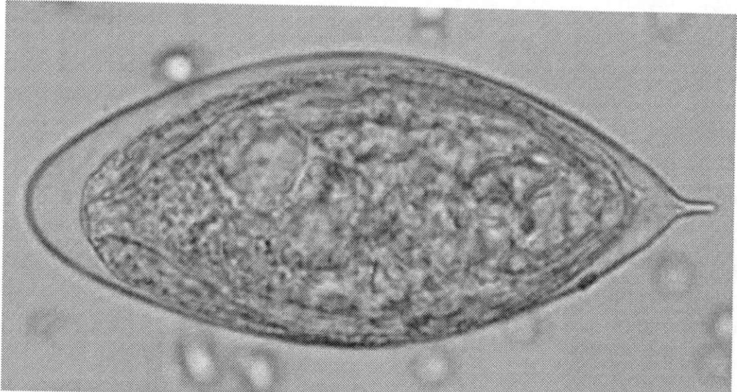

Figure 19.4. *S. haematobium* egg in urine sediment. Note characteristic terminal spine at right. Reproduced from: Centers for Disease Control and Prevention (CDC), Atlanta, GA (http://www.dpd.cdc.gov/dpdx).

Parasite eggs may persist in human tissues for a number of years, even following effective therapy. Because of this, egg viability testing (by vital staining or miracidial hatching) may be necessary to establish the presence of active infection. Serologic testing for *S. haematobium* is sensitive, but not very specific due to the fact that antiparasite antibodies persist after infection resolves. In consequence, the specificity is too low to be used as a diagnostic tool for patients in endemic areas. Radiologic testing remains yet another option that can provide useful diagnostic information. For example, an ultrasound may demonstrate evidence of obstructive uropathy with findings of bladder wall thickening and characteristic granulomas, calcification, hydroureter and hydronephrosis.

Treatment

For patients with *S. haematobium* infection who present with cercarial dermatitis, this initial presentation is usually self-limited and responds to local topical care. Treatment of acute schistosomiasis, as described earlier, focuses on control of inflammation first, followed by specific antiparasite therapy. It is important to note that immature schistosome parasites are relatively resistant to the effects of the standard antischistosomal drug, praziquantel. To fully eradicate infection, it may be necessary to provide retreatment of patients with acute schistosomiasis at a point 2-3 months after last exposure.

For chronic schistosomiasis haematobia, the treatment of choice is praziquantel, 40 mg/kg, as a single oral dose or in two divided doses. Praziquantel is formulated as scored, breakable 600 mg pills, so that dose can be appropriately adjusted to body weight (see Table 19.1).

A single treatment is 80-90% effective with limited side effects. Praziquantel side effects include sedation, malaise, headache, dizziness, abdominal discomfort and nausea. The minority of patients who fail to clear their S. *haematobium* infection after praziquantel therapy nevertheless typically have a >95% decrease in their egg burden, which indicates significant reduction of infection. If egg excretion persists for more than one month, an additional round of praziquantel therapy (40 mg/kg, PO, once) should be given. There are some patients that have an acute exacerbation of symptoms during praziquantel treatment and these individuals may need a short course of anti-inflammatory medications such as antihistamines or corticosteroids. Metrifonate is an alternative treatment of choice for patients with *S. haematobium* infection, but may not be commercially available.

Table 19.1. Praziquantel doses to treat schistosomiasis

Body Weight (kg)	Height	No. of Praziquantel Tablets (600 mg tablet)
10.0-14.9	(94-110 cm)	1
15.0-22.4	(110-125 cm)	1½
22.5-29.9	(125-138 cm)	2
30.0-37.4	(138-150 cm)	2½
37.5-44.9	(150-160 cm)	3
45.0-59.9	(160-178 cm)	4
60.0-75.0	(>178 cm)	5

Prevention and Control

Disease due to *S. haematobium* infection remains the most prevalent form of urinary schistosomiasis in sub-Saharan Africa. The WHO has established the goal to treat 75% of school-age children with infection and at risk of morbidity by 2010. Morbidity is particularly associated with cases with *S. haematobium* infection who develop immunopathological sequelae, such as hepatosplenomegaly, bladder deformity and hydronephrosis. Endoscopic/surgical interventions may be needed for hematuria, urinary obstruction, bladder cancer and CNS disease. For persons living in high transmission areas, regular treatment may prevent *S. haematobium* infection from developing into serious disease.

Even though an estimated 150,000 individuals die from schistosomiasis infection on a yearly basis, there is still no form of prophylaxis for patients with risk of exposure. However, there are many important ways to improve upon current risk for transmission. It is vital to avoid fresh immersion in any freshwater in contaminated areas. Because transmission is highly focal and sporadic, one should avoid cutaneous exposure to all ponds, lakes, rivers and streams in endemic areas. Bath water should be heated for at least 5 minutes at 150 degrees F, or let stand in a cistern or container (known to be completely free of snails) for 3 days.

For residents of endemic areas, sanitation, socioeconomic development and health education are key components.

Suggested Reading

1. Brouwer KC, Ndhlovu PD, Wagatsuma Y et al. Urinary tract pathology attributed to Schistosoma haematobium: does parasite genetics play a role? Am J Trop Med Hyg 2003; 68:456-62.
2. Centers for Disease Control. Schistosomiasis among river rafters—Ethiopia. MMWR 1983; 32:585-6.
3. Gonzalez E. Schistosomiasis, cercarial dermatitis and marine dermatitis. Dermatol Clin 1989; 7:291-300.
4. King CH, Mahmoud AA. Drugs five years later: praziquantel. Ann Intern Med 1989; 110:290-6.
5. King CH. Disease in Schistosomiasis Haematobia. In: Mahmoud AAF, ed. Schistosomiasis. London: Imperial College Press, 2001:265-96.
6. King CH. Ultrasound monitoring of structural urinary tract disease in S. haematobium infection. Memorias do Instituto Oswaldo Cruz 2002; 97:149-52.
7. King CL. Initiation and regulation of disease in schistosomiasis. In: Mahmoud AAF, ed. Schistosomiasis. London: Imperial College Press, 2001:213-64.
8. Smith JH, Christie JD. The pathobiology of Schistosoma haematobium infection in humans. Hum Pathol 1986; 17:333-45.
9. Peters PAS, Kazura JW. Update on diagnostic methods for schistosomiasis. In: Mahmoud AAF, ed. Balliere's Clinical Tropical Medicine and Communicable Diseases, Schistosomiasis, Vol. 2. London: Bailliere Tindall, 1987:419-33.
9. van der Werf MJ, de Vlas SJ, Brooker S et al. Quantification of clinical morbidity associated with schistosome infection in sub-Saharan Africa. Acta Trop 2003; 86:125-39.
10. Visser LG, Polderman AM, Stuiver PC. Outbreak of schistosomiasis among travelers returning from Mali, West Africa. Clin Infect Dis 1995; 20:280-5.
11. World Health Organization. IARC Monographs on the Evaluation of Carcinogenic Risks to Humans. Schistosomes, Liver Flukes and Helicobacter pylori. Vol. 61 Geneva: World Health Organization, 1994.
12. World Health Organization. Prevention and control of schistosomiasis and soil-transmitted helminthiasis: Report of a WHO expert committee. Technical Report Series 912 2002. Report No. 912.

SECTION III

Cestodes

CHAPTER 20

Taeniasis and Cyticercosis

Hannah Cummings, Luis I. Terrazas and Abhay R. Satoskar

Background

Taeniasis and cysticercosis are diseases resulting from infection with parasitic tapeworms belonging to *Taenia species*. Approximately 45 species of *Taenia* have been identified; however, the two most commonly responsible for human infection are the pork tapeworm *Taenia solium* and the beef tapeworm *Taenia saginata*. Parasitic tapeworm infections occur worldwide, causing sickness, malnutrition and often resulting in the death of their host. Infection with adult tapeworms of either *T. solium* or *T. saginata* cause taeniasis in humans. The metacestode, or larval stage, of *Taenia solium* causes the tissue infection, cysticercosis. Clinical manifestations associated with the tapeworm infection can vary greatly and may range from mild forms where patients exhibit little to no symptoms, to severe life-threatening forms which are often fatal.

Geographic Distribution and Transmission

Taenia infections are estimated to affect 100 million people worldwide, with major endemic areas located primarily in the developing countries of South America, Africa, India, China and Southeast Asia. The ingestion of cysticerci from raw or undercooked meat facilitates the transmission of *T. solium* from pigs to humans and is presumably responsible for the high prevalence of human cysticerosis in these regions. It is estimated that anywhere between 5-40% of individuals carrying the adult tapeworm will develop cysticercosis. *Taenia* infections are less common in North America; however neurocysticercosis has been recognized as an important health problem in California. Although this disease is mainly seen in migrant workers from Latin American, it has also been reported in US residents who have not traveled to endemic countries.

Life-Cycle

The complete life-cycle of *Taenia solium* involves two hosts: the pig and the human, whereas that of *Taenia saginata* involves the cow and the human (Fig. 20.1). Humans act as the definitive host and harbor the adult tapeworm in the small intestine. Infection is acquired either through the accidental ingestion of embryonated eggs passed in the feces of an individual infected with the adult tapeworm, or through the consumption of raw or poorly cooked meat containing cysticerci. The cysticerca develops into an adult worm in the gut; these worms can survive up to 25 years. Depending on the species of *Taenia,* an adult worm can reach lengths between 2-25 meters and may produce as many as 300,000 eggs

Medical Parasitology, edited by Abhay R. Satoskar, Gary L. Simon, Peter J. Hotez and Moriya Tsuji.

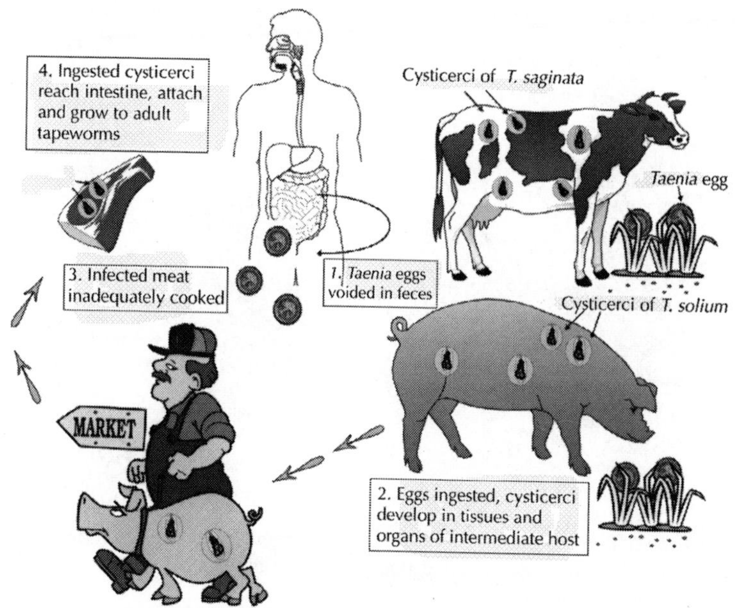

Figure 20.1. Life cycles of the beef tapeworm, *Taenia saginata* and the pork tapeworm, *T. solium*. Reproduced from: Nappi AJ, Vass E, eds. Parasites of Medical Importance. Austin: Landes Bioscience, 2002:61.

20

per day. The morphology of the adult worm consists of a scolex and a strombila. The scolex acts as the organ of attachment and consists of four suckers equipped with hooklets. The strombila consists of several segments (proglottids) with the gravid or egg-carrying proglottids located toward the posterior end of the worm (Fig. 20.2). Individual proglottids may contain as many as 40,000 eggs in *T. solium* or as many as 100,000 eggs in *T. saginata*.

Both the proglottids and the eggs are released with the feces of infected individuals and serve as a source of infection for pigs and cattle, which act as intermediate hosts for these parasites. Following the ingestion of eggs, mature larvae (onchospheres) are released in the gut. These onchospheres enter the blood stream by penetrating the small intestine and migrate to skeletal and cardiac muscles where they develop into cysticerci. Cysticerci may survive in the host tissues for several years causing cysticercosis (Fig. 20.3). The consumption of raw or undercooked meat containing cysticerci facilitates the spread of infection from pigs to humans. In humans, cysticerci transform into adult tapeworms which persist in the small intestines for years causing taeniasis. The time between initial infection and the development of the adult worm occurs over a period of approximately 2 months. In some instances, an infected individual harboring the adult worm can become auto-infected through the accidental ingestion of eggs released in the feces.

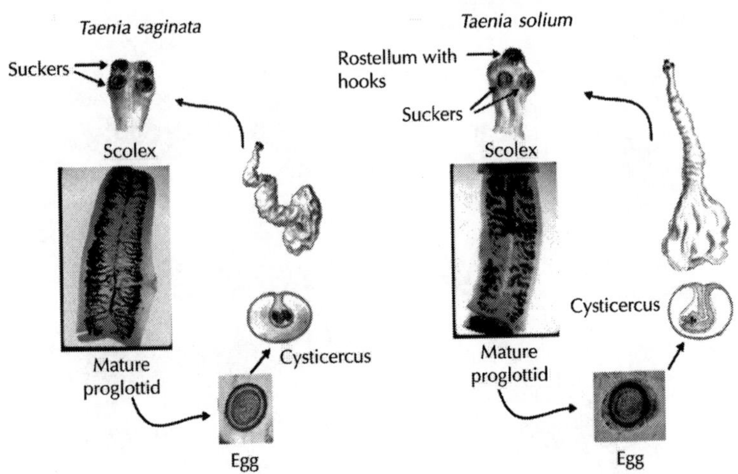

Figure 20.2. Morphology of *Taenia saginata* and *T. solium*. Reproduced from: Nappi AJ, Vass E, eds. Parasites of Medical Importance. Austin: Landes Bioscience, 2002:62.

Immunobiology

Infection with the adult tapeworm occurs in the small intestine of the human host and has been shown to induce a Th2-type immune response characterized by high levels of IL-4 and IL-10 expression and an increase in immunoglobulin production, primarily IgG. Antibodies produced in response to parasite antigens appear to be somewhat effective in the destruction of the early larval form, but offer little to no protection against cysticerci present within the tissues.

Viable cysticerci produce little to no inflammation within the surrounding tissues and their ability to suppress the host inflammatory response undoubtedly plays a major role in their ability to survive within the host for extended periods of time. In contrast, the death or destruction of cysticerci within host tissues has been shown to induce a strong Th1-type cell-mediated inflammatory response, characterized by high levels of interferon-gamma and the formation of granulomas containing lymphocytes, eosinophils, granulocytes and plasma cells. Experimental data using a mouse model suggest that the development of a Th1 cell-mediated inflammatory response controls parasite growth, whereas a Th2-type response increases levels of susceptibility to chronic infection.

These parasites have developed numerous methods for evading the host immune response. Although the ingested oncospheres which are capable of penetrating the intestinal mucosa are susceptible to destruction by host compliment and antibody responses, the time required to generate these antibodies allows the oncosphere to transform into the highly resistant metacestode form. The metacestode form, resistant to complement-mediated destruction, produces a variety of molecules effective in evading the host immune response. The serine-threonine protease

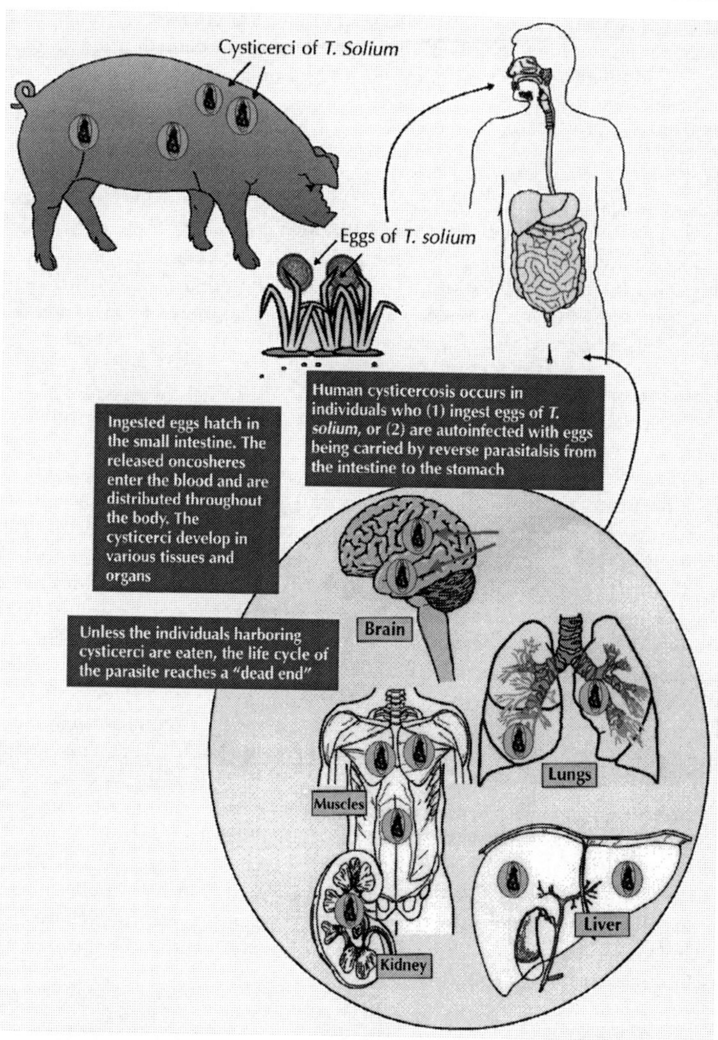

Figure 20.3. Development of cysticercosis in humans. Reproduced from: Nappi AJ, Vass E, eds. Parasites of Medical Importance. Austin: Landes Bioscience, 2002:63.

inhibitor, Taeniastatin, inhibits complement activation, blocks cytokine production and interferes with neutrophil function. Paramyosin renders parasite killing by the host complement cascade ineffective, primarily through inhibiting the activity of C1q. Activated complement is directed away from the parasite by the production of sulfated polysaccharides. Antibodies produced by the host bind the metacestode form through Fc receptors and are degraded, possibly functioning as a source of amino acids for the parasite. Glutathione S-transferase and other small molecules produced by the cyst form are involved in the detoxification of toxic oxygen intermediates and the suppression of host inflammation.

Signs and Symptoms

Taeniasis

Taeniasis is an infection with the adult tapeworm which usually remains confined to the small intestine. Most often, such infection results in minor gastrointestinal irritation and is frequently accompanied by nausea, diarrhea, constipation, hunger pains, chronic indigestion and passage of proglottids in the feces. Although these symptoms are usually milder when the infection is caused by *T. solium*, the risk of developing cysticercosis remains high.

Cysticercosis

Cysticercosis refers to the tissue infection caused by the metacestode, or larval stage, of *Taenia solium* and is acquired by the accidental ingestion of eggs. The clinical manifestations associated with cysticercosis are a direct result of the inflammatory response induced to control parasite growth and may occur months to years after initial infection. Manifestations of disease are dependent upon a variety of factors including the site of infection as well as the number of cysticerci present within the tissues, which most often localize to sites within the eyes, skeletal muscles and brain. Cysticercosis is the most common intra-orbital parasitic infection and is observed in 13-46% of infected individuals. Infection may involve the sub-retinal space (intra-ocular) or the extraocular muscles, eyelid and/or lachrymal glands (extra-ocular) surrounding the eye(s). Patients suffering from ocular infection frequently experience pain in the eyes accompanied by blurriness and partial or complete loss of vision. In extreme cases, infection may cause complete detachment of the retina.

Patients infected with cysticerci in the skeletal muscles and/or subcutaneous tissues are usually asymptomatic. In most cases, multiple cysts are present within the tissues, although solitary cysts may also be detected. Cysts range from 10-15 mm in length and arrange themselves in the same orientation as the muscle fibers. Leakage of fluid into the tissues, or death of the parasite, can trigger a strong inflammatory response, resulting in sterile abscess formation accompanied by localized pain and swelling.

Neurocysticercosis

Neurocysticercosis is the most common parasitic infection of the human central nervous system and is observed in 60-90% of infected patients. Cysts localized within the brain may range anywhere from 4-20 mm in length, but most commonly average between 8-10 mm. As with cysts localized in skeletal muscles and subcutaneous tissues, the destruction of parasites induces an inflammatory response, granulomas and fibrosis which may result in a subacute encephalitis.

Seizures are the most common symptom reported in patients with neurocysticercosis and occur in 70-90% of infected patients. Other commonly associated clinical manifestations include headache, dizziness, involuntary muscle movement, intercranial hypertension and dementia. Not all patients with neurocysticercosis are symptomatic; a certain percentage of patients with neurocysticercosis never develop any symptoms and these infections are often self-resolving.

Diagnosis

Diagnosis is often difficult due to the nonspecific nature of symptoms associated with cysticercosis. Therefore, proper diagnosis of the diseases is most often based on a combination of clinical, serological and epidemiological data.

MRI and CAT scans are considered to be the most sensitive methods of detection of neurocysticercosis and are useful in establishing diagnosis. However, the high costs associated with these radiologic methods greatly restrict the availability and/or accessibility of these tests in most underdeveloped countries where the disease is endemic.

Serological methods of detection most often include the ELISA (enzyme-linked immunoassays) and the EITB (enzyme-linked immunoelectrotransfer blot) and involve the detection of antibodies against cysticerci. EITB is highly sensitive and is considered to be the best immunological diagnostic test available. However, EITB is not effective in the detection of antibodies when only one cyst is present. The ELISA, while not as sensitive, is technically simpler and is therefore used extensively in clinical settings. It should be noted, however, that detection of anticysticercal antibodies may simply indicate previous exposure or infection and is not an exclusive indication of a current, active infection within the host. Other methods of detection include compliment fixation and indirect haemagglutination assays.

Treatment

Praziquantel and albendazole are the two anticysticercal drugs used to treat patients diagnosed with cysticercosis in the brain and skeletal muscles. Treatment with praziqauntel (50-100 mg/kg/d × 30 d) and albendazole (400 mg bid for 8-30 d) has been shown to completely eliminate cysts in 80% of treated patients, with an additional 10% of patients experiencing a significant reduction in the number of cysts present. Some investigators recommend 100 mg/kg/d in three divided doses × 1 day and then 50 mg/kg/d in 3 doses for 29 days of praziquantel. Neither drug is toxic; however, a percentage of patients undergoing therapy experience adverse side effects such as headache, nausea, vomiting, dizziness and increased pressure on the brain. These effects are most likely a result of the host immune response resulting from the massive destruction of parasites and therefore, treatment with either praziquantel or albendazole is often administered concomitantly with corticosteroids in order to prevent excessive inflammation. Dexamethasone is the steroid most often administered in conjunction with either praziquantel or albendazole. Prednisone may be used as a replacement in patients when long-term therapy is required. Antiepileptic drugs may be necessary adjuncts for treatment of seizures in patients being treated for neurocysticercosis.

Surgical removal of cysts from infected tissues is possible and, prior to the development of anticysticercal drugs, was the primary means of treatment. However, the invasiveness and high risk of complications associated with surgery makes this method less favorable to treatment with chemotherapeutic agents.

Prevention and Prophylaxis

The most effective means of preventing infection is to ensure that meats are cooked thoroughly prior to consumption. Good hygiene and sanitation are highly effective in decreasing the risk of infection associated with fecal-oral transmission.

The costs associated with chemotherapy and other medical resources, as well as losses in production, are enormous and efforts to prevent and/or eliminate disease have been a primary concern for public health systems in endemic countries for a long time. More recently, an increase in the number of imported cysticercoses in developed countries has made the eradication of the diseases a primary health concern worldwide.

Improvements in sanitation and public health care are essential for preventing the further spread of disease. Altering the infrastructure to keep pigs from roaming freely and contacting human feces will help reduce human-to-pig transmission. Effective measures to control and regulate meat inspection at slaughterhouses has been extremely effective in Europe and North America; however, programs to ensure proper compensation for the loss of infected livestock must be developed in order to discourage the underground trafficking of livestock by local farmers in endemic regions.

Vaccines aimed at preventing infection in pigs may play a role in efforts to control the spread of disease. Due to their typically short-life span (approximately one year), pigs do not require long-term immunity; therefore, vaccines which provide only short term resistance may be sufficient to prevent the spread of infection to humans. Additionally, the vaccination, rather than the confiscation, of pigs is often a more favorable alternative to local farmers.

To date, the most effective vaccines have involved the expression of recombinant oncosphere antigens TSOL18 and TSOL45 in *E. coli*. TSOL18 appears to be more effective, inducing greater than 99% protection in the five vaccine trials undertaken thus far. Current efforts are focused on developing the methods necessary to make the vaccine widely available and successful on a practical scale. The use of recombinant vaccines in pigs, combined with anticysticercal chemotherapy in humans, seems to be the most effective approach in the battle against cysticercosis and appears to have potential to control and/or eradicate the disease.

Concluding Remarks

Cysticercosis and taeniasis resulting from tapeworm infections currently affect millions of people worldwide and continue to exert increasing pressure on public health care systems in endemic countries and non-endemic countries alike. The high prevalence of the diseases in endemic countries as well as increasing incidences of these diseases in non-endemic regions has grabbed the attention of health officials worldwide. Further research to elucidate the mechanisms of the host immune response to parasitic infection, including the mechanisms by which parasites are able to evade destruction by the host, will likely facilitate the development of effective vaccines to control the further spread of disease. Successful programs to eradicate the diseases will require the combined efforts of scientists and physicians as well as the development of social and economic programs geared towards improving public education and the quality of life in many impoverished, underdeveloped countries in which Taenia infections are endemic.

Suggested Reading

1. Carpio A. Neurocysticercosis: an update. The Lancet Infectious Diseases 2002; 2:751-62.
2. Hoberg EP. Phylogeny of Taenia: species definitions and origins of human parasites. Parasitol Int 2006; 50::S23-30.
3. Singh G, Prabhakar S. Taenia solium Cysticercosis: From Basic to Clinical Science. New York: CABI Publishing, 2002.
4. Becker H. Out of Africa: The origins of the tapeworms. Agricultural Research 2001; 49:16-8.
5. Sciutto E, Fragoso G, Fleury A et al. Taenia solium disease in humans and pigs: an ancient parasitosis disease rooted in developing countries and emerging as a major health problem of global dimensions. Microbes and Infection 2000; 2:1875-90.
5. Wandra T, Ito A, Yamasaki H et al. Taenia solium Cysticercosis, Irian Jaya, Indonesia. Emerg Infect Dis 2003; 9:884-5.
6. White AC Jr, Robinson P, Kuhn RE. Taenia solium cysticercosis: host-parasite interactions and the immune response. Chem Immunol 1997; 66:209-30.
7. Rahalkar MD, Shetty DD, Kelkar AB et al. The Many Faces of Cysticercosis. Clin Radiol 2000; 55:668-74.
8. Sloan L, Schneider S, Rosenblatt J. Evaluation of Enzyme-Linked Immunoassay for Serological Diagnosis of Cysticercosis. J Clin Microbiol 1995; 33:3124-8.
9. Garcia H, Evans C, Nash TE et al. Current Consensus Guidelines for Treatment of Neurocysticercosis. Clin Microbiol Rev 2002; 15:747-56.
10. Garg RK. Drug treatment of neurocysticercosis. Natl Med J India 1997; 10:173-77.
11. The Medical Letter (Drugs for Parasitic Infections) 2004; 46:e1-e12.

Chapter 21

Hydatid Disease

Hannah Cummings, Miriam Rodriguez-Sosa and Abhay R. Satoskar

Background

Hydatid disease, also called hydatidosis or echinococcosis, is a cyst-forming disease resulting from an infection with the metacestode, or larval form, of parasitic dog tapeworms from the genus *Echinococcus*. To date, five species of *Echinococcus* have been characterized. The vast majority of human diseases are from *Echinococcus granulosus* and *Echinococcus multioccularis* which cause cystic echinococcosis and alveolar echinococcosis, respectively. Millions of people worldwide are affected by human hydatid disease and as a result, the diagnosis, treatment and prevention of the disease has become a serious concern for public health care systems around the world.

Geographic Distribution

Echinococcus infections are estimated to affect between 2-3 million people worldwide with endemics located primarily in regions of North and South America, Europe, Africa and Asia associated with the widespread raising of sheep and other livestock.

Life Cycle

Hydatid disease is caused by infection with the larval form of *E. granulosus* (and/or *E. multiocularis)* and results in the formation of cysts within various host tissues. The complete life cycle of *Echinococcus granulosus* requires two hosts (Fig. 21.1). Domestic dogs act as the primary definitive host of the mature adult worms and a single infected dog may harbor millions of adult worms within its intestines. Other canines such as wild dogs, wolves, coyotes, foxes and jackals may also act as a definitive host harboring the adult tapeworms. Intermediate hosts become infected with the larval form of the parasite and include a wide range of herbivorous animals, primarily sheep, cattle, pigs, goats and horses. The life cycle is completed by the ingestion of one or more cysts and its contents by the canine host through the consumption of infected viscera of sheep and and/or other livestock. Protoscoleces released in the small intestine attach to the intestinal wall through the action of four suckers and a row of hooks and within two months mature into adult worms capable of producing infective eggs.

Humans may become infected though the ingestion of food and/or water contaminated with infective eggs released in the feces of dogs harboring the adult

Medical Parasitology, edited by Abhay R. Satoskar, Gary L. Simon, Peter J. Hotez and Moriya Tsuji.

Figure 21.1. Life cycle of *Echinococcus*. Reproduced from: Nappi AJ, Vass E, eds. Parasites of Medical Importance. Austin: Landes Bioscience, 2002:65.

tapeworm(s). Once ingested, the eggs release oncospheres capable of actively penetrating the intestinal mucosa. These oncospheres gain access to the blood stream via the hepatic portal vein and migrate to various internal organs where they develop into cysts. Hydatid cysts most often localize within the liver and the lungs; however, cysts may also form in the bones, brain, skeletal muscles, kidney and spleen. The clinical manifestations of hydatid disease vary depending on a variety of factors including the location, size and number of cysts present within the infected tissues.

Similar to *E. granulosus*, the complete life cycle of *E. multiocularis* also requires two hosts. The primary definitive host for *E. multiocularis* is the fox, although the parasite may also infect wild and domesticated dogs and occasionally cats. Rodents such as field mice, voles and ground squirrels act as natural intermediate hosts and acquire infection by ingesting infective eggs released into the environment.

Immunobiology

The development of an immune response to infection with the larval form of the parasite is generally divided into two broad phases: the preencystment phase and the postencystment phase. Both cellular and humoral immunity are induced during each phase; however, neither response is sufficient to eliminate the parasite.

Early stages of a primary infection with *E. granulosus* are characterized by the substantial activation of a cell-mediated type immune reaction against the parasite. The release of oncospheres promotes an increase in leukocytosis, primarily by eosinophils, lymphocytes and macrophages. Host complement pathways contribute to the host inflammatory response and are activated by both living organisms as well as by material derived from dead parasites. Intense, dense granulomas form around the cyst and are responsible for much of the tissue destruction and subsequent clinical pathology associated with the disease.

Parasite-specific antibodies can be detected in the sera of patients shortly after infection and include IgG, IgA and IgM. Studies suggest that early oncospheres may be killed through antibody-dependent cell-mediated cytotoxicity reactions involving neutrophils. A certain percentage of patients develop an immediate-type hypersensitivity reaction to larval antigens, characterized by the nonspecific degranulation of basophils and increased levels of circulating IgE. Anaphylaxis-type reactions may occur and are often induced by the rupture of a cyst or the leakage of hydatid cyst fluid within the tissues.

The postencystment phase of infection is marked by an increase in the levels of IgG, IgM and IgE. The infiltration of eosinophils, neutrophils, macrophages and fibrocytes initiated early in infection persists throughout the later phases of cyst development; however, the presence of mature cysts within the tissues does not result in an intense inflammatory response.

Cytokine profiles of infected patients suggest the development of both a Th1- and Th2-type immune response to infection. Live parasites have been shown to actively induce Th1 cytokines, suggesting that the development of a Th2-type response is involved in host susceptibility to infection. In addition, Th2 cytokines are the predominant cytokines detected in sera from patients with active or transitional cysts. In contrast, patients with inactive cysts or undergoing effective chemotherapy exhibit a strong Th1-type response. This Th1 response dominates the Th2 response and suggests that a predominant Th1 response induced late in infection may be responsible for the successful resolution of infection.

Signs and Symptoms

Echinococcus granulosus and *Echinococcus multiocularis* are the two species most often identified in human hydatid disease. Cystic echinococcosis, caused by *E. granulosus*, is the most common and accounts for approximately 95% of all

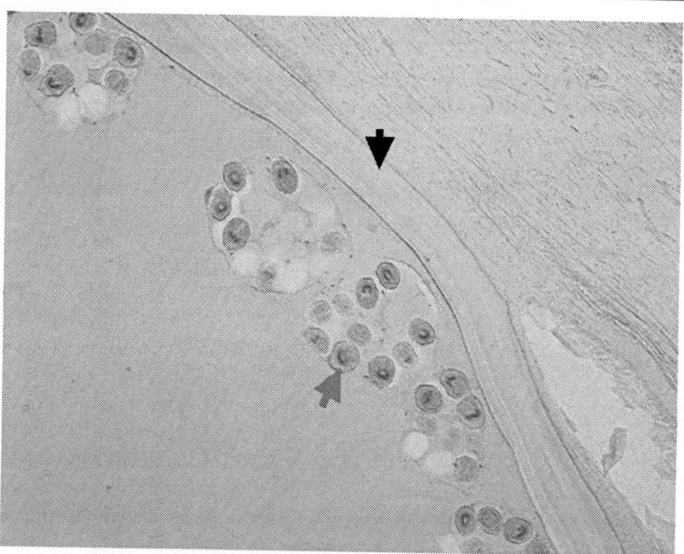

Figure 21.2. Photomicrograph of a hydatid cyst from the liver. Note the hyaline membrane (black arrow) and the protoscolex in the brood capsules (gray arrow).

global cases. Cystic echinococcosis may affect people of all ages, but hydatid cysts are most often present in patients between 15-35 years of age.

Infection with *E. granulosus* results in the rapid growth of large, uniocular cysts filled with fluid (Fig. 21.2). Most cysts develop within the tissues of the liver and lung, with 55-75% of cysts found in the liver and 10-30% of cysts found in the lungs. Cysts may survive in the liver for several years and often do not cause any symptoms in the infected host. Symptoms arise when the cysts become large enough to be palpable and/or cause visual abdominal swelling and pressure. Patients frequently experience abdominal pain in the right upper quadrant, often accompanied by nausea and vomiting. The rupture or leakage of cysts within the tissue can result in anaphylactic shock and facilitate the spread of secondary cysts through the release and dissemination of germinal elements. Biliary tract disease and portal hypertension may complicate liver involvement and postobstructive infection due to erosion of cysts into the biliary tract may further complicate echinococcal infection. Pulmonary cystic echinococcosis is acquired early during childhood, but the clinical manifestations associated with the disease do not typically appear until the third or fourth decade of life.

Cysts residing within the lung tissue often remain silent producing little to no symptoms. Problems arise when cysts grow large enough to obstruct or erode a bronchus, often causing the rupture of cysts and the dissemination of cystic fluids. Patients infected with pulmonary cysts frequently experience chronic dry cough, chest pain and hemoptysis often accompanied by headache, sweating, fever and malaise.

Alveolar echinococcosis affects between 0.3-0.5 million people and is usually caused by *Echinococcus multiocularis*. It is characterized by the formation of multiocular hydatid cysts which contain little to no fluid. These cysts lack both the hyaline membrane and the brood capsules which facilitate the widespread metastasis of larvae into the surrounding tissues. These larvae invade adjacent tissues and proliferate indefinitely causing extensive and progressive tissue necrosis and eventual death in 70% of infected patients.

Hydatid disease can affect a wide range of organs including the bones, central nervous system, heart, spleen, kidneys, muscles and eyes. Patients diagnosed with the disease should be screened for the presence of multiple cysts in various tissues.

Diagnosis

Proper diagnosis and treatment of hydatid disease is difficult. Individuals often remain asymptomatic for several years after initial infection, allowing time for the growth of large, debilitating cysts. Various imaging techniques are used to visually detect cysts present within host tissues. CT scans and MRIs are used extensively in clinical settings and are useful in the detection of developing, dying or dead cysts. Typical features include thick cyst walls, detached germinal membranes, internal septae and/or the presence of daughter cysts. X-ray, ultrasound and scintillography may also be useful in the detection of hydatid cysts and in diagnosis of the disease.

Numerous serological assays are currently available and are useful in the detection and diagnosis of hydatid disease. Common detection methods include indirect hemagglutination assays (IHA), indirect immunofluorescence, counter-current immunoelectrophoresis (CIEP), enzyme-linked immunoassays (ELISA) and enzyme-linked immunotransfer blots (EITB). Most serological assays involve the detection of specific serum antibodies, primarily the detection of IgG to hydatid cyst fluid-derived or recombinant antigen B subunits. Although high levels of sensitivity have been achieved (92.2%), complications may arise due to cross-reactivity between hydatid disease and cysticercosis.

Detection of mitochondrial DNA using molecular techniques like PCR is extremely useful and is often used to analyze genotypic variations between species and/or strains.

Treatment

Surgery remains the treatment of choice for the removal of cysts. Patients diagnosed with multiple cysts often require numerous staged operations. Complete excision of the cysts is difficult: surgical removal may cause the rupture or leakage of cysts/cystic fluid resulting in the release and dissemination of infective protoscoleces.

Albendazole is frequently used to treat patients with hydatid disease. Patients typically receive 10 mg/kg/d or 400 mg orally twice per day for 1-6 months. Although neither regimen has been proven to be effective in resolving the disease alone, the use of drug therapy in conjunction with surgical treatment has shown to greatly reduce the risk of development of new cysts and is currently the therapy of choice.

PAIR, or percutaneous aspiration, followed by injection of 95% ethanol or another scolicidal agent and then reaspiration, may sometimes be used as an alternative to therapy, especially for the treatment of inoperable cysts.

Prevention and Prophylaxis

The most effective means to control hydatid disease in humans and eliminate the consequences of *Echinococcus* infections in livestock is through the broad- range education of people living in endemic regions. Education to prevent the feeding of infected viscera to dogs is essential for controlling the spread of infection from livestock to dogs. Most human infections are due to close contact with infected dogs. Deliberate actions aimed at reducing the rate of dog infection in endemic regions will undoubtedly reduce the number of human infections. In addition, the reduction and removal of stray and unwanted dogs, as well as the regular treatment of dogs with anthelminthic drugs, will facilitate the widespread efforts geared towards controlling disease transmission.

The development of vaccines designed to prevent infection of either or both the definitive and intermediate host(s) offers the greatest possibility of success in the control and eradication of hydatid disease in both the livestock and human populations. EG95 is a 16.5 kDa recombinant GST fusion protein derived from *E. granulosus* oncospheres and functions as a highly effective vaccine for grazing livestock. EG95, which induces immunity through complement-fixing antibodies, has been shown to induce high levels of protection (96-98%) against the development of hydatid cysts.

Concluding Remarks

Human hydatid disease affects millions of people and has attracted the attention of health professionals around the world. The treatment of echinococcus infections within the domestic animal population would likely result in a reduction in the number of human cases of hydatid disease and, therefore, has become the focus of many studies aimed at the development of effective vaccines to control the spread of disease. Although vaccines are an invaluable tool for the control and eradication of disease, increasing public education and awareness of the effects of infection and the mode of transmission will be essential for control within remote areas where the disease is endemic.

Suggested Reading

1. Thompson RCA. The Biology of Echinococcus and Hydatid Disease. London: George Allen & Unwin Ltd, 1986:85.
2. Zhang W, Li J, McManus DP. Concepts in immunology and diagnosis of hydatid disease. Clin Microbiol Rev 2003; 16:18-36.
3. Craig PS, McManus DP, Lightlowlers MW et al. Prevention and control of cystic echinococcosis. Lancet Infect Dis 2007; 7:385-94.
4. Sturton SD. Geographic distribution of hydatid disease. Chest 1968; 54:78.
5. Ceran S, Sunam GS, Gormus N et al. Cost-effective and time-saving surgical treatment of pulmonary hydatid cysts with multiple localization. Surg Today 2002; 32:573-6.
6. Jenkins DJ, Power K. Human hydatidosis in New South Wales and the Australian Capital Territory, 1987-1992. Med J Aust 1996; 164:14-7.

7. Goldsmith RS. 35 Infectious diseases: protozoal and helminthic. Current Medical Diagnosis and Treatment 2007; 46:
8. Dickson DD, Gwadz RW, Hotez PJ. Parasitic Diseases, 3rd Edition. New York: Springer-Verlag, 1995:93-8.
9. Parija SC. Text Book of Medical Parasitology: Protozoology and Helminthology. Chennai: All India Publishers & Distributors, 2001: 214-9.
10. Arora DR, Arora B. Medical Parasitology. New Delhi: CBS Publishers and Distributors, 2002:120-3.
11. Moro P, Schantz PM. Cystic echinococcosis in the Americas. Parasitol Internat 2005; 55:S181-6.
12. Magambo J, Njoroge E, Zeyhle E. Epidemiology and control of echinococcosis in sub-Sahara Africa. Parasitol Internat 2006; 55:S193-5.
13. Torgerson PR, Oguljahan B, Muminov AE et al. Present situation of cystic echinococcosis in Central Asia. Parasitol Internat 2006; 55:S207-12.
14. Shaikenov BS, Vaganov TF, Torgerson PR. Cystic Echinococcosis in Kazakhstan: An emerging disease since independence from the Soviet Union. Parasitol Today 1999; 15:173-4.
15. Romig T, Dinkel A. Mackenstedt. The present situation of echinococcosis in Europe. Parasitol Internat 2006; 55:S187-91.
16. Baz A, Ettlin GM, Dematteis S. Complexity and function of cytokine responses in experimental infection by Echinococcus granulosus. Immunobiology 2006; 211:3-9.
17. Warren KS. Immunology and Molecular Biology of Parasitic Infections, 3rd Edition. Chelsea: Blackwell Scientific Publications, 1993:438-48.
18. Ferreira M, Irigoin F, Breijo M et al. How echinococcus granulosus deals with compliment. Parasitol Today 2000; 16:168-72.
19. Zhang W, You H, Zhang Z et al. Further studies on an intermediate host murine model showing that a primary Echinococcus granulosus infection is protective against subsequent oncospheral challenge. Parasitol Internat 2001; 50:279-83.
20. Rosenzvit M, Camicia F, Kamenetzky L et al. Identification and intra-specific variability analysis of secreted and membrane-bound proteins from Echinococcus granulosus. Parasitol Internat 2006; 55:S63-7.
21. Kizaki T, Kobayashi S, Ogasawara K et al. Immune Suppression Induced by Protoscoleces of Echinococcus multiocularis in Mice: Evidence for the Presence of $CD8^+$ dull Suppressor Cells in Spleens of Mice Intraperitoneally Infected with E. multiocularis. J Immunol 1991; 147:1659-66.
22. Markel EK, John DT, Krotoski WA. Markell and Voge's Medical Parasitology, 8th Edition. Philadelphia: W.B. Saunders Company, 1999:254-60.
23. Tor M, Atasalihi, Altuntas N et al. Review of cases with cystic hydatid lung disease in a tertiary referral hospital in an endemic region: a 10 Years' experience. Respiration 2000; 67:539-42.
24. Elton C, Lewis M, Jourdan MH. Unusual site of hydatid disease. Lancet 2000; 355:2132.
25. Bahloul K, Ghorbel M, Boudouara MZ et al. Primary vertebral echinococcosis: four case reports and review of literature. Br J Neurosurg 2006; 20:320-3.
26. Todorov T, Mechkov G, Vutova K et al. Benzimidazoles in the treatment of abdominal hydatid disease: a comparative evaluation. Parasitol Internat 1998; 47:105-31.
27. Heath D, Yang W, Tiaoying L et al. Control of hydatidosis. Parasitol Internat 2006; 55:S247-52.
28. Parija SC. A review of some simple immunoassays in the serodiagnosis of cystic hydatid disease. Acta Tropica 1998; 70:17-24.

Section IV

Protozoans

Chapter 22

American Trypanosomiasis (Chagas Disease)

Bradford S. McGwire and David M. Engman

Introduction

American trypanosomiasis is a vector-borne infection caused by the protozoan parasite *Trypanosoma cruzi*. Also called Chagas disease, named after the Brazilian physician Carlos Chagas who described the infection in 1909, it is found only on the American continent. The parasite alternately infects triatomine insects (reduviid, assassin or "kissing" bugs) and a wide range of vertebrate hosts in a complex lifecycle. Human infection results in a myriad clinical syndromes resulting from localized and disseminated infection arising from the initial deposition of infective parasites during feeding of the blood sucking triatomine. Chagas disease is an important public health concern, being widespread in Central and South America and chronic infection is the leading cause of heart failure in these regions. Transmission via transfusion of blood products and organ transplantation is a matter of concern, even in North America. This review will cover the lifecycle and epidemiology, pathogenesis, clinical diagnosis, management and prevention of *T. cruzi* infection.

Epidemiology of *T. cruzi* Infection

The triatomine insects that transmit *T. cruzi* are present throughout the Americas, spanning vast regions from the central United States throughout Central and South America, extending to the south-central portions of Chile and Argentina. *T. cruzi* infection is primarily a zoonosis and humans are only incidental hosts; thus, natural transmission occurs primarily in rural areas where insects are abundant. The incidence of human infection is increasing in these regions due to deforestation for farming, which has caused the insects to migrate to the rudimentary human dwellings made of mud and thatch, wood or stone. Despite the presence of *T. cruzi*-infected insects in the United States, the low incidence of acute Chagas disease in this country is thought to be due to the relatively high quality of housing. The World Health Organization currently estimates that 13 million people are infected with *T. cruzi*, with 200,000 new infections occurring annually in 15 countries. In addition to insect-borne disease, *T. cruzi* can also be transmitted congenitally or by blood transfusion or organ transplantation. Transmission of *T. cruzi* infection by blood transfusion is increasing in the US due to the increasing influx of infected immigrants who donate blood. Thus, there is a pressing need to implement widespread screening of blood products for the presence of *T. cruzi*.

Medical Parasitology, edited by Abhay R. Satoskar, Gary L. Simon, Peter J. Hotez and Moriya Tsuji.

T. cruzi Life Cycle and Transmission

Trypanosoma cruzi is a eukaryote possessing a membrane bound nucleus and mitochondrion. The mitochondrial DNA is a complex structure which resides in a specialized region (kinetoplast) adjacent to the base of the flagellum (Fig. 22.1). *T. cruzi* has four distinct life cycle stages (Fig. 22.2). Within the midgut of the reduviid bug, parasites replicate as flagellated epimastigotes (epi). As epis replicate and increase in number they migrate to the hindgut of the bug where they differentiate into infective metacyclic trypomastigotes (meta). Metas are discharged in the feces of the bug as they take a blood meal. Infection results from the contamination of the insect bite or open wounds, mucous membranes or conjunctiva with parasite laden bug feces. Once in the vertebrate host, the meta, which is unable to replicate, must invade host cell within which it can differentiate into the replicating amastigote (ama). During invasion the meta is initially present within a membrane bound vacuole, but it escapes this vacuole and differentiates into the aflagellated ama, which divides in the cytoplasm. After a number of rounds of replication, the amas fill the cytoplasm and differentiate into motile trypomastigotes (tryp), which lyse the infected cell and escape to infect adjacent cells or disseminate throughout the body via the bloodstream and lymphatics. Tryps, like metas, cannot replicate and must invade host cells and differentiate into amas to survive. Alternatively, they may be taken up by a triatomine insect during a blood meal and differentiate into epis in the insect midgut, thereby completing the life cycle. Within the vertebrate host, parasites can infect any nucleated cell, but have a predilection for muscle, particularly of the heart and gastrointestinal tract. This tissue tropism ultimately leads to the two predominant clinical forms of chronic *T. cruzi* infection: cardiomyopathy and megacolon/megaesophagus.

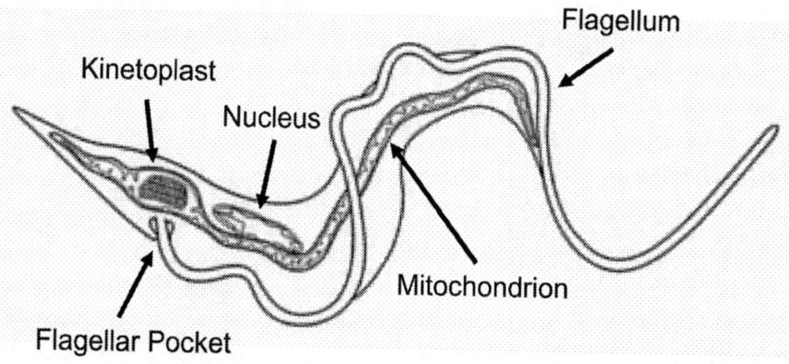

Figure 22.1. Cellular features of a *Trypanosoma cruzi* trypomastigote. *Trypanosoma cruzi* is a protozoan parasite, possessing the organelles of all eukaryotes including a membrane bound nucleus and mitochondrion. A single membrane-bound flagellum emerges from the trypanosome's flagellar pocket and runs the length of the cell, attached to the cell body membrane via a desmosome-like adhesive junction. At its origin, the flagellum is physically connected to the mitochondrial DNA, which resides in a specialized region of the mitochondrion termed the kinetoplast.

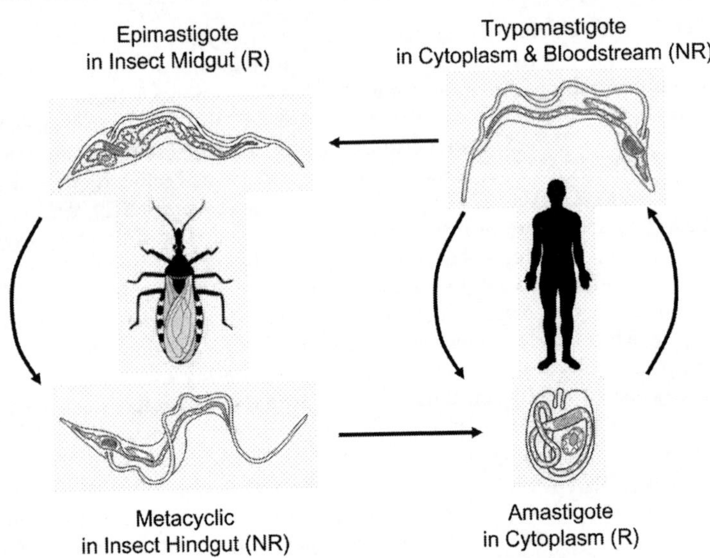

Figure 22.2. Life cycle of *Trypanosoma cruzi*. *T. cruzi* possesses four basic life cycle stages. In the insect, noninfectious epimastigotes replicate (R) in the midgut and differentiate into infectious but nonreplicating (NR) metacyclics as they migrate to hindgut. The fecal material of the insect, which contains metacyclics is deposited on the skin during a bloodmeal and infection occurs when this material contaminates the insect bite or a mucous membrane, which the trypomastigotes can penetrate. Within the human host, metacyclics invade host cells and differentiate into amastigotes, which replicate, burst out of the cell and either invade other cells or are taken up by another insect. Within the insect gut, the trypomastigotes differentiate into replicating epimastigotes, thus completing the cycle.

Pathogenesis of Chagas Disease

Acute *T. cruzi* infection results from the contamination of wounds or mucous membranes with insect feces containing expelled infective parasites. Locally deposited parasites bind to and invade host tissue and transform into and replicate as intracellular amastigotes. Infection leads to the formation of parasite "pseudocysts," so named because the amastigote nests are intracellular. This stimulates a localized inflammatory response mediated predominantly by lymphocytes and macrophages. Lymphatic drainage of the infected area into regional lymph nodes results in activation and proliferation of cells, resulting in regional lymphadenopathy. As the process continues, the amas transform into trypomastigotes, escape host cells and disseminate throughout the body. Infection and lysis of liver cells results in transient increases in serum liver enzyme levels. In chronic infection, tissue parasites are difficult to detect but significant interstitial fibrosis occurs, damaging the affected tissue. The molecular pathogenesis of Chagas disease is not completely understood, but likely results from (i) parasite-mediated tissue destruction, (ii) inflammation and fibrosis resulting from immune responses generated to parasites and residual parasite antigen, (iii) parasite-induced microvascular spasm and ischemic damage and/or (iv) autoimmune responses triggered by release of self-antigen during parasite lysis of host cells. Because there are many outcomes of chronic *T. cruzi* infection (see below), it is likely that each of these mechanisms occurs in isolation or in combination in a given individual, depending on the specific pathogenic potential of the strain of parasite (tissue tropism, replication rate, etc.) and the immunogenetic susceptibility of the infected individual.

Clinical Syndromes of Chagas Disease

Acute infection by *T. cruzi* is marked by the development of localized swelling and erythema at the site of the insect bite, which is termed a chagoma. This is a result of the local replication of parasites and the influx of fluid and inflammatory cells into the infected area. Infection through the conjunctiva can result in periorbital swelling, termed Romaña's sign (Fig. 22.3D). As parasites disseminate patients experience nonspecific symptoms such as fever, malaise and anorexia. Parasite infestation of peripheral tissues can give rise to hepatosplenomegaly and, in some cases, meningeal signs. Initial infection of heart tissue can lead to acute myocarditis and cardiac sudden death due to parasitization of the cardiac conduction system. The signs and symptoms of acute *T. cruzi* infection can last from days to weeks but are often unrecognized due to their nonspecific nature. The disease then proceeds to a quiescent phase lasting months to years and often decades, prior to the onset of chronic disease. It should be noted that the majority of *T. cruzi*-infected individuals do not develop any parasite-related disease and simply harbor low levels of parasites for life. Less than one-third of infected people develop chronic Chagas disease. The two hallmarks, usually mutually exclusive, disorders that occur in chronically infected patients are cardiomyopathy and megaorgan syndromes (Fig. 22.3E and G. respectively).

Cardiac involvement is heralded by the development of fibrosis within the heart muscle (Fig. 22.3F) and conduction system which leads to arrhythmias and heart failure, that latter being predominantly right-sided. Loss of ventricular muscle leads to wall thinning which can be associated with the development of apical aneurysms and subsequent formation of thrombi, which may have serious thromboembolic consequences (Fig. 22.3E). In the gastrointestinal tract, chronic infection leads to parasympathetic denervation, resulting in massive dilatation of the esophagus and/or colon. Esophageal involvement results in achalasia, associated odynophagia, dysphagia and esophageal dysmotility, often resulting in aspiration pneumonia. Colonic involvement results in abdominal pain, constipation, obstruction with perforation and secondary intrabdominal infection. Immunosuppression of patients with chronic Chagas disease, regardless of the mechanism (HIV infection, usage of immunosuppressive drugs in organ transplantation) can lead to recrudescence of parasite replication, massive parasitosis and death. Clinical disease in this setting is often fulminant with more extensive involvement of the central nervous system.

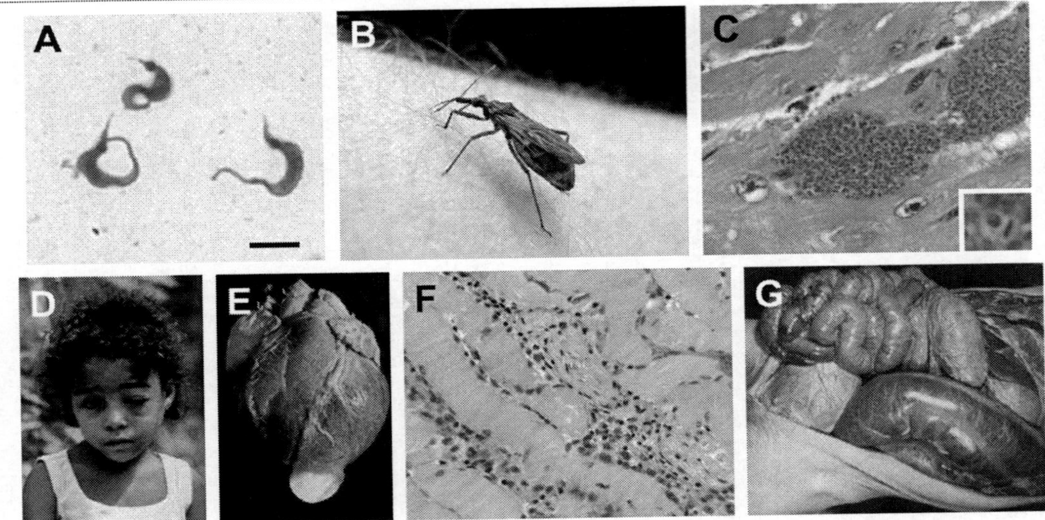

Figure 22.3. Various aspects of *Trypanosoma cruzi* biology and Chagas disease. A) *T. cruzi* trypomastigotes stained with Giemsa (bar = 5 μm). Note the prominent darkly-stained kinetoplast DNA. B) Reduviid bug. C) Nests of amastigotes in heart tissue, often termed "pseudocysts" since they are intracellular collections of parasites. The inset shows an amastigote with clearly visible nucleus (round structure) and kinetoplast (bar-like structure). D) Romaña's sign. E) Apical aneurysm (illuminated by light bulb) can occur after chronic fibrosis and weakening of the apical wall of left ventricle. F) Histopathology of Chagas heart disease: myofibrillar swelling and degeneration, mononuclear cell infiltration, fibrosis and edema in the absence of parasites are typical. G) Megacolon: a serious sequela of infection that is poorly understood. Photographs are courtesy of Cheryl Olson (A,C,F), Dr. Chris Beard (B), Dr. Michael Miles (D), Prof. F. Köberle (E,F,G).

Diagnosis of *T. cruzi* Infection

The diagnosis of *T. cruzi* infection initially requires a high degree of clinical suspicion. History of potential exposure to *T. cruzi* is important to document. Patients with a history of travel to or having had blood transfusion within endemic areas are at increased risk of *T. cruzi* infection. The presence of, or recent history of a chagoma or Romaña's sign are indicators of recent infection. The mainstay of diagnosis is detection of trypomastigotes in the blood or the presence of *T. cruzi*-specific antibodies in serum to indicate acute or chronic infection, respectively. Direct detection of parasites in blood is easier in immunocompromised patients in whom the immunologic control of parasites is not as efficient. Heavy parasite burdens in the tissues of such patients can permit diagnosis via direct examination of tissue (lymph nodes, or bone marrow) or fluids (cerebrospinal or pericardial fluid). In addition these specimens can be cultured in vitro in liquid medium or by growth within uninfected insect vectors (xenodiagnosis). During chronic infection parasites are frequently not detectable in the blood, and the presence of *T. cruzi* IgG, using commercial immunoassays, ELISA, complement fixation, or hemagglutination based tests, establishes the diagnosis. Direct detection of parasites using PCR based testing has been demonstrated but is not yet available for routine laboratory diagnosis. Potential blood donors throughout the Americas are asked questions related to risk factors of *T. cruzi* infection, but transfusion-associated disease remains a serious problem. As a result, the blood in much of South and Central America is screened for *T. cruzi*-specific antibodies, and many feel that the United States blood supply will be screened beginning within a few years.

Treatment of *T. cruzi* Infection

Benznidazole, an imidazole (trade name *Rochagan*, produced by Roche in Brazil) and Nifurtimox, a nitrofuran (trade name *Lampit*, produced by Bayer in Germany), are the two agents approved for treatment of Chagas disease and are available in the United States through contact with the Centers for Disease Control in Atlanta, Georgia. These agents have similar efficacy but have many adverse effects. Benznidazole is given orally for 1-3 months at a dose of 5-7 mg/kg/d in two divided doses. The side effects of this medication include rash and peripheral neuropathy but can also include bone marrow suppression. In adults, Nifurtimox is given for 120 days at a dose of 8-10 mg/kg/d in four divided doses. In children the drug is given for 90 days in four divided doses but the amount is based on age: 11-16 years (12.5-15 mg/kg/d); and under 11 years (15-20 mg/kg/d). Gastrointestinal maladies (nausea, vomiting, abdominal pain) are the predominant side effects of this medication but up to 30% of patients can also experience central nervous system effects such as polyneuritis, confusion or focal or generalized seizures. Skin rash can also develop in some patients. Individuals with glucose-6-phosphate dehydrogenase deficiency can experience drug-induced hemolytic anemia. Treatment is undertaken in cases of acute or congenital infection natural infection or in cases of accidental laboratory inoculation. Recent systematic reviews of clinical trials of trypanocidal therapy in patients with chronic *T. cruzi* infection suggest that treatment of asymptomatic immunocompetent patients may result in a reduction of progression to chronic disease (development of megaorgan syndromes, cardiomyopathy and arrhythmia). In contrast, there is no convincing data that support the use of trypanocidal therapy in patients who have already manifested end-organ damage as a result of chronic *T. cruzi* infection. It is clear that randomized controlled trials are necessary to truly understand the clinical benefit of trypanocidal therapy in chronic Chagas disease. The management of *T. cruzi* induced cardiac failure, achalasia and megacolon are approached in the same way that these end-organ problems are approached due to other causes.

Prevention of *T. cruzi* Infection

Limiting exposure to *T. cruzi* infected insects and blood is the mainstay of the prevention of Chagas disease. Persons living in or traveling to areas endemic for *T. cruzi* should avoid residing in substandard housing frequented by reduviid bugs. The use of bed nets and insect repellent are also recommended for this purpose. Barrier protection for those working with *T. cruzi* in the laboratory setting, such as protective clothing, gloves and eyewear is a must. Since the incidence of transfusion- and transplantation-associated *T. cruzi* infection is increasing in the Americas, serologic screening of donated blood seems advisable. Such is the practice in endemic countries within South America. As the number of potentially-infected immigrants to the United States increases, this will likely increase the number of transfusion-associated *T. cruzi* infections despite the presence of blood bank questionnaires.

Suggested Reading

1. Engman DM, Leon JS. Pathogenesis of Chagas heart disease: role of autoimmunity. Acta Trop 2002; 81:123-32.
2. Kirchhoff LV. Trypanosoma Species (American Trypanosomiasis, Chagas Disease): Biology of Trypanosomes. In: Mandell GL, Douglas RG, Bennett JE, eds. Principles and Practice of Infectious Diseases, 6th Edition. New York: Churchill Livingstone, 2005:
3. Mascola L, Kubak B, Radhakrishna S et al. Chagas disease after organ transplantation—Los Angeles, California. MMWR Morb Mortal Wkly Rep 2006; 55:789-800.
4. Villar JC, Marin-Neto JA, Ebrahim S et al. Trypanocidal drugs for chronic asymptomatic Trypanosoma cruzi infection. Cochrane Database Syst Rev 2002; CD003463.
5. Tyler KM, Miles MA. American Trypanosomiasis. Norwell: Kluwer Academic Publishers, 2003.

CHAPTER 23

African Trypanosomiasis

Guy Caljon, Patrick De Baetselier and Stefan Magez

Abstract

African trypanosomiasis is a vector-born disease that severely affects a broad range of vertebrate hosts, including humans, on the sub-Saharan African continent. The infection is caused by flagellated unicellular parasites (*Trypanosoma sp.*) and is lethal without treatment. Disease manifestations are pleotropic and are dependent on the host and infection-stage. Currently available diagnostic tests are adapted for field usage but have a low specificity, while an accurate differential diagnosis of human pathogenic *Trypanosoma* subspecies and correct determination of the infection stage is essential for appropriate treatment. For treatment of human African trypanosomiasis (HAT), four drugs with significant side-effects are currently available, with only one of them being registered in the last 50 years. This chapter will introduce the disease, its diagnosis, treatment and prospects for new therapeutic approaches.

Introduction

African trypanosomes are extracellular protozoan parasites that cause lethal infections in humans and livestock in large parts of sub-Saharan Africa. The responsible flagellated parasite (*Trypanosoma sp.*) is approximately twice the size of erythrocytes (15-30 μm, Fig. 23.1A) and relies on tsetse flies for its transmission (Fig. 23.1B). These arthropods are obligate bloodsucking insects (genus *Glossina*), that get infected through feeding on a parasitized host and accommodate the trypanosome during their entire lifespan. Engorged trypanosomes colonize the midgut, proliferate and undergo differentiation while directionally migrating towards the insect salivary glands. The vertebrate-infective metacyclic form of the parasite resides in the salivary glands or mouthparts of the fly, using the bloodfeeding behaviour for its transmission to a new host. Upon transmission to the vertebrate host, trypanosomes will transform into actively proliferating (long slender) forms to allow a systemic colonization of the host. Eventually, trypanosomes in the bloodstream become quiescent (short stumpy) and pre-adapt to uptake and subsequent survival in the tsetse fly. During the complex life cycle (Fig. 23.1C) of the parasite in the insect and vertebrate host, trypanosomes undergo several metabolic changes for the acquisition of free-energy from different available sources and modify mechanisms for the uptake of host nutrients, such as iron complexed with transferrin. In the fly, trypanosomes utilize amino acids (e.g., proline) as primary energy sources while trypanosomes in the vertebrate hosts metabolize glucose via glycolysis in a unique organelle, the glycosome. Only two subspecies,

Medical Parasitology, edited by Abhay R. Satoskar, Gary L. Simon, Peter J. Hotez and Moriya Tsuji.

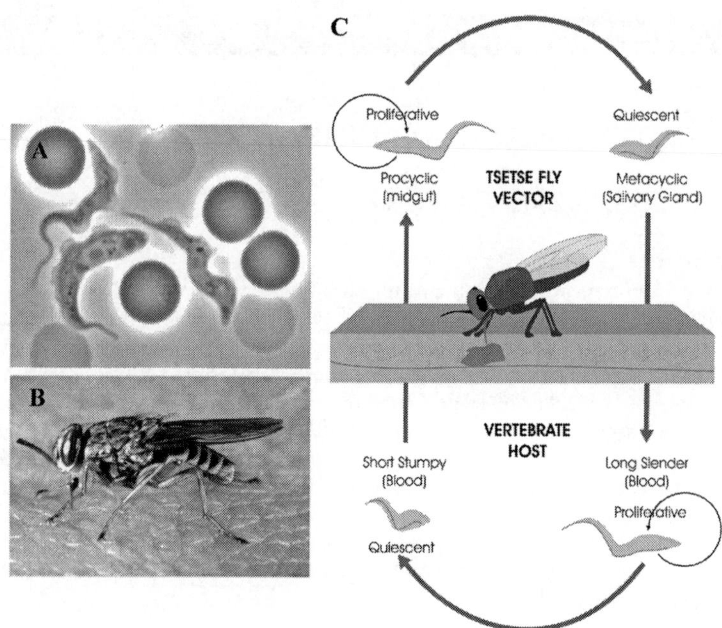

Figure 23.1. The trypanosome's lifecycle: A) a microscopic image (obtained with permission from Dr. David Pérez-Morga) of the causative agent of HAT, the trypanosome; B) a photograph of the vector of the disease, the tsetse fly (*Glossina morsitans*) (obtained with permission from Jan Van Den Abbeele); C) the lifecycle in the mammalian and insect host, indicating a proliferative stage for host colonization and a quiescent form, pre-adapted to survival in a new host (obtained with permission from Guy Caljon).

23

Trypanosoma brucei rhodesiense and *Trypanosoma brucei gambiense*, have the additional feature of resistance to normal human serum (NHS) that is trypanolytic for strictly livestock-threatening trypanosomes. Although both subspecies are pathogenic to human, they differ significantly in virulence and geographical occurrence (Fig. 23.2). *T. b. gambiense* causes chronic infections in West and Central Africa which can persist up to 10 years while *T. b. rhodesiense* is more prevalent in Eastern Africa and mostly results in acute human infections that can be lethal within a few months. The diseases caused by both subspecies are categorized under human African trypanosomiasis, better known as sleeping sickness and are responsible for an estimated 50,000 deaths a year. In contrast, *T. congolense, T. vivax* and *T. brucei brucei* are trypanosomes species that cause the majority of livestock infections with an estimated loss in agriculture of more than 1 billion $ per annum. Although sleeping sickness was largely controlled by the early 1960s, the disruption of health infrastructures and population displacement, as well as the lack of human and financial resources for disease control, led to a current epidemic scale of the disease in specific regions of Africa. Moreover, the development of protec-

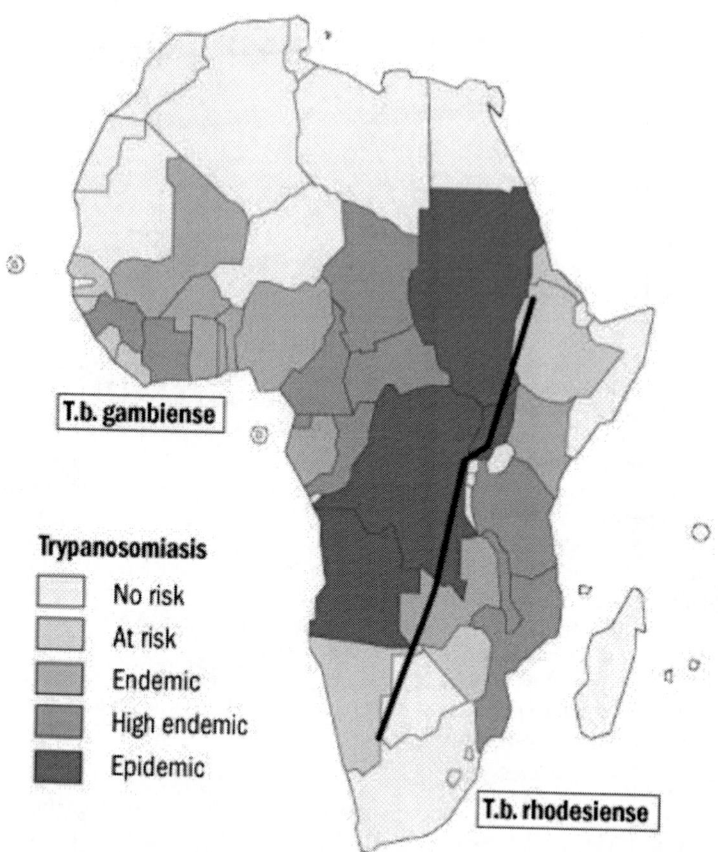

Figure 23.2. Distribution of HAT in sub-Sahara Africa: predominant occurrence of *T. b. gambiense* in Central and West Africa and *T. b. rhodesiense* in East Africa. The colour-scale indicates the incidence of HAT in the different countries. This figure was reproduced with permission from the World Health Organization (http://www.who.int/en/).

tive vaccines has been unsuccessful until now, mainly due to the ability of African trypanosomes to escape adaptive immune responses by antigenic variation of the most abundant surface glycoprotein VSG (variant-specific surface glycoprotein). These VSG molecules form a densely packed coat of up to 10^7 identical copies per cell, making invariant epitopes as potential immune targets inaccessible to conventional antibodies. There are about 1,000 genes present in the trypanosome genome that encode for these VSGs, but, due to a strictly controlled gene expression, only one gene at a time is translated to construct the actual outer protein coat. As the vertebrate host mounts an efficient antibody response against the VSG leading to partial parasite clearance of the major variant antigenic type (VAT), a minor part

23

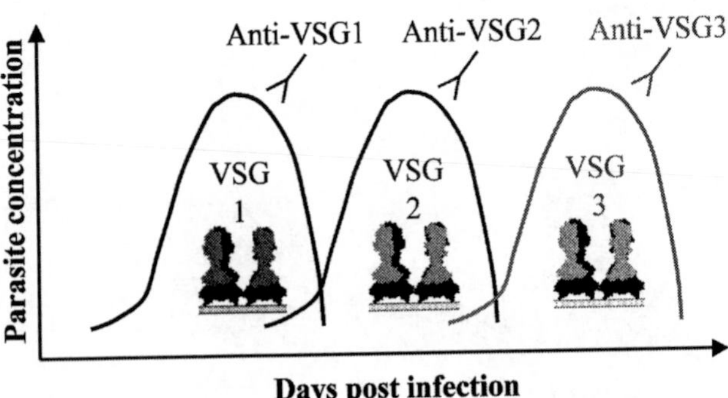

Figure 23.3. Antigenic variation of VSG: illustration of the escape of the trypanosome from specific host antibody responses. Each parasitemia wave represents a population that expresses another VSG-molecule or variant antigenic type (VAT) that escaped the host antiparasite immunity. Obtained with permission from Guy Caljon.

of the parasite population will initiate transcription from another VSG gene or undergo genetic rearrangements, resulting in the expression of a new VSG type. As such, a new wave of trypanosomes will emerge, expressing a VSG that is not recognized by the raised anti-VSG immune response (Fig. 23.3). Together with the modulation of functions of antigen presenting cells (e.g., macrophages) and T-lymphocytes, the trypanosome is able to avoid elimination by the immune system and to maintain a well controlled growth in a broad range of hosts.

Pathology

23

African trypanosomiasis is lethal unless the parasite is completely eliminated from the body of the infected individual by drug treatment. However, death in infected hosts is rarely due to an uncontrolled parasite expansion. In human infections, mortality results from neurological complications after penetration of the parasite into the central nervous system, while cattle succumb from infection-induced severe anemia or complications associated with secondary opportunistic infections.

Human African trypanosomiasis (HAT) is characterized by two disease stages. During the first (haemolymphatic) stage of the infection, parasites will proliferate in the blood and the lymphatic circulation. Symptoms at this stage are pleotropic and nonspecific and include fever, lymphadenopathies, splenomegaly and endrocrine disorders. Systemic inflammation finally leads to increased blood-brain barrier (BBB) permeability allowing parasites to penetrate the central nervous system and cerebrospinal fluid, ushering in the second (encephalitic) stage of HAT. The symptoms of this stage include sensory, motoric and psychic disturbances, neuroendocrine abnormalities and disturbed circardian rhythms, eventually resulting in coma and death. The disturbed day-night cycles in the late stage of infection are characteristic for "sleeping sickness".

Experimental models for trypanosomiasis indicated that the transition from the first to second stage of HAT is dependent on inflammatory cytokine secretion by macrophages and microglial cells, which in turn can activate matrix metalloproteinases that selectively cleave basement membrane components from BBB endothelial cells. This could facilitate the migration of leukocytes and trypanosomes across the BBB. Tumor necrosis factor (TNF) seems to be especially involved in all stages of the inflammatory pathology of trypanosomiasis. In this context, TNF was first identified as cachexin, the causative agent of the tremendous weight loss (cachexia) observed during cattle trypanosomiasis. Circulating serum TNF concentrations and disease severity are correlated for HAT, cattle trypanosomiasis and experimental mouse infections. Interestingly, TNF was also demonstrated to exert an effector function in parasitemia control, attributed to a direct trypanolytic effect of this cytokine. A $TNF^{-/-}$ mouse model confirmed the involvement of TNF in both parasite control and immune pathology, as knockout mice show significantly less signs of morbidity as compared to wildtype mice although having 20-fold higher parasite concentrations in the blood (10^9/ml versus 5×10^7/ml).

Another aspect of trypanosomiasis-associated pathology, causing extensive morbidity during animal trypanosomiasis, is anemia. The mechanism underlying trypanosomiasis-elicited anemia was proposed to rely on (i) the release of trypanosome components lytic to red blood cells (RBCs), (ii) antibody-mediated lysis and/or phagocytosis of opsonized erythrocytes and (iii) suppression of RBC replenishment by erythropoiesis. Detailed analysis of anemia has recently uncovered a major role of T-cells, found to be a major source of IFNγ during infection, and responsible for excessive macrophage activation and TNF production. As such, activated macrophages and TNF are proposed to play key roles in both the induction of pathology and the control of parasitemia.

Beside the pathology occurring during infection, treatment with a trypanocidal drug during second stage HAT can prove fatal due to several complications including cardiomyopathy, hepatitis and especially posttreatment reactive encephalopathy (PTRE). To study PTRE, several mouse models were generated mainly relying on sub-curative drug treatment. This treatment clears parasites from the circulation, but not from the central nervous system, eventually leading to encephalitic shock. Immunological analysis revealed that severe PTRE was associated with astrocyte activation and increased IL (interleukin)-1α, -4, -6, MIP-1 (macrophage inflammatory protein-1) and TNF mRNA levels in the brain. This indicates that inflammatory responses in the central nervous system are associated with the occurrence of encephalopathy after treatment. A profound continued analysis of PTRE is mandatory in order to develop appropriate treatment protocols that will reduce fatalities during HAT treatment.

Diagnosis

The main bottlenecks in HAT diagnosis are the identification of the infecting trypanosome subspecies as well as the determination of the disease stage. Correct differential diagnosis of *T. b. gambiense* and *T. b. rhodesiense* infections will determine the applied treatment, as *T. b. rhodesiense* is acute and refractory to two of the four trypanocidal drugs. The disease stage will determine whether BBB-crossing drugs that are generally more toxic, should be administered. To date, diagnosis of first stage HAT mainly relies on microscopic detection of trypanosomes in blood smears and lymph node aspirates. Second stage HAT diagnosis is based on parasite detection or lymphocyte counting in the cerebrospinal fluid (CSF) taken by lumbar puncture. So far, molecular and serological tools cannot substitute for the classical parasitological procedures. For field conditions, the card agglutination test for trypanosomiasis (CATT) is the preferred first-line serological detection method for *T. b. gambiense*, but must be followed by parasitological confirmation and stage determination. The assay relies on the detection of anti-VSG antibodies in an agglutination reaction. However, since this diagnostic test is based on the recognition of one variable antigen type (VAT), LiTat 1.3, antigenic variation of the parasite population in a given foci can result in the disappearance of this VAT and false negative CATT assay results. In addition, a variable percentage of CATT seropositive individuals shows no clinical signs of infection and cannot be confirmed by parasitological detection. In order to complement the CATT assay, a *T. b. gambiense* specific polymerase chain reaction (PCR) was recently developed, based on the presence of a *T. b. gambiense*-specific gene, i.e., *tgsGP*. This PCR was proven to be unaffected by antigenic variation of the VSG and able to detect infections in individuals that scored negative in the antibody-based CATT test. In addition, false positive CATT results could be excluded by this PCR-based technique. Finally, since trypanosome-specific antibodies remain in the circulation after curative HAT treatment, the serological CATT assay cannot be used to detect relapses or re-infection. Here, the further development of a TgsGP reversed transcription PCR (RT-PCR) for the detection of mRNA from living parasites could provide a discriminative diagnosis in previously infected individuals.

For *T. b. rhodesiense*, infections can only be diagnosed by microscopic analysis as no serological field test is available. Based on advances in molecular parasitology, a new *T. b. rhodesiense* diagnostic PCR method has been developed based on the restricted presence of the serum resistance antigen (SRA) gene. The SRA-based PCR could be appropriate for diagnosis and to delineate the distribution pattern of *T. b. rhodesiense* in livestock, an issue that will become crucial for correct HAT management.

In addition to PCR diagnostics, rapid and low cost diagnostic approaches are developed and validated with a focus on increased sensitivity and specificity (overview Fig. 23.4). As an alternative for PCR, a new DNA amplification method under isothermal conditions, the loop-mediated isothermal amplification (LAMP) for the detection of African trypanosomes has been developed for easier field use. As alternative serological tests, the immunofluorescent antibody test (IFAT) and enzyme-linked immunosorbent assay (ELISA) methods have been proposed. However, due to the simplicity and rapidity of the CATT, it remains the most efficient field serological test. As an alternative for the CATT, the LATEX/*T. b. gambiense* has been developed. This test is, similarly to the CATT test, based on an agglutination reaction, using latex particles coated with three purified variable surface antigens, LiTat 1.3, 1.5 and 1.6. In recent field studies conducted in several West and Central African countries, LATEX/*T. b. gambiense* showed a higher specificity (96 to 99%) but a lower or similar sensitivity (71 to 100%) as compared to the CATT. Further evaluation of this test is ongoing before it can be recommended for routine field use.

	Detection			
	Microscopical	Enhanced microsopical analysis	Serological	Molecular
T. b. rhodesiense	1st stage: blood, lymph node 2nd stage: CSF (trypanosome, lymphocyte counting)	mHCT QBC mAECT PNA-FISH		SRA-PCR
T. b. gambiense			CATT/*T.b.gambiense* LATEX/*T.b.gambiense* IFAT ELISA	TgsGP-PCR

Figure 23.4. Overview of the major diagnostic tests based on microscopy, serology and molecular techniques for field diagnosis of *T. b. rhodesiense* and *T. b. gambiense* infections. Obtained with permission from Guy Caljon.

Another approach is to improve the microscopic detection of trypanosomes by including a parasite concentration step. Three techniques have been proposed for this purpose: (i) the microhematocrit centrifugation technique (mHCT), (ii) the quantitative buffy coat (QBC) and (iii) the mini-anion-exchange centrifugation technique (mAECT). The mHCT and QBC techniques are based on the concentration of parasites in the white blood cell fraction of total blood by high speed centrifugation in capillary tubes. Both techniques enhance the detection limit significantly (<100-500 trypanosomes/ml), but remain quite labor intensive. The mAECT allows the separation of trypanosomes from blood cells, based on the differences in surface electrical charge. Validation of a newly produced mAECT version for field usage is ongoing. Recently, spotting and methanol-fixing 23
of blood samples after erythrocyte lysis on microscopy slides followed by specific trypanosome-detection by fluorescence in situ hybridization (FISH) with peptide nucleic acid (PNA) probes was proposed as alternative approach with improved detection limits (5 trypanosomes/ml). PNA probes are pseudopeptides that are resistant to nucleases and proteases and hybridize specifically to a complementary nucleic acid target (DNA or RNA). Using *Trypanozoon* 18S ribosomal DNA sequences that are not affected by the mechanisms of antigenic variation, specific probes were generated for batch hybridization assays. Together, detection limits of microscopic diagnosis can be improved by several techniques but make the procedure more labor intensive and require mobile teams to be equipped with additional apparatus for field diagnosis.

Treatment

In *T. b. gambiense* infections, the human reservoir is the primary source for new HAT cases as the disease is chronic and might take years to result in fatal outcome. Treatment relies on suramine, pentamidine, melarsoprol and eflornitine. Suramine is a polysulphonated symmetrical naphthalene derivative first used to treat HAT in 1922. The drug is administered through slow intravenous injection, typically

100-200 mg (test dose), then 1 g IV on days 1, 3, 7, 14 and 21. Severe side effects have often been reported, including anaphylactic shock, severe cutaneous reactions, neurotoxic signs and renal failure. The exact trypanocidal mode of action of suramine remains to be elucidated. However, as suramine does not cross the BBB, its application is limited to the treatment of the haemolymphatic stage HAT.

Compared to suramine, pentamidine is better tolerated and has been in use since 1940. It is an aromatic diamidine that exerts a direct trypanocidal activity but also does not cross the BBB. The typical administration protocol is a regime of seven intramuscular doses of 4 mg/kg per injection given daily or every alternate day. Hypotension and hypoglycemia are the most common side effects. As a polycation, pentamidine interacts with polyamines and circular DNA molecules in the mitochondrion upon uptake into the parasite by a specific receptor/transporter (P2 amino-purine transporter). As such, specific point-mutations or loss of expression of this transporter can render parasites resistant to pentamidine treatment. Interestingly, loss of this transporter function also renders trypanosomes resistant to melarsoprol, the main drug for treatment of second stage HAT.

The melaminophenyl arsenical melarsoprol is a trivalent organo-arsenical compound that was first used in HAT treatment in 1949. The drug is water-insoluble and is dissolved in 3.6% propylene glycol. Generally, melarsoprol treatment is preceded by one or two injections with suramine or pentamidine to clear the parasites from the bloodstream. The most common treatment protocol consists of 3 to 4 series of intravenous injections separated by rest periods of at least 1 week (8-10 days), as melarsoprol is a highly toxic drug that penetrates the central nervous system. Each series consists of one intravenous injection of 2-3.6 mg/kg/d on 3 consecutive days. Although this arsenical derivate very efficiently lyses trypanosomes, posttreatment reactive encephalopathy occurs as an adverse drug reaction in up to 12% of the cases. A potential target of melarsoprol is trypanothione (N^1,N^8-bis-glutathionylspermidine), a low molecular weight thiol comprising
23 two glutathione molecules conjugated with spermidine. In trypanosomatids, trypanothione fulfills most of the roles carried out by glutathione as the major redox reactive metabolite in mammalian cells. As arsenic is documented to interact very stably with thiols, complexation with trypanothione could account for a complete disturbance of the redox balance and a rapid trypanotoxic effect.

The fourth drug used to treat *T. b. gambiense* is eflornithine (DFMO or DL-alpha-difluoromethylornithine). This is the only new molecule registered for HAT treatment in the last 50 years and was first used in 1981. DFMO is difficult to administer as it requires to be given at 400 mg/kg/d in 4 doses for 14 days. While it is better tolerated than melarsoprol for the treatment of second stage HAT, it still can cause pancytopenia, diarrhea, convulsions and hallucinations. DFMO is an analogue of ornithine, which acts as a specific inhibitor of ornithine decarboxylase (ODC) resulting in a suppression of the trypanothione and polyamine biosynthesis. Specificity for parasite killing results from a several orders of magnitude faster turnover of ODC in mammalian cells as compared to trypanosomes.

In contrast to *T. b. gambiense*, *T. b. rhodesiense* is a zoonotic parasite that mainly infects livestock and wild animals. Infections in humans are acute and require a fast and accurate diagnosis to initiate an appropriate treatment. Treatment relies only on suramine (first stage) and melarsoprol (second stage) as *T. b. rhodesiense* is refractory to pentamidine and DFMO.

In the context of novel chemotherapies against HAT, ongoing preclinical and clinical trials are focusing on combinational treatment strategies in order to increase cure rates by lower dosages and milder treatment schedules. An example of such approach is the combination of eflornitine with nifurtimox (5-nitrofuran), an orally administered drug that is used for the treatment of Chagas disease and sometimes for the encephalitic stage of *T. b. gambiense* HAT if eflornitine or melarsoprol are ineffective. Combination of eflornitine (DFMO) and suramine is also in trial for treatment of second stage *T. b. rhodesiense* infection. Other compounds are being tested for targeting several biochemical pathways in the trypanosome including, e.g., the polyamine biosynthesis, trypanothione reductase and glycolytic enzymes. In that context, DB289, an aromatic diamidine (pentamidine analog) and prodrug of the active metabolite diphenyl furan diamidine (DB75), is currently in Phase III clinical trials as a new orally administered candidate drug to treat 1st stage HAT.

A novel immunotherapeutic approach in the preclinical phase is based on the generation of a 15 kD VSG-recognizing molecule derived from nonconventional heavy chain camel antibodies. In contrast to most other mammals, camelids have a separate class of single chain antibodies that enable the engineering of small antigen-specific moieties (nanobodies) through a relatively simple procedure of cloning and affinity panning. The generated nanobody was shown to be a promising tool for targeting effector molecules to the trypanosome membrane as it is able to penetrate into the VSG coat and bind to conserved trypanosome surface epitopes that are inaccessible to lager conventional antibodies. In the further development of immunotoxins for trypanosomiasis therapy, trypanosome-specific nanobodies might be coupled to conventional drugs or new trypanocidal molecules. Recently, a toxin was generated from apolipoprotein L-1 (ApoL-1), a naturally occurring trypanolytic component in normal human serum and coupled to a VSG recognizing nanobody for in vivo use.

Conclusion

African trypanosomiasis is a devastating disease that is making a fast comeback in sub-Saharan Africa. Limited local resources for trypanosomiasis prevention and control have made this disease a major humanitarian and economic disaster affecting more than 10 million Km^2 of the African continent. Currently, no serological field test for *T. b. rhodesiense* is available, while differential diagnosis of the two human-pathogenic subspecies relies on relatively sophisticated molecular-based (PCR) tests. Moreover, available trypanocidal drugs have considerable levels of toxicity and are generally used for a specific disease stage. Combined efforts for new drug design approaches will be needed to combat this disease in the future. Beside toxicity, the rise of drug-resistance in trypanosomes is an important issue to be taken into account, urging that mechanisms of resistance need to be elucidated. One of those mechanisms is dependent on the reduced uptake of drugs through the P2 amino-purine transporter leading to resistance of trypanosomes to pentamidine, melaminophenyl arsenicals and potentially all new drug variants derived hereof. While preclinical and clinical studies on new chemotherapeutic and immunotherapeutic approaches seem to offer alternative approaches, further research is needed to evaluate pharmacological properties and applicability in the field. Other lines of research that might lead to the discovery of new low-toxicity antitrypanosomal agents will probably emerge through unraveling unique biochemical pathways utilized by the parasite. For instance, the specific compartmentalization of the glycolysis in glycosomes or unique metabolic features of the trypanosome could yield new antiparasite drug-targets. In this context, the full sequencing of the trypanosome genome will probably also contribute to the identification of new potential drug targets.

Suggested Reading

1. Vanhamme L, Lecordier L, Pays E. Control and function of the bloodstream variant surface glycoprotein expression sites in Trypanosoma brucei. Int J Parasitol 2001; 31:523-31.
2. Magez S, Radwanska M, Beschin A et al. Tumor necrosis factor alpha is a key mediator in the regulation of experimental Trypanosoma brucei infections. Infect Immun 1999; 67:3128-32.
3. De Baetselier P, Namangala B, Noel W et al. Alternative versus classical macrophage activation during experimental African trypanosomosis. Int J Parasitol 2001; 31:575-87.
4. Pays E, Vanhollebeke B, Vanhamme L et al. The trypanolytic factor of human serum. Nat Rev Microbiol 2006; 4:477-86.
5. Chappuis F, Loutan L, Simarro P et al. Options for field diagnosis of human African trypanosomiasis. Clin Microbiol Rev 2005; 18:133-46.
6. Fairlamb AH. Chemotherapy of human African trypanosomiasis: current and future prospects. Trends Parasitol 2003; 19:488-94.

CHAPTER 24

Visceral Leishmaniasis (Kala-Azar)

Ambar Haleem and Mary E. Wilson

Abstract

Amongst the many clinical forms taken by leishmaniasis, visceral leishmaniasis is the form that most often leads to a fatal outcome. Ninety percent of cases world-wide occur in three regions: northeast India/Bangladesh/Nepal, the Sudan and northeast Brazil. The disease is most often caused by *L. donovani* or *L. infantum* in the Old World, or by *L. chagasi* in the New World. The clinical presentation differs somewhat in different geographic regions, with humans serving as a main reservoir of infection in India due to the high incidence of PKDL, but dogs serving as a major reservoir in Brazil and around the Mediterranean. Treatment of visceral leishmaniasis is complicated by a need to administer standard therapy parenterally, toxicity of therapeutic agents and emerging parasite resistance to standard medications.

Introduction

Leishmaniasis is a vector-borne disease caused by obligate, intracellular protozoa belonging to the genus *Leishmania*; 21 out of the 30 mammalian-infecting species of *Leishmania* cause disease in humans.[5,13,14] The etiologic parasite was discovered in 1903, when Leishman and Donovan separately described the protozoan now called *Leishmania donovani* in splenic tissue from patients in India. They correctly identified this as the causative agent of the life threatening disease visceral leishmaniasis.[5] The insect vector is a female phlebotomine sand fly which acquires the parasite while feeding on an infected mammalian host. A total of about 30 sand fly species have been identified as vectors transmitting the different *Leishmania* species, although not all sand flies are capable of hosting all *Leishmania* species.[5] The disease leishmaniasis refers to several clinical syndromes. The most common are visceral (VL), cutaneous (CL) and mucocutaneous (MCL) leishmaniasis, which result from pathological changes in the reticuloendothelial organs, dermis and naso-oropharynx, respectively. The following chapter will include a discussion of the epidemiology, pathogenesis, clinical features and diagnostic and therapeutic approaches employed in visceral leishmaniasis.

Leishmaniasis is caused by a large number of *Leishmania* species that lead to characteristic clinical syndromes, with some overlap between species. The distribution of the different Leishmania infections is very regional and treatment is challenging. With the spread of AIDS, visceral leishmaniasis (VL) has become recognized as an opportunistic co-infection in HIV-infected people particularly in the Iberian Peninsula. The high morbidity and mortality associated with visceral

Medical Parasitology, edited by Abhay R. Satoskar, Gary L. Simon, Peter J. Hotez and Moriya Tsuji.

leishmaniasis, lack of available and affordable diagnostic and therapeutic modalities and increasing drug resistance in the developing world continue to pose major challenges in eradication of this infection.

Epidemiology

Leishmania donovani and *Leishmania infantum* in the Old World (particularly India, Nepal, Bangladesh and Sudan) and *Leishmania chagasi* in the New World (Latin America) are responsible for most of the cases of visceral leishmaniasis worldwide. Some of the *Leishmania* spp. that are commonly associated with CL (*L. amazonensis* in Latin America and *L. tropica* in Middle East and Africa) are on occasion, also isolated from patients with visceral disease.[5,13,14] Molecular techniques have revealed that most likely *L. chagasi* and *L. infantum* are the same organism causing disease in diverse geographic locations.[7]

Leishmaniasis has been reported from 88 countries around the world. Approximately 90% of the estimated 500,000 new annual cases of visceral disease occur in rural areas of India, Nepal, Bangladesh, Sudan and Brazil. (http://www.cdc.gov/ncidod/dpd/parasites/leishmania/). Disease is transmitted primarily through the bite of a sand fly, although rarely disease is transmitted by the congenital route, blood transfusions, accidental needle stick injuries in the laboratory or by sharing of leishmania-contaminated needles by intravenous drug users.[13]

Visceral leishmaniasis encompasses a broad range of clinical manifestations. Infection can assume an asymptomatic/subclinical and self-resolving form, or follow an aggressive, systemic course of illness (classic kala-azar or black fever). The disseminated form of infection is fatal if untreated and has resulted in mass epidemics in India and Sudan.

Most leishmania infections are zoonotic, with dogs or rodents as reservoir hosts. Only two species can maintain an anthroponotic cycle (human reservoir).[14] These two species are *L. donovani*, responsible for VL in the Indian subcontinent (particularly Bihar and Assam states) and East Africa, as well as *L. tropica* that causes CL in the Old World. Particularly in East Africa, people affected by post-kala-azar dermal leishmaniasis (PKDL) may serve as a reservoir for visceral disease.[13]

Pathogenesis

Leishmania spp. parasites exist in two stages, the promastigote and the amastigote. The promastigote is a15-20 μm × 1.5-3.5 μm flagellated form found in the gut of sand flies. The amastigote is a nonflagellated, intracellular form measuring 2-4 μm in diameter that replicates in macrophage phagosomes. Amastigotes are the only form present in mammalian hosts.

After inoculation into skin by a sand fly, promastigotes are phagocytosed by dermal macrophages, where they convert to amastigotes and multiply within acidic parasitophorous vacuoles. Additional mononuclear phagocytes are attracted to the site of the initial lesion and become infected (Fig. 24.1). Amastigotes then disseminate through regional lymphatics and the vascular system to infect mononuclear phagocytes throughout the reticuloendothelial system (Fig. 24.1). Progressive recruitment of amastigote-infected mononuclear phagocytes and inflammatory cells within organs results in distortion of the native tissue architecture and often, massive hepatosplenic enlargement. Parasitized reticuloendothelial cells can be found in bone marrow, lymph nodes, skin and other organs.

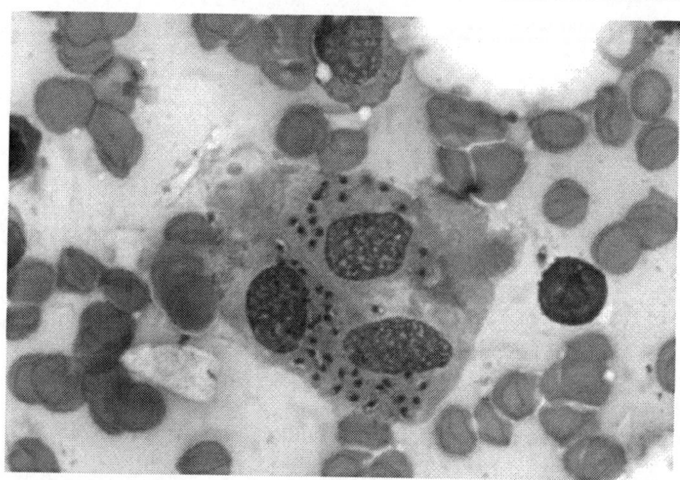

Figure 24.1. Bone marrow aspirate from a patient suffering from VL showing amastigotes in the macrophages. Photo kindly provided by Selma Jeronimo, MD, PhD, Universidade Federal do Rio Grande do Norte, Natal RN Brazil.

Why the infection follows a self-resolving course in certain human hosts and progresses to overwhelming, life-threatening disease in others remains an area of intense research. Mice self-cure infection and thus are a better model of asymptomatic infection than disease. Although murine models cannot completely explain the unique milieu present during human infection, murine models have illuminated the cytokines and chemokines that play key roles in determining whether the parasite replicates within quiescent macrophages or is killed by activated macrophages. Some mouse strains are inherently susceptible or resistant to *Leishmania* spp. infections. Similarly, genetically determined human immune responses influence the manifestations of leishmania infection in the human host.[5,12,20] 24

Early in infection of genetically resistant mouse strains, expansion of leishmania-specific $CD4^+$ T-cells of the Th1 type that secrete interferon gamma (IFN-γ) and interleukin 2 (IL-2) confers resistance to disease progression.[20] In contrast, expansion of Th2-type $CD4^+$ cells producing IL-4, IL-10 and IL-13 leads to progression of infection caused by *L. major* or other species inducing CL in mice. Transforming growth factor β (TGF-β) in the absence of a Th2 response promotes progressive murine infection due to the visceralizing *Leishmania* species. IL-2 enables differentiation of Th1 cells and production of IFN-γ, which then activates murine macrophages to kill amastigotes largely through nitric-oxide dependent mechanisms.[20]

Similar to mice, humans who have either had self-limiting infection with *L. donovani* or *L. infantum/L. chagasi*, or who have been successfully treated for symptomatic VL, develop protective Type 1 immunity against the same parasite. Leishmania-specific Type 1 responses are lacking in human hosts during progressive VL, although there is not often a clear expansion of Type 2 or TGF-β response during progressive infection.[20] Nonetheless, antileishmanial antibodies from polyconal B-cell activation are produced in high titer during progressive VL, but are not protective, similar to the murine Th2 response. Humans can develop reactivation of disease in the setting of immune suppression such as occurs during HIV-1 co-infection.

Clinical Manifestations of Kala Azar

Infection with the *Leishmania* species causing visceral leishmaniasis can manifest as a progressive fatal disease or as an asymptomatic form. The incubation period typically varies from 3 to 8 months, but can be weeks or years. Typical, symptomatic VL is associated with heavily infected mononuclear phagocytes throughout the reticuloendothelial system and suppressed cellular immune responses. VL can be fatal if left untreated.

The onset of disease is insidious in most cases and marked by the progressive development of fever, weakness, anorexia, weight loss and abdominal enlargement from hepatosplenomegaly. Fever, accompanied by chills is usually intermittent or remittent with twice-daily temperature spikes. During the less common acute cases, fever can be of abrupt onset and have a periodicity similar to that of malaria.

Progressive and massive hepatosplenomegaly is characteristic of VL. Infected individuals in the Sudan often also develop lymphadenopathy (Fig. 24.2) and in India patients with VL commonly develop hyperpigmentation of extremities, face and abdomen. Hemorrhage can occur from various sites. Severe cachexia is a prominent feature of VL, driven in part by high levels of TNF-α. Death from VL occurs either from the primary, multisystem disease causing malnutrition and bone marrow suppression and/or from secondary bacterial infections such as tuberculosis, dysentery, pneumonia and measles.[13]

Important laboratory findings in advanced visceral disease include profound pancytopenia, eosinopenia, hypoalbuminemia and hypergammaglobulinemia (mainly IgG). The erythrocyte sedimentation rate is usually elevated. Kidneys may show evidence of immune complex deposition, but renal failure is rare.

Several infectious and hematologic diseases can mimic visceral leishmaniasis. These include malaria, schistosomiasis, miliary tuberculosis, African trypanosomiasis, typhoid fever, brucellosis, histoplasmosis, bacterial endocarditis, lymphoma and leukemia.

Coinfection with HIV-1

Reactivation (or newly acquired) visceral leishmaniasis is a recognized opportunistic infection in T-cell impaired/deficient persons. Examples include individuals with HIV-1 infection, neoplasm, or receiving steroids, cancer chemotherapy or antirejection agents in organ transplantation. The leishmania parasite may be a cofactor in the pathogenesis of HIV infection. A major surface molecule, the lipophosphoglycan of *L. donovani*, induces transcription of HIV in $CD4^+$ cells.[4]

Most of the data on HIV co-infected persons with VL is derived from three countries in southern Europe, in particular from Spanish patient cohorts. Based on these studies, it appears that most HIV-infected patients manifest VL late in the course of HIV infection (CD4 cell count <200 cells/mm^3 in 90% of patients). The clinical presentation can be atypical.[1,5,6,13] Splenomegaly may be absent, whereas the gastrointestinal tract and oro-mucosal surfaces are commonly involved. Visceral leishmaniasis usually follows a chronic and relapsing course in HIV-positive

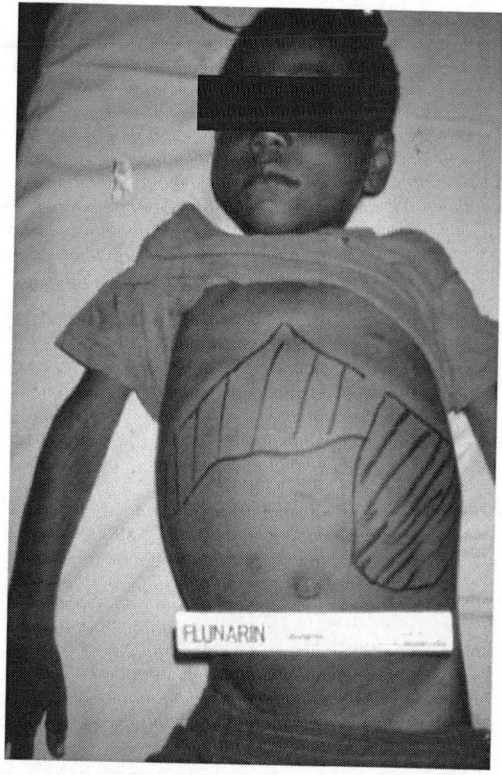

Figure 24.2. Hepatosplenomegaly in patients suffering from VL. Photo kindly provided by Dr. John David, Harvard School of Public Health, Boston.

patients. Initial responses to traditional VL therapy are lower in these hosts and adverse drug reactions are frequent. 50-70% of HIV-infected patients relapse within 12 months after discontinuing treatment.[10,13]

Post-Kala-Azar Dermal Leishmaniasis

PKDL is a syndrome encountered after completion or premature cessation of treatment for visceral leishmaniasis due to *L. donovani*. PKDL occurs in 5-10% of persons with VL in India and approximately 50% of those in Sudan.[13] PKDL may also occur in some HIV-coinfected people. The clinical presentation of PKDL in India and Sudan is similar, although the onset and duration of skin lesions differs between these two patient populations. In India, skin lesions typically appear 1 to 2 years after therapy and can persist for as long as 20 years, whereas the timing of appearance and persistence of lesions is much shorter in Sudan. PKDL lesions presumably serve as a source of leishmania infection for sand flies and the long duration of PKDL in Indian patients helps explain the fact that humans serve as the major reservoir of disease in this country.

PKDL is generally asymptomatic other than widespread skin lesions on the face, trunk, extremities, oral mucosa or genitalia. These can vary from hyperpigmented macules to overt nodules. Lesions may resemble the lesions of leprosy clinically and pathologically.

Diagnosis

The clinical features of visceral leishmaniasis are highly suggestive of, but not specific for this disease. Particularly in developing countries the differential is wide, including leukemia and a variety of tropical infections such as malaria, schistosomiasis and tuberculosis amongst others. Moreover, people from non-endemic areas or those with HIV co-infection can manifest atypical manifestations of VL. Hence, diagnosis must be confirmed by demonstration of the parasite in tissues.[5]

Tissue Diagnosis

Reliable diagnostic methods for leishmaniasis primarily involve invasive procedures with visualization of amastigotes in Wright-Giemsa stained smears of tissues, or by culture of promastigotes from human samples.[5,14] Splenic, liver or bone marrow biopsy, lymph node aspirates (particularly in Sudan) or buffy coat of peripheral blood can be utilized to look for the parasite microscopically. Splenic aspiration, although incurring a risk of hemorrhage, is the most sensitive means (95%) for diagnosing leishmaniasis.[5,14] Bone marrow biopsy demonstrates amastigotes in approximately two-thirds of patients. Liver biopsy is less sensitive than either splenic or bone marrow specimens.[13]

Syndromes such as VL or PKDL are characterized by many parasites in tissues, whereas lesions of mucosal leishmaniasis characteristically exhibit an exuberant inflammatory infiltrate with few parasites present. Logically, the abundance of parasites in tissues of the reticuloendothelial system during visceral leishmaniasis enables relatively easy demonstration of parasites from tissue smears as compared to diagnosis in cutaneous and mucosal syndromes of leishmaniasis. Furthermore, in HIV co-infected individuals parasites may be isolated and cultured from a multitude of sites, often atypical. These include bronchoalveolar lavage and pleural fluid, biopsies of the gastrointestinal tract or peripheral blood smears. The Giemsa-stained peripheral blood smear has a sensitivity of about 50% and parasite culture of a buffy coat preparation, about 70% for diagnosis of VL in HIV-positive patients.[5,13] In post-kala-azar dermal leishmaniasis syndrome, diagnosis is primarily clinical although amastigotes can be readily visualized in dermal macrophages in 80% of Sudanese patients.

In order to make an accurate diagnosis of leishmaniasis, amastigotes should be visualized by light microscopy under oil immersion. Identifying features are the parasite size (2-4 μm in diameter), shape (round to oval) and morphologic characteristics (nucleus and kinetoplast).[5] The kinetoplast is a rod-shaped mitochondrial structure that contains the extranuclear mitochondrial DNA.[14]

In vitro culture of promastigotes from tissue aspirates should be performed in concert with microscopic demonstration of amastigotes. However, it can take several weeks to achieve a detectable concentration of parasites in culture. The standard method for *Leishmania* species identification is by isoenzyme analysis of cultured promastigotes. Various molecular methods are promising tools but require further assessment for field application. These can employ polymerase chain reaction (PCR) to detect leishmanial DNA and RNA targets.[14]

Serology

Serological assays are sensitive tools for the diagnosis of VL in immuncompetent persons. Leishmania parasite-specific antibody responses peak during active VL and decline after treatment or spontaneous resolution of infection, although a period of seropositivity after recovery can confound the interpretation of a positive response. Antibodies do not correlate with protective immunity and do not remain positive for life and as such they cannot be used for studies of prevalence or presumed immunity to reinfection. They are, however, extremely useful for the diagnosis of VL. Factors that preclude widespread use of serological and immunodiagnostic methods are test costs, availability of a full-equipped laboratory, adaptability to field/rural conditions in developing countries and conflicting results in immunocompromised patients. Antibodies produced during several other infections (leprosy, cutaneous leishmaniasis, Chagas disease) can yield false-positive results on tests for VL, whereas patients with AIDS may have false-negative results owing to absence of an antileishmanial humoral response.

Recent advances that have increased the feasibility of serodiagnosis include the discovery that humoral responses to recombinant K39 antigen of *L. chagasi/L. infantum* are highly correlated with acute disease and the development of an rK39 dipstick test for use in the field. Nonetheless, it is sometimes difficult to distinguish between remote and recent infection. There is no one diagnostic method that is perfect.[5,13,14]

Commonly employed antigen-based tests are the enzyme linked immunosorbent assay (ELISA) using whole parasite lysate, the rK39 ELISA and the direct agglutination test (DAT). Indirect fluorescent antibody detection test (IFAT) is the only one that has been used on a limited scale.[14] The DAT entails agglutination of Coomassie blue-stained promastigotes in serum dilutions and is simple and reliable for field use. The ELISA assay is highly sensitive (80-100%) but the specificity varies with the antigen used, from 80-94% with whole parasite lysate to or 100% with rK39. Studies have found that dipstick tests using a recombinant kinesin-like antigen, rK39 from *L. infantum/L. chagasi* have 100% sensitivity and specificity for diagnosis of VL and that the test has good predictive value in detecting VL in immunocompromised patients.[5,13,14]

Antigen-impregnated nitrocellulose paper strips with rK39 antigen are being used successfully on the field for VL diagnosis. The DAT assay has been found to be 91-100% sensitive and 72-100% specific in various studies.

Skin Testing

It should be mentioned that the delayed-type hypersensitivity (DTH) leishmania antigen skin test (Montenegro test) is generally negative during active VL and may be negative during post-kala-azar dermal leishmaniasis. The test becomes positive in the majority of people in whom infection spontaneously resolves or who have undergone successful therapy. However, the lag between clinical recovery and DTH development can be months or longer. DTH testing is very useful in the diagnosis of cutaneous leishmaniasis. It is utilized primarily as an epidemiological tool and has little role in establishment of a diagnosis of acute VL.[13]

Treatment

The decision to treat leishmaniasis depends on the clinical syndrome, the infecting *Leishmania* spp., the immunologic status of the host and drug availability and cost. The goal of antileishmanial therapy is to prevent death from visceral disease and limit morbidity from cutaneous and mucocutaneous syndromes.

Therapeutic strategies for treatment of VL have evolved since 1912 when the first drugs against *Leishmania* species, the organic pentavalent antimonials were developed. Two commonly used pentavalent antimony preparations are stibogluconate sodium and meglumine antimoniate. However, the recent emergence of parasite strains resistant to pentavalent antimony (Sb) preparations has limited the utility of this mainstay of thereapy, primarily in the Indian subcontinent which harbors the bulk of VL cases globally. In the early 1990s, failure rates of Sb reached 65% in the endemic Indian state of Bihar. In spite of treatment advances in VL, all currently approved drugs are parenteral, toxic and must be administered for long durations. The high cost of alternative medications has been a major drawback to the use and development of newer agents in developing countries. Furthermore, for HIV co-infected patients where relapse of visceral disease is common after therapy there are few, good treatment options.

Outside of India, pentavalent antimonials are still the mainstay of therapy for visceral disease in both children and adults. The recommended dosage is 20 mg/kg/d for 28-30 days with usual cure rates of ~90%.[10] Inadequate response to the initial course of therapy or infection relapse may require a second course of treatment with the same agent. Common side effects are mainly gastrointestinal and include abdominal pain, nausea, vomiting and anorexia. Chemical hepatitis, chemical pancreatitis, arthralgias and myalgias can also occur. These drugs should be used cautiously in the elderly and in persons with heart disease, since dose-dependent arrhythmias and sudden death have been reported. This is a particular risk with doses >20 mg of Sb/kg/d. Renal failure can rarely occur.

Over the past 10 years practitioners in Sb-resistant areas of India have turned to amphotericin B for first-line therapy. The target of amphotericin B is ergosterol-like sterols, the major membrane sterols of *Leishmania* spp., leading to parasite killing through the creation of membrane pores. At a dose of 0.75-1 mg/kg for 15 infusions on alternate days, the drug cures ~95-97% of patients.[8,10,19] Occasional relapses (~1%) with amphotericin B can be successfully treated with the same drug.[14] Limitations to the use of amphotercin B in developing countries include the need for parenteral administration, cost and serious adverse effects such as nephrotoxicity.

Lipid formulations of amphotericin B are effective and less toxic than amphotericin B deoxycholate. In these preparations, various liposomes or other lipid components have been used to replace the usual carrier deoxycholate. Liposomes target the drug toward macrophages of the reticuloendothelial system and thus are ideal for treatment of VL in which the parasite resides in these same cells. Renal toxicity is a less frequent complication compared to conventional amphotercin B because lipid drug formulations are specifically targeted toward macrophage-rich organs and not to the kidney. This also allows for higher daily doses and shorter courses of amphotercin therapy to be delivered.[2,15] Lipid formulations of amphotericin B include liposomal amphotericin B (L-Amb: Ambisome), amphotericin B colloidal dispersion (ABCD: Amphocil) and amphotericin B lipid complex (ABL: Abelecet). The US Food and Drug Administration licensed liposomal amphotericin B in 1999 for treatment of VL and has recommended treating immunocompetent persons with 3 mg/kg daily on days 1-5,14 and 21 (total 21 mg/kg) and immunosuppressed patients with 4 mg/kg daily on days 1-5, 10, 17, 24, 31 and 38 (total 40 mg/kg).[3] An alternative recommendation for immunocompetent persons is treatment on days 1-5 and day 10 with 3-4 mg/kg daily in Europe and Brazil, 3 mg/kg daily in Africa and 2-3 mg/kg daily in India.[3] Although the reason is not known, the continent-specific therapeutic requirements could reflect different causative *Leishmania* species and/ or differences in the susceptible populations. Ambisome seems best-suited pharmacologically for ultra-short-course of therapy. Indeed, a trial in India showed the treatment duration can be compressed to a single day in India in adults and children (one infusion of 5 or 7.5 mg/kg)[17] and to two days in children in Greece (two infusions of 10 mg/kg each)[18] with some loss of cure rate but much improved cost and compliance. Relapses can occur after treatment with the lipid formulations in persons with AIDS and repeat courses may be necessary. However, the high cost, even with lower total doses and short courses of therapy, often proves prohibitive for use outside the developed world.

Another approved agent for treatment of visceral leishmaniasis is parenteral pentamidine. Promising investigational treatment agents include parenteral paromomycin and the only oral drug, Miltefosine, which has proven effective for Indian VL but is under investigation in other countries. Pentamidine has considerable toxicity and is not effective for Indian VL. Paromomycin is an inexpensive aminoglycoside that has appeared promising in limited studies. Miltefosine is an exciting oral option for therapy for VL in naive and Sb-refractory patients across the world. It was registered for use in India for both children and adults in 2002[10] at a dose of 50-100 mg (~2.5 mg/kg) for 4 weeks.[16] The drug is still undergoing trials for therapy of VL in other regions of the world. Major limitations to the use of miltefosine include its potential for teratogenicity and its long median half-life of 154 hrs, which could encourage emergence of drug resistance.[14] Transient, minor side effects are anorexia, diarrhea and nausea.

The imidazoles, ketoconazole and itraconazole, have been used successfully in some cases of cutaneous leishmaniasis, but primary failures occur in visceral leishmaniasis and they are not recommended for general use. Combination drug therapy can be an attractive approach for enabling cure of VL. At present, the only potential combination therapy is miltefosine and paromomycin but adequate clinical trials are lacking in this area.

Persons co-infected with HIV-1 pose a major challenge to effective therapy because of frequent relapses and reduced treatment responses. Most HIV-positive patients are given one of the standard regimens for VL. Concomitant HAART therapy and resulting increase in CD4 count >200 cells/mm^3 would be expected to enhance the efficacy of antileishmanial therapy and reduce relapse rates, but no randomized controlled trials have been conducted to investigate this hypothesis. Moreover, there is no general consensus about long-term maintenance therapy for VL in HIV-coinfected patients.[9]

Antileishmanial treatment is indicated in Indian PKDL, and treatment duration often spans 4 months. Most PKDL lesions of the Sudanese form self-resolve. When therapy is necessary, stibogluconate sodium 20 mg/kg/d for 2 months is given.

Most patients with visceral leishmaniasis become afebrile during the first week of treatment.[5] Splenomegaly and biochemical abnormalities often take weeks to resolve. Freedom from clinical relapse for at least 6 months is usually indicative of cure.[5] Even after symptomatic cure, parasites can remain in tissues indefinitely and do not necessarily represent an indication for retreatment. However, if symptoms relapse or do not resolve, a re-evaluation of the diagnosis is warranted followed sometimes by retreatment. Relapses usually occur within 6 months of therapy and are more common in persons with AIDS than in immunocompetent persons.

References

1. Albrecht H, Sobottka I, Emminger C et al. Visceral leishmaniasis emerging as an important opportunistic infection in HIV-infected persons living in areas nonendemic for Leishmania donovani. Arch Pathol Lab Med 1996; 120:189-98.
2. Berman JD. Human leishmaniasis:clinical, diagnostic and chemotherapeutic developments in the last 10 years. Clin Infect Dis 1997; 24:684-703.
3. Berman JD. US Food and Drug Administration approval of AmBisome (liposomal amphotericin B) for treatment of visceral leishmaniasis. Clin Infect Dis 1999; 28:49-51.
4. Bernier R, Barbeau B, Tremblay MJ et al. The lipophosphoglycan of Leishmania donovani up-regulates HIV-1 transcription in T-cells through the nuclear factor-kappaB elements. J Immunol 1998; 160:2881-8.
5. Herwaldt BL. Leishmaniasis. Lancet 1999; 354:1191-9.
6. Lopez-Velez R. The impact of highly active antiretroviral therapy (HAART) on visceral leishmaniasis in Spanish patients who are co-infected with HIV. Ann Trop Med Parasitol 2003; 97:143-7.
7. Maurício IL, Stothard JR, Miles MA. The strange case of Leishmania chagasi. Parasitol Today 2000; 16:188-9.
8. Mishra M, Biswas UK, Jha DN et al. Amphotericin versus pentamidine in antimony-unresponsive kala-azar. Lancet 1992; 340:1256-7.

24

9. Murray HW. Kala-azar—progress against a neglected disease. N Engl J Med 2002; 347:1793-4.
10. Murray HW. Progress in the treatment of a neglected infectious disease:visceral leishmaniasis. Expert Rev Anti Infect Ther 2004; 2:279-92.
11. Murray HW. Treatment of visceral leishmaniasis in 2004. Am J Trop Med Hyg 2004; 71:787-94.
12. Reed SG, Scott P. T-cell and cytokine responses in leishmaniasis. Curr Opin Immunol 1993; 5:524-31.
13. Selma MB, Jeronimo AD, Richard QS et al. Leishmania species: Visceral (Kala-Azar), cutaneous and mucocutaneous leishmaniasis. In: Mandell GL, Bennett JE, Dolin R, eds. Principles and Practice of Infectious Diseases. London: Churchill Livingstone, 2004.
14. Singh RK, Pandey HP, Sundar S. Visceral leishmaniasis (kala-azar): challenges ahead. Indian J Med Res 2006; 123:331-44.
15. Sundar S, Goyal AK, More DK et al. Treatment of antimony-unresponsive Indian visceral leishmaniasis with ultra-short courses of amphotericin-B-lipid complex. Ann Trop Med Parasitol 1998; 92:755-64.
16. Sundar S, Jha TK, Thakur CP et al. Oral miltefosine for Indian visceral leishmaniasis. N Engl J Med 2002; 347:1739-46.

17. Sundar S, Jha TK, Thakur CP et al. Single-dose liposomal amphotericin B in the treatment of visceral leishmaniasis in India: a multicenter study. Clin Infect Dis 2003; 37:800-4.
18. Syriopoulou V, Daikos GL, Theodoridou M et al. Two doses of a lipid formulation of amphotericin B for the treatment of Mediterranean visceral leishmaniasis. Clin Infect Dis 2003; 36:560-6.
19. Thakur CP, Kumar M, Pandey AK. Comparison of regimes of treatment of antimony-resistant kala-azar patients: a randomized study. Am J Trop Med Hyg 1991; 45:435-41.
20. Wilson ME, Jeronimo SM, Pearson RD. Immunopathogenesis of infection with the visceralizing Leishmania species. Microb Pathog 2005; 38:147-60.

Chapter 25

Cutaneous Leishmaniasis

Claudio M. Lezama-Davila, John R. David and Abhay R. Satoskar

Background

The leishmaniases comprise several diseases that are caused by *Leishmania* species, intracellular protozoan parasites which lead to a wide spectrum of clinical manifestations. Over 12 million people currently are infected with *Leishmania* and approximately 2 million are infected annually, making it a major health problem in parts of Asia, Africa, the Middle East, Southern Europe and Latin America. The clinical manifestations depend on the parasite species. Visceral leishmaniasis (VL or kala-azar) is caused by *Leishmania donovani* or *Leishmania infantum* (the latter called *chagasi* in Latin America). Cutaneous leishmaniasis (CL) is caused by *Leishmania mexicana* and *Leishmania brasiliensis* complexes in the New World and by *Leishmania major*, *Leishmania tropica* and *L. aethiopica* complexes in the Old World. *Leishmania* parasites are transmitted by about 50 different species of new (*Lutzomia*) and old (*Phlebotomus*) world sand flies. Most cases of human CL are due to a zoonotic mode of transmission (animal to man), but *L. tropica* infections are due to an anthroponotic mode of transmission (human to human).

Life Cycle

The female sand flies inoculate infective *Leishmania* metacyclic promastigotes during blood meals taken for egg production. Macrophages migrate to the inoculation site and phagocytose parasites which then transform into amastigotes. Amastigotes multiply in infected cells and skin and depending in part on the *Leishmania* species, migrate to other organs. This is the basis of the diverse clinical manifestations of leishmaniasis. Then, more sand flies become infected when they ingest infected macrophages full of amastigotes while taking blood meals. In the sand fly's midgut, amastigotes differentiate into promastigotes, which multiply and migrate to the proboscis transforming into the infective metacyclic stage. For life cycle details, see Figure 25.1.

Geographical Distribution

The disease is found in 88 tropical and subtropical countries around the world. Approximately 350 million people live in these areas. Each year, an estimated 1-1.5 million children and adults develop symptomatic cutaneous leishmaniasis. The settings in which cutaneous leishmaniasis is found range from rain forest and other vegetative sites in Mexico, Central and South America to desert-like areas in West Asia and Africa.

Medical Parasitology, edited by Abhay R. Satoskar, Gary L. Simon, Peter J. Hotez and Moriya Tsuji.

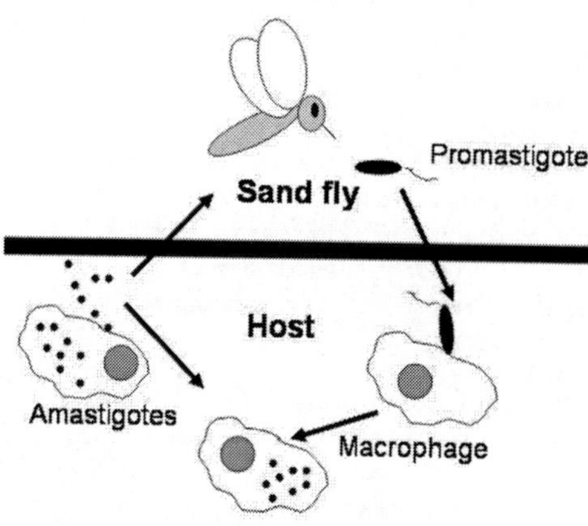

Figure 25.1. Life cycle of *Leishmania*.

New World cutaneous leishmaniasis in the Americas spreads from Northern Argentina to Southern Texas, although most affected places are in Mexico, Central America (Belize, Guatemala, Nicaragua, Honduras, El Salvador and Panama) and South America (Guyana, Brazil, Colombia, Venezuela, Bolivia, Peru, Ecuador and Argentina. The United States of America presents some endemic foci in Texas. Old World cutaneous leishmaniasis in Africa spreads more abundantly in East and North Africa, with some cases elsewhere. In Asia and the Middle East the affected countries include Afghanistan, Pakistan, India, Iran, Iraq, Saudi Arabia, Palestine and Israel. More than 90% of the world's cases of cutaneous leishmaniasis occur in Afghanistan, Algeria, Brazil, Iran, Iraq, Peru, Saudi Arabia and Syria.

Immunobiology

Macrophages and other monocytic lineage cells are the primary target cells for *Leishmania*, but they are also the principle effector cells that kill intracellular parasites via nitric oxide (NO)-dependent and-independent mechanisms when activated by the cytokines IFN-γ, MIF and TNF-α. Activated macrophages also function as antigen presenting cells (APCs) and secrete cytokines such as IL-12, IL-18 and TNF-α, all of which regulate innate and acquired immunity during CL. The dendritic cells (DCs) also play a critical role in induction of acquired immunity against CL. The disease protective role of DCs has been attributed to their ability to present parasite antigens to $CD4^+$ T-cells, as well as produce cytokines such as IL-12, which is required for NK cell activation and subsequent differentiation of $CD4^+$ T-cells into a Th1 subset. These lymphocytes produce IFN-γ, which not only activates macrophages to kill *Leishmania* but also facilitates Th1 differentiation of $CD4^+$ T-cells by signaling via the STAT1-mediated pathway. In contrast, $CD4^+$ Th2 cells, which produce IL-4, IL-10 and IL-13, are believed to play a role in disease exacerbation in CL. Genetic factors also influence outcomes of CL. It is important to note that particular alleles at the TNF-α and TNF-β genetic loci and overproduction of TNF-α are associated with a higher risk of CL in studies with Venezuelan, Brazilian and Mexican patients. However, the overall group of genes associated with resistance or susceptibility of different forms of cutaneous leishmaniasis remains to be defined.

Signs and Symptoms

Skin-lesions appear within weeks or months after the sand fly bite. Lesions are normally painless clean ulcers, but can be nodular or plaque-like. Lesions may self heal, but may take months or even years to resolve without treatment. They frequently can become secondarily infected. Some lesions may also grow and leave disfiguring scars. They rarely spread to the mouth or nose unless they are caused by *L. brasiliensis* or *L. aethiopica.*

Localized Cutaneous Leishmaniasis (LCL)

This disease is caused by *L. tropica, L. major* and *L. aethiopica* in the Old World and by *L. (V) brasiliensis, L. (V) guyanensis, L. (V) panamensis, L. (V) peruviana, L. (L) mexicana, L. (L) amazonensis* and *L. (L) garhani* in the New World. The incubation period of the disease varies between 1 week to 3 months, depending on the size of the inoculum and parasite species, but it may be shorter in travelers from non-endemic countries who are infected in endemic regions. The infection caused by most of these *Leishmania* species commonly manifests as painless ulcerative skin lesion(s) at the site of parasite inoculation, which resolve after antimonial treatment in most patients or self-heal (Fig. 25.2). However, *L. tropica* infection can produce disfiguring scars on the face in some patients and *L. aethiopica* and *L. amazonensis* can disseminate to other skin sites.

25

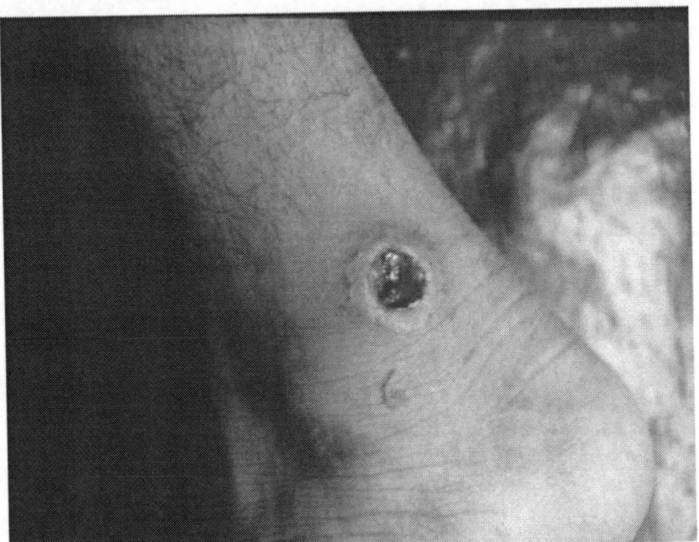

Figure 25.2. A typical ulcerating skin lesion in a patient suffering from localized cutaneous leishmaniasis (LCL). Courtesy Dr. J.R. David.

Lesions that resolve within a year are considered acute cutaneous leishmaniasis patients. On the other hand, patients with persistent lesions for more than a year (1-2 years) are considered chronic CL patients. These patients present a higher morbidity since they develop chronic, large lesions which may be difficult to diagnose because they contain few or undetectable parasites.

Mucocutaneous Leishmaniasis (MCL)

This disease is caused most commonly by *L. (V) brasiliensis,* but occasionally by *L. (V) panamensis* and *L. (V) guayanensis.* It begins as a single lesion that disseminates and produces new metastases affecting the mucosa, mouth, palate and nose (Fig. 25.3). The extensive mucocutaneous damage associated with MCL is mediated by an exaggerated host immune response against the parasite. MCL can be highly disfiguring if not promptly treated and is difficult to treat in some patients. Severe MCL can diminish the ability to eat and can be fatal.

Disseminated Cutaneous Leishmaniasis (DCL)

This form of the disease is normally produced by *L. aethiopica* and *L. amazonensis,* although some cases of disseminated cutaneous leishmaniasis have been reported where the causative agent is *L. major* or *L. mexicana* and is characterized by an initial lesion that disseminates throughout the skin (Fig. 25.4). This disease is usually associated with the failure of the host to mount an antigen-specific T-cell-response against the parasite. Lesions in patients with DCL present as nodules and erythematous plaques with variable degrees of verrucous changes, scaling and scarring. Microscopically, these lesions show abundance of parasites with minimal inflammation. DCL has been described in patients co-infected by *L. braziliensis* and HIV.

Leishmaniasis Recidiva Cutis (also known as Lupoid Leishmaniasis and Leishmaniasis recidivans)

This clinical form of the disease is rare and peculiar form provoked by *L. tropica* in the Old World and *L. (V) brasiliensis* in the New World. The disease is associated with the development of new lesions within the scar of a healed acute lesion, mimicking lupus vulgaris. Lesions appear as scaly erythematous papules that may evolve before the classic ulcer has healed or develop afterwards.

Post-Kala-Azar Dermal Leishmaniasis

This is a manifestation that happens in 5 and 20% of patients recovering from visceral leishmaniasis in Africa and India, respectively. It is less prevalent in the New World. The eruption is papular and lasts for months in African patients, whereas in India the lesions usually start as erythematous and hypopigmented macules that enlarge into patches. Later these asymptomatic patches become nonulcerative erythematous nodules that are particularly disfiguring.

Cutaneous Leishmaniasis and Immunodeficiency Virus (HIV) Co-Infection

With the increased transmission of human immunodeficiency virus infection (HIV) from urban and peri-urban locations to rural areas where leishmaniasis is endemic, incidence of CL as a co-infection in patients infected with HIV is increasing. Most cases (90%) of coinfections have been described in Southern Europe (Spain,

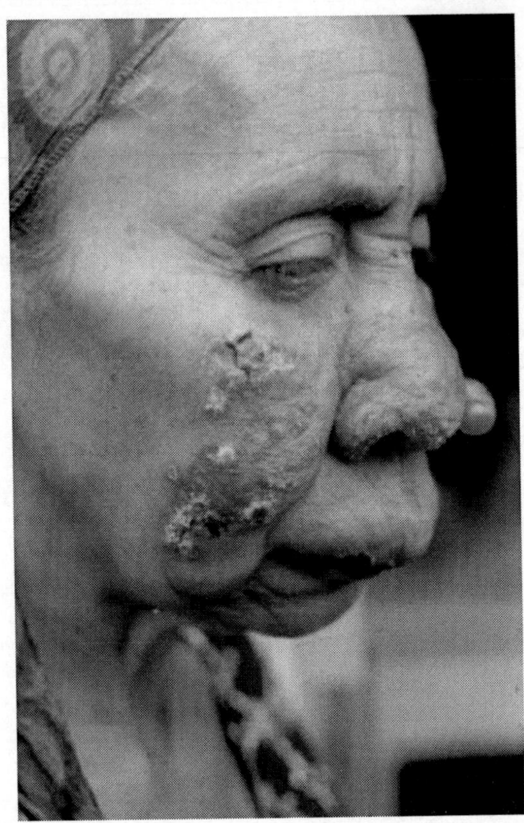

Figure 25.3. Clinical manifestations in a patient suffering from mucocutaneous leishmaniasis caused by *L. brasiliensis*. Courtesy Dr. P. Marsden.

France and Italy) and it is suspected that the mode of transmission of parasites is by shared contaminated needles among intravenous drug abusers. It has been described in Brazil in patients co-infected with *L. braziliensis* and HIV. Clinically, cutaneous lesions in these patients manifest as papulonodular, ulcerative, diffuse, kaposi-like and psoriasis-like forms, with spread to the mucosa in some patients.

Diagnosis and Treatment

Microscopic examination of Giemsa-stained imprint smears, hematoxylin eosin-stained punch biopsy specimens, or needle aspirations from suspected lesions are the most commonly used techniques for detecting amastigote forms of the parasites. Additionally, parasites can also be isolated from the ulcer by culturing a lesion biopsy or aspirate in tissue culture media (e.g., Schneider's medium, RPMI-1640, M199 medium) supplemented with 20% fetal bovine serum and penicillin/streptomycin or by inoculation into laboratory animals such as hamsters. This method not only allows parasitologic confirmation of the diagnosis, but also

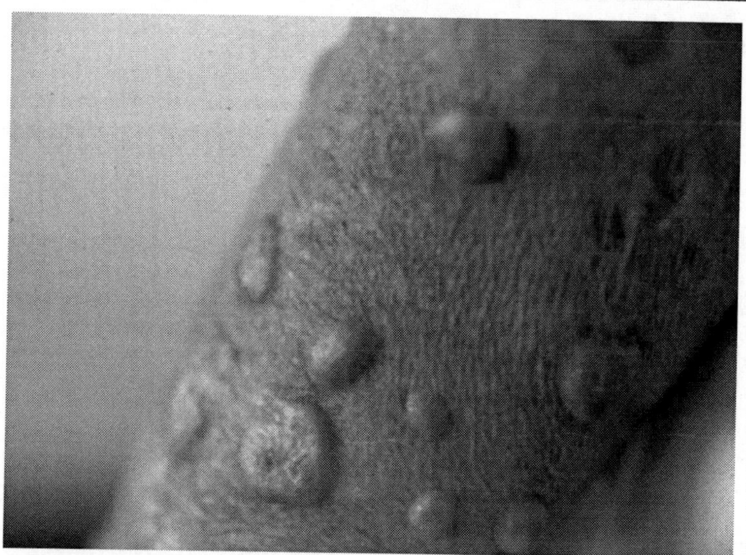

Figure 25.4. Disseminated cutaneous leishmaniasis (DCL) caused by *L. amazonensis*. This infection clinically manifests as multiple skin lesions. Courtesy of Dr. J.R. David.

provides sufficient number of promastigotes for further investigations (e.g., identification of *Leishmania* species by isoenzyme analysis). Other diagnostic techniques exist that allow parasite detection and/or species identification using immunologic (monoclonal antibodies and immunoassays) and molecular (PCR) approaches. Such techniques, however, are expensive and may not be readily available in general diagnostic laboratories in some developing countries or in the field.

Pentavalent antimony (Sb^v) drugs such as sodium stibogluconate (Pentostam™) or meglumine antimoniate (Glucantime™) have been used for more than 50 years to treat cutaneous leishmaniasis. Both these drugs are required to be administered daily via intravenous or intramuscular route for 20 days and sometimes several (20 mg/kg/d) courses are needed. These drugs, however, are not approved by the Food and Drug Administration, although they are approved and produced in Great Britain (Pentostam™) and France, (Glucantime™). They must be administered under an Investigational New Drug (IND) protocol. Pentostam™ is available under an IND protocol from the CDC Drug Service to physicians. Paramomycin can also be used topically for the treatment of CL caused by *Leishmania* species that have low potential to spread to mucosa and Imiquimod in combination with Glucantime™ has been used to treat CL patients in the New World that are refractory to Glucantime™ alone.

Drugs of choice for treating MCL are Pentostam™, Glucantime™ or Amphotericin B. Both Pentostam™ and Glucantime™ can be administered (20 mg/kg IV or IM) daily for 28 days whereas as Amphotericin B (0.5-1 mg/kg) is administered IV daily or every second day for up to 8 wks. New formulations of Amphotericin B encapsulated into liposomes are being investigated. Oral Miltefosine (2.5 mg/kg/d oral) × 28 d was also found to be effective in treatment of *L. panamensis*, but not *L. brazilliensis* or *L. mexicana*. Physicians may consult the CDC to obtain information on how to treat leishmaniasis. Walter Reed Army Medical Center (WRAMC) in Washington and Brooke Army Medical Centre Medical at Fort Sam Houston, Texas offer treatment to military beneficiaries, including reservists no longer on active duty. An alternative treatment for CL is radiowave-induced heat by a small portable battery-driven instrument. The lesions are anesthetized and a small applicator producing heat to 50°C is applied for 30 seconds across the lesions. One treatment lasting less than 5 minutes is usually sufficient. Cryotherapy using liquid nitrogen has also been used but this requires numerous applications. For *L. major* infections it is suggested that fluconazole may decrease the time of healing and, once the parasite species has been determined as the causative agent, a course of 6 weeks treatment is advised only for the clinical form associated to this particular species. Also under investigation are extracts from certain plants that have been used successfully for the treatment of CL in Mexico.

Prevention and Prophylaxis

The best prevention for leishmaniasis is to avoid sand flies bites. Some useful measures are as follows:

a. Stay in air-conditioned rooms from dusk to dawn or at least in a properly inspected closed tents free of sand flies.
b. Wear long sleeved shirts, long pants and socks when going outside. Tuck undershirts into pants and pants into socks. Insect repellent should be used liberally on the face, under the ends of the sleeves and pant legs. Clothing should be treated with permethrine in aerosol spray and retreated every six washes.
c. People not having access to air-conditioned rooms or screened tents should use thin mess nets with at least 18 holes per inch and cover and tuck it into the mattress. The bed and the mattress should be treated with permethrin, since sometimes sand flies can fly through the mesh.
d. Domestic pets such as dogs and rodents should be avoided near the sleeping area.

25 Presently, there are no commercial vaccines that protect humans against any type of leishmaniasis. Several experimental vaccines, however, are currently under investigation. One of them under study in human populations contains a mixture of parasite antigens and is known as Leish 111f. Another strategy is the use of immunotherapy-based vaccines in humans that include BCG (Bacille Calmette-Guerin) and a mixture of killed parasites. This preparation is useful to improve chemotherapy, thus reducing effective doses of antimonial treatments.

Concluding Remarks

Although each of the *Leishmania* species may have its own manifestations and areas of endemicity, none of the clinical forms of this disease are unique to a particular species because of considerable clinical diversity and overlap. Thus, the clinical picture is dependent on determinants associated with parasite and host interactions, such as parasites-virulence, genetic susceptibility and immune responses. New chemotherapeutic contributions, such as development of liposomal amphotericin B and miltefosine to treat resistant visceral leishmaniasis and the development of new vaccines for human application are needed to establish better control of this disease in endemic areas around the world.

Suggested Reading

1. Murray WH, Berman JD, Davis CR et al. Advances in leishmaniasis. Lancet 2005; 366:1561-77.
2. Mott KE, Nuttalli I, Desjeux P et al. New geographical approaches to control of some parasitic zoonoses. Bull World Health Organ 1995; 73:247-57.
3. Alexander J, Bryson K. T helper (h)1/Th2 and Leishmania: paradox rather than paradigm. Immunol Lett 2005; 99:17-23.
4. Berman JD. Human leishmaniasis: clinical, diagnostic and chemotherapeutic developments in the last 10 years. Clin Infect Dis 1997; 24:684-703.
5. Malla N, Nahajan RC. Pathophysiology of visceral leishmaniasis—some recent concepts. Indian J Med Res 2006; 123:267-74.
6. Schwartz E, Hatz C, Blum J. New World cutaneous leishmaniasis in travellers. Lancet Infect Dis 2006; 6:342-9.
7. Zijlstra EE, Musa AM, Khalil EAG et al. Post-kala-azar dermal leishmaniasis. Lancet Infect Dis 2003; 3:87-98.
8. Singh S. New developments in diagnosis of leishmaniasis. Indian J Med Res 2006; 123:311-30.
9. Roscoe M. Leishmaniasis: early diagnosis is key. JAAPA 2005; 18:47-50,53-4.
10. Vega-Lopez F. Diagnosis of cutaneous leishmaniasis. Curr Opin Infect Dis 2003; 16:97-101.
11. Blum J, Desjeux P, Schwartz E et al. Treatment of cutaneous leishmaniasis among travellers. J Antimicrob Chemother 2004; 53:158-66.
12. Croft SL, Coombs GH. Leishmaniasis—current chemotherapy and recent advances in the search for novel drugs. Trends Parasitol 2003; 19:502-8.
13. Magill AJ. Cutaneous leishmaniasis in the returning traveler. Infect Dis Clin North Am 2005; 19:241-66.
14. Davies CR , Kaye P, Croft SL et al. Leishmaniasis: new approaches to disease control. BMJ 2003; 326:377-82.
15. Bogdan C, Guessner A, Solbach W et al. Invasion, control and persistence of Leishmania parasites. Curr Opin Immunol 1996; 8:517-25.
16. Convit J. Leishmaniasis: Immunological and clinical aspects and vaccines in Venezuela. Clin Dermatol 1996; 14:479-87.
17. Khamesipour A, Rafati S, Davoudi N et al. Leishmaniasis vaccine candidates for development: a global overview. Indian J Med Res 2006; 123:423-38.
18. Reithinger R, Mohsen M, Wahid M et al. Efficacy of thermotherapy to treat cutaneous leishmaniasis caused by Leishmania tropica in Kabul, Afghanistan: a randomized, controlled trial. Clin Infect Dis 2005; 40:1148-55.
19. Lobo IMF. Soares MBP, Correia TM et al. Heat therapy for cutaneous leishmaniasis elicits a systemic cytokine response similar to that of antimonial (Glucantime) therapy.Trans Roy Soc Trop Med Hyg 2006; 100:642-9.
20. Drugs for parasitic infections. The Medical Letter 2004; 46:e5.

Chapter 26

Toxoplasmosis

Sandhya Vasan and Moriya Tsuji

Definition

Toxoplasmosis is an infection caused by the protozoan obligate intracellular parasite *Toxoplasma gondii.*

Overview and Incidence

The incidence of toxoplasmosis varies greatly by country and by age, but may affect up to one-third of the global human population. The majority of immunocompetent adults, pregnant women and children infected with *Toxoplasma gondii* experience no or mild symptoms during acute infection. Infants of women who seroconvert during pregnancy may develop congenital toxoplasmosis. This incidence ranges from one to ten per 10,000 live births. Immunocompromised individuals are at risk for reactivation of latent infection, including potentially fatal encephalitis.

Causes and Risk Factors

The house cat and other members of the family Felidae serve as definitive hosts in which the sexual stages of the parasite develop. The life cycle of *Toxoplasma gondii* begins when a cat ingests toxoplasma-infected tissue from an intermediate host, usually a rodent. Tissue cysts within the muscle fibers or brain are digested in the cat's digestive tract. The parasite then undergoes sexual development, multiplies in the intestine of the cat and is eventually shed in cat feces, mainly into litter boxes and garden soil. A human may become infected in one of the following ways:

1. By accidentally ingesting oocysts passed in cat feces through contaminated soil or handling of cat litter.
2. By ingesting tissue cysts within raw or undercooked meat (lamb, pork and beef), drinking unpasteurized milk, contaminated water, or unwashed fruits or vegetables.
3. By direct transmission of tachyzoites from mother to fetus through the placenta (congenital infection) or, rarely, by blood transfusion or solid organ transplantation from a positive donor to a previously uninfected host.

Parasite Life Cycle (Fig. 26.1)

Oocysts: Infective stages transmitted via cat feces after sexual development in cat.

Tachyzoites: Infective stages that infect macrophages and are carried throughout the human body via macrophages, causing pathology.

Medical Parasitology, edited by Abhay R. Satoskar, Gary L. Simon, Peter J. Hotez and Moriya Tsuji.

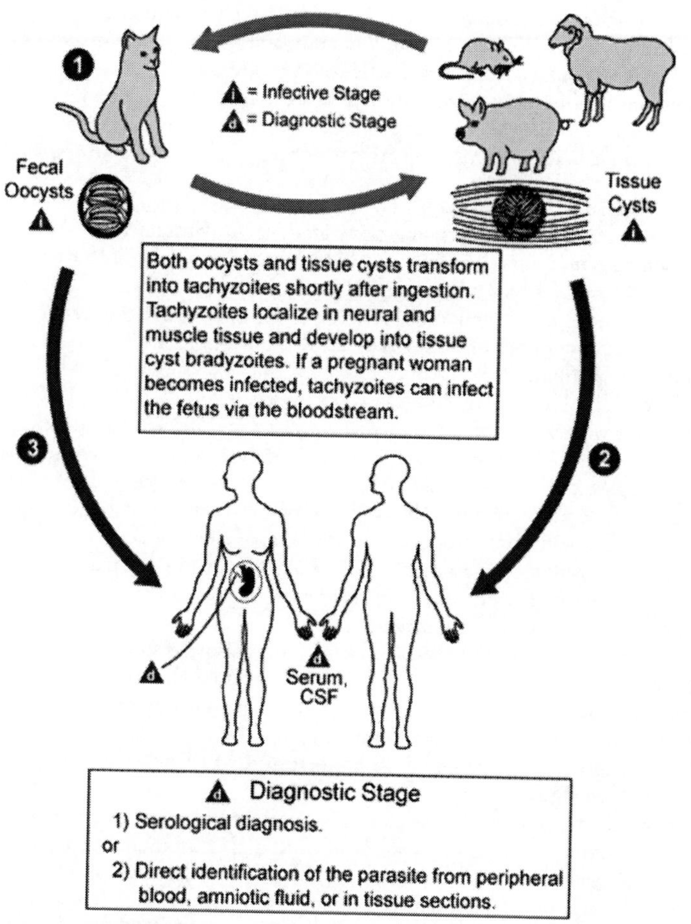

Figure 26.1. Toxoplasmosis Life Cycle. Image reproduced with permission of the Centers for Disease Control and Prevention (CDC) (http://www.cdc.gov).

Tissue cysts (pseudocysts): Large cyst-like forms that become quiescent in response to host adaptive immune responses.

Bradyzoites: Slowly developing forms within tissue cyst.

Pathology

T. gondii invades numerous organs, infecting a broad spectrum of cell types. Tachyzoites infect macrophages and are disseminated through the blood to many organs, where they invade, asexually multiply and cause cellular disruption, leading to cell death. The resulting necrosis attracts inflammatory host cells, such as lymphocytes and monocytes. It is this inflammatory response that causes the major pathology in infected individuals.

As host resistance develops, usually around 3 weeks post infection, tissue cysts may form in many organs, primarily in brain and muscle. These quiescent cysts enable *Toxoplasma gondii* to evade the adaptive host immune. As tissue cysts periodically rupture, the released bradyzoites are killed by the host immune system. If immune surveillance becomes compromised, such as due to chemotherapy or AIDS, these bradyzoites develop into tachyzoites, causing active toxoplasmosis.

Clinical Manifestations

Over 80-90% of primary infections produce no symptoms. The incubation period for symptoms is 1 to 2 weeks. Mild symptoms of primary infection include localized, painless cervical or occipital lymphadenopathy, usually persisting 4-6 weeks, or nonspecific symptoms including myalgia, headache, rash or sore throat that persist for one month or longer. Recently, newer more virulent strains causing severe symptoms in immunocompetent individuals have been reported.

Congenital toxoplasmosis is caused by infection with *Toxoplasma gondii* in a pregnant woman. Infants born to women who were infected before conception do not develop disease due to protection by maternal antibodies. In contrast, new infections with detectable maternal parasitemia are associated with up to a 50% transmission rate to the fetus. The likelihood of transmission and severity of disease in the fetus are inversely proportional. Mothers who develop acute toxoplasmosis in the first trimester have a much lower fetal transmission rate than in the third trimester, but fetuses exposed early are at much higher risk for severe symptoms or death and spontaneous abortion.

Up to 85% of newborns with congenital toxoplasmosis show no initial symptoms. Infants may show signs of central nervous system disorders such as hydrocephalus, microcephaly, or mental retardation. Hepatomegaly, splenomegaly, rash, fever, jaundice, anemia may also be present.

The most common pathology is chorioretinitis, which may result in strabismus or blindness. The age of onset ranges from 1-2 months to several years. A characteristic residual pigmented scar is left on the retina after resolution of infection.

Severe toxoplasmosis occurs most often in immunocompromised adults that develop either acute infection or reactivation from quiescent tissue cysts. In these cases, the disease may affect the brain, lung, heart, eyes, or liver. Brain lesions are associated with fever, headache, confusion, seizures and abnormal neurological findings. Systemic manifestations include myocarditis, pneumonitis and chorioretinitis, although immunocompetent individuals may also experience ocular damage from toxoplasmosis. The hallmark finding of chorioretinitis is white focal retinal lesions accompanied by vitreous inflammaion.

Diagnosis

Diagnosis relies on either indirect serological tests or direct detection of the organism. Serologic tests, indicating recent or past infection, are most effective in immunocompetent adults who are able to mount a humoral response to the parasite. These include ELISA, IFA, complement fixation and in the past, the Sabin-Feldman dye test to detect IgG antibodies. IgG antibodies develop 1-2 weeks post infection and then persist and therefore will not distinguish between recent and past infection. Rising titers on serial examination may be indicative of active or ongoing infection. The presence of a high IgM titer in the absence of a significant IgG titers indicates early stages of primary infection. A negative IgM titer is helpful for ruling out recent infection. However, due to considerable variability in tests and a high false positive rate, a positive IgM test should be verified in a reference laboratory (e.g., the US Centers for Disease Control or Toxoplasmosis Serology Lab, Palo Alto Medical Foundation).

Direct diagnosis can be made by PCR on several bodily fluids and tissues, including blood, bronchioalveolar lavage, vitreous fluid, amniotic fluid after 18 weeks gestation and brain biopsies. It is particularly useful in cases of advanced immunosuppression, where antibody titers may be less reliable. Diagnosis is occasionally made from identification of trophozoites in infected tissue. Histologic examination of lymph nodes may reveal a characteristic histiocytic hyperplasia. Additional indirect diagnostic tests include cranial MRI or CT scan for cysts in brain tissue and ocular slit lamp examination for characteristic retinal lesions.

In infants, it is important to distinguish between maternal antibodies found in a non-infected infant versus a high titer of antibodies being produced by an infected infant. In this regard, the simultaneous measurement of maternal and infant IgG antibodies specific for the parasites is critical. An infant: maternal IgG ratio of four or higher is indicative of new infection. This test should be repeated in the infant at 4 months of age. In addition, the presence of high titers of specific IgM antibodies in the infant's serum is diagnostic. Current procedures allow diagnosis of an active infection in the fetus in utero by means of PCR of amniotic fluid obtained by amniocentesis. The majority of infants will appear normal on prenatal ultrasound although findings may include intracranial calcifications, ventricular dilatation, hepatic enlargement, increased placental thickness and ascites.

Treatment

The immunocompetent, nonpregnant individual over 5 years of age with acute toxoplasmosis is treated with specific medications only if signs and symptoms are severe or persistent, or in the case of active chorioretinitis. Immunocompromised patients must be treated, often for 4 to 6 weeks after cessation of symptoms. Treatment of new infections in pregnant women is controversial because of the toxicity of the medications, but treatment is still advocated. Congenitally-infected newborns are treated aggressively.

Medications to treat the infection include: pyrimethamine (25-100 mg/d × 3-4 wks) plus either trisulfapyrimidines or sulfadiazine (1-1.5 gm qid 3-4 wks). Pregnant women are usually treated with spiramycin. The length of treatment varies widely between countries and may be followed by a course of pyrimethadine/sulfadiazine. Additional medications include sulfonamide drugs, folinic acid, clindamycin and trimethoprim-sulfamethoxazole. In immunocompetent adults and newborns with chorioretinitis, corticosteroids may help suppress the acute vitreous inflammatory response. Treatment in AIDS patients is continued as long as the individual is considered immunocompromised, usually until the $CD4^+$ T-cell count is above 200 cells/µL, to prevent reactivation of the disease.

Prognosis

Toxoplasmosis in immunocompetent adults carries a good prognosis without long term sequellae. In contrast, infection in immunocompromised adults can be fatal if not treated early, often due to neurological complications. Congenital toxoplasmosis can result in permanent disability in infants due to ocular and neurological involvement, including blindness, learning disorders and mental retardation.

Prevention

General

- Protect children's play areas from cat and dog feces. Cover sandboxes when not in use to avoid cat defecation.
- Wash hands thoroughly after contact with soil that may be contaminated with animal feces.
- Control flies and cockroaches as much as possible. They can spread contaminated soil or cat feces onto food.
- Avoid rubbing eyes or face when preparing food, especially raw meat or poultry. After food preparation, wash hands thoroughly with soap and water and clean the counter.
- Avoid ingesting raw or undercooked meat or poultry, raw eggs and unpasteurized milk. Fruits and vegetables should be peeled or thoroughly washed.

During Pregnancy

- Pregnant women should have their serologic testing for toxoplasma antibodies that may be repeated several times during the pregnancy depending upon initial results.
- Avoid exposure to cat feces by having other family members change the cat litter box. If the litter box must be changed, wear rubber gloves to avoid contact with the litter. Wash hands thoroughly with soap and water afterwards.
- Use work gloves when gardening and wash hands afterwards.

HIV-Infected Individuals

Patients with HIV disease should have toxoplasma antibody titers checked. If the results are positive and if the $CD4^+$ T-cell count is less than 200 cells/μL, patients should be given prophylactic antibiotics, such as trimethoprim-sulfamethoxazole, in conjunction with antiretroviral therapy until the $CD4^+$ T-cell count has risen.

Selected Internet Resources

http://www.cdc.gov/ncidod/dpd/parasites/toxoplasmosis/default.htm—CDC Division of Parasitic Diseases.

www.cfsph.iastate.edu/Factsheets/pdfs/toxoplasmosis.pdf—Center for Food Security and Public Health, Institute for International Cooperation in Animal Biologics, Iowa State university, 2005.

Suggested Reading

1. Lopez A, Dietz VJ, Wilson M et al. Preventing congenital toxoplasmosis. MMWR 2000; 49:57-75.
2. Olliaro P. Congenital toxoplasmosis. Clin Evid 2004; 12:1058-61.
3. Kravetz JD, Federman DG. Toxoplasmosis in pregnancy. Am J Med 2005; 118:212-6.
4. Montoya JG, Liesenfield O. Toxoplasmosis. Lancet 2004; 363:1965-1976.
5. Bossi P, Bricaire F. Severe acute disseminated toxoplasmosis. Lancet 2004; 364:579.

Chapter 27

Giardiasis

Photini Sinnis

Introduction

Giardia intestinalis, also called *Giardia lamblia* and *Giardia duodenalis*, is one of the most common intestinal parasites in the world, occurring in both industrialized and developing countries with an estimated 2.8 million new cases annually. First observed by Anton Van Leuwenhoek in 1681 in a sample of his own diarrheal stool, and later described in greater detail by Vilem Lamble, *Giardia* was initially thought to be a commensal and has only been recognized as a pathogen since the mid 1900s. In this chapter, salient features of the parasite and the disease it causes are described.

Life Cycle and Structure

This one-celled flagellated protozoan has a simple life cycle consisting of two stages: trophozoite and cyst (Fig. 27.1). Cysts are the transmission stage and are excreted in the feces of infected individuals into the environment where they can survive for weeks. When ingested, exposure to the low pH of the stomach and pancreatic enzymes induces excystation, with two trophozoites developing from each cyst. Trophozoites attach to epithelial cells of the upper intestine, primarily the jejeunum but also the duodenum, where they grow and divide. Attachment is required to prevent being swept away by peristalsis and is mediated by the ventral disk of the trophozoite as well as adhesins on the parasite surface. As the intestinal epithelial cell surface is renewed, trophozoites move and reattach to other epithelial cells. In some cases, the detached trophozoite is carried down the intestinal tract where exposure to bile salts, which occurs when the trophozoite is no longer protected by the mucous layer of the epithelium, and cholesterol starvation induce encystation.

The structure of trophozoite and cyst are shown in Figure 27.2. Trophozoites have two nuclei and each nucleus contains a prominent karyosome, giving the parasite its characteristic face-like appearance. In addition it has four pairs of flagella, an axostyle (a microtubule-containing organelle to which the flagella attach), a ventral disk and two median bodies, organelles whose function is not known. Cysts, which are slightly smaller than trophozoites, have a carbohydrate-rich cell wall which likely protects them from the environment and two to four nuclei. *Giardia* is an aerotolerant anaerobe that metabolizes glucose and scavenges cholesterol, phospholipids, purines, pyrimidines and amino acids. *Giardia* does not contain classical mitochondria but does make ATP in double-membraned organelles called mitosomes, which may represent degenerate mitochondria. The lack of

Medical Parasitology, edited by Abhay R. Satoskar, Gary L. Simon, Peter J. Hotez and Moriya Tsuji. ©2009 Landes Bioscience.

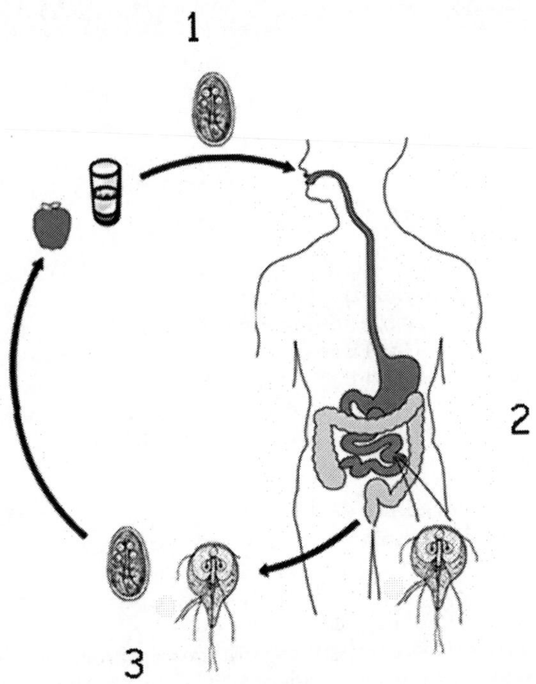

Figure 27.1. Lifecycle of *Giardia intestinalis*. 1) Ingestion of cysts either directly or via fecally-contaminated water or food. 2) Excystation of trophozoites in small intestine where they grow and multiply. 3) Encystation as the trophozoite leaves the small intestine and excretion of cysts into the environment. Some trophozoites may be found in feces but are not adapted for survival in the environment. Adapted from: Centers for Disease Control and Prevention (CDC) (http://www.dpd.cdc.gov/dpdx/images/ParasiteImages).

mitochondria originally led people to classify these parasites as early eukaryotes; however, the recent sequencing of the *Giardia* genome has identified genes that can be traced to the common prokaryotic ancestor of the mitochondria raising the possibility that mitosomes have evolved from mitochondria as an adaptation to the microaerophilic environment of the gut and that *Giardia* is an ancestor of an aerobic mitochondria-containing flagellate.

Epidemiology

Giardia transmission occurs by the fecal-oral route, either directly, via person to person contact or indirectly, via contamination of surface water or food The salient features of *Giardia* cysts that influence disease transmission include their stability in the environment, their immediate infectivity upon leaving the host and the small number of cysts required to cause infection.

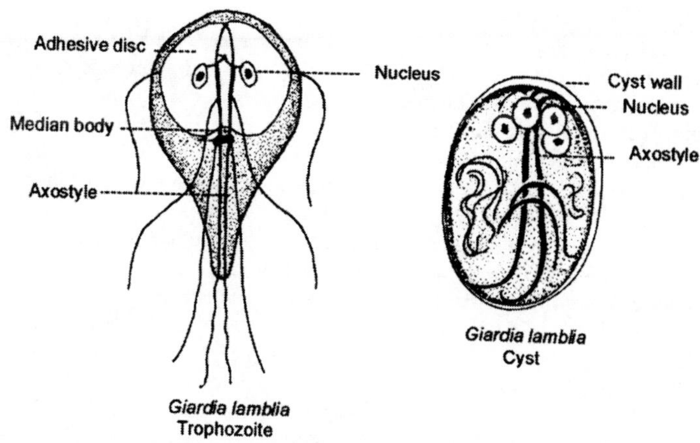

Figure 27.2. Structure of trophozoite (10 - 20 microns by 5 - 15 microns) is shown on the left and the slightly smaller cyst (7 - 10 microns by 8 - 12 microns) is on the right. Reproduced from: Nappi AJ, Vass E, eds. Parasites of Medical Importance. Austin: Landes Bioscience, 2002:16.

The stability of cysts in the environment means that fecal contamination of both food and surface water can lead to disease transmission. Common source outbreaks of giardiasis are primarily the result of contaminated water, although foodborne outbreaks have been reported. In the United States, *Giardia* is the most common cause of waterborne diarrheal disease. Risk factors for *Giardia* infection via waterborne transmission include drinking tap water, swallowing water while swimming, contact with recreational fresh water and eating lettuce, the last factor likely reflecting the use of contaminated water or fertilizer in farming. *Giardia* cysts can also be transmitted directly from person to person and this commonly occurs in daycare centers, custodial institutions and among men who have sex with men. In developing countries inadequate means of disposing of human waste leads to very high rates of infection, likely via both contamination of drinking water and direct person to person contact.

There are five species of *Giardia*, only one of which, *G. intestinalis*, infects humans; the others infect a variety of domestic and wild animals. *G. intestinalis*, can be further subdivided into six genetically distinct groups or Assemblages, only two of which, Assemblages A and B, are associated with human infection. Further molecular typing may find that these genotypes actually represent distinct species. The two human-infecting genotypes have a broad host-range that includes aquatic mammals such as seals, domestic animals and beavers. Although it has been suggested that these animals serve as reservoirs for human infection, recent studies using molecular tools that allow for precise genotyping of isolates have suggested that, on the contrary, other animals are indicator species for environmental contamination with human waste. More studies that incorporate genotyping of *Giardia* isolates are required before we fully understand the zoonotic potential of the *Giardia* parasites that infect humans.

Clinical Presentation

In the early 1920s, Rendtorff gave human volunteers *Giardia* cysts in gelatin capsules and found that although infection can occur after ingestion of as few as 10 cysts, it more reliably occurs upon ingestion of 100 or more cysts. More recent studies demonstrate that ingestion of cysts results in one of three outcomes: no infection, asymptomatic infection with excretion of cysts, or symptomatic infection. Analysis of infection rates during outbreaks indicates that 35 to 70% of exposed people remain uninfected, without symptoms or any trace of infection, possibly in part due to a low innoculum, 5 to 15% of people remain completely asymptomatic but pass cysts in their feces and 15 to 60% of people become symptomatic. The majority of symptomatic people have an acute self-limited diarrhea, lasting 7 to 15 days while a small proportion of symptomatic patients develop chronic diarrhea accompanied by signs of malabsorption and significant weight loss.

Symptoms usually begin 7 to 14 days after cyst ingestion but can begin as late as 4 weeks later. The onset is usually acute with diarrhea accompanied by diffuse abdominal cramping and discomfort, bloating, flatulence and fatigue. Nausea can be present but vomiting is rare. Fever, when present, is low-grade and seen early in the course of infection. Initially stools are usually profuse and watery but later in the course of the disease can become greasy and malodorous. Stools usually do not contain blood or white blood cells, the peripheral white blood cell count is not elevated and there is no increase in the absolute eosinophil count. Weight loss is common and it is not unusual for patients to present with a greater than 10 lb weight loss. Very rarely, extraintestinal symptoms such as urticaria and polyarthritis are seen and there are a few case reports in which *Giardia* infection extended into the biliary tract. Two distinguishing features of *Giardia* infection are its duration, usually in the range of 2 to 4 weeks, and the associated weight loss. Although the majority of symptomatic *Giardia* infections are self-limiting, symptoms can nonetheless be quite severe and occasionally lead to hospitalization, usually secondary to volume depletion. In an outbreak in New Hampshire, it was estimated that about 13% of symptomatic individuals required hospitalization.

A small percentage of symptomatic individuals will have chronic infection, lasting months or longer. These patients can have chronic diarrhea or a waxing and waning course characterized by intermittent symptoms of diffuse abdominal discomfort, malaise and diarrhea. Chronic giardiasis is frequently accompanied by weight loss, which can be significant, and malabsorption. Malabsorption of fats, vitamins A and B12, disacharrides, especially lactose, and protein are observed, with malabsorption of fats and lactose being most common. Lactose malabsorption is likely due to damage to the brush border which affects the disaccharidase-containing microvilli. All of these absorption deficits resolve with treatment of the infection. However, it can take several weeks for full restoration of the disaccharidase activity of the brush border so that lactose intolerance can persist even after the infection is eradicated.

More severe and prolonged symptoms are observed in patients with immunodeficiencies, especially those affecting the humoral arm of the immune system. Adults with common variable immunodeficiency (T and B cell deficits resulting in impaired production of most immunoglobulins), selective IgA deficiency and

X-linked agammaglobulinemia have an increased incidence of symptomatic disease with a prolonged course. HIV patients do not have more severe disease, with the exception of a subset of AIDS patients with low CD4 counts who have severe symptoms and are difficult to cure. Exactly how this group of AIDs patients differ from most is not clear.

One group of patients worth particular mention is children living in poor communities where baseline nutritional status is borderline and infection with intestinal helminths and protozoans is common. Many studies in poor communities in Central and South America, the Middle East and other places have found that *Giardia* infection in children is strongly associated with growth stunting. In children harboring several intestinal parasites, *Giardia* infection is most robustly associated with decreased growth. In one study, it was found that more than one episode per year of giardiasis was associated with a decrease in cognitive function. In contrast to these findings, growth of children in a daycare setting in Israel, was not affected by *Giardia* infection, despite the fact that prevalence of infection was high (20 to 40%). It is likely that underlying conditions, such as protein-energy malnutrition, affect the ability of the child to mount an effective immune response, resulting in more severe, chronic infection and leading to malabsorption of essential vitamins and fats. This situation is compounded by the fact that malabsorption in a child with borderline nutritional status will have more far-ranging and deleterious effects than it would in otherwise healthy children.

It is not yet known why there is such a large range of outcomes after ingestion of *Giardia* cysts. Infectious dose does not seem to play a role in outcome other than the fact that very low infectious doses may not result in infection at all. Likely both host and parasite factors are involved. The immune status of the host is clearly important with nonimmune and immunocompromised hosts more likely to become ill or severely ill. In addition, some studies suggest that different strains of *Giardia* differ in their capacity to cause disease. These types of studies, however, are still in their infancy and there is not yet consensus as to which strains or Assemblages of *Giardia* are more virulent. Undoubtedly as studies are performed with more markers, the role of strain variation in virulence will become clear.

Immunity

There is immunity to *Giardia* infection but it is not complete and reinfection is common. The existence of immunity is supported by several clinical observations: First, in developing countries where the prevalence of *Giardia* infection is high, the incidence of giardiasis is higher in younger age groups, suggesting immunity develops with exposure. Second, infection is typically more severe and of longer duration in people with immunodeficiencies, especially those with deficiencies in the humoral immune response.

In mouse models, using either *G. lamblia* or the rodent parasite *G. muris*, it has been shown that mice clear the infection and become resistant to reinfection, indicating that there is an immune response to the parasite in mice. Immunodeficient mice such as nude mice or thymectomized mice cannot clear the parasite and have chronic persistant infections. More detailed studies have found that depletion of B-cells leads to chronic giardiasis, indicating a role for antibodies in parasite clearance. IgA knockout mice cannot clear the infection although they control the infection better than B-cell knockout mice, suggesting that IgA antibodies are critical for parasite clearance but that other isotypes play a role in controlling infection or can partly compensate for the deficiency in IgA. Other experiments using rodent models found that the cytokine IL-6 is important in the acute phase of infection. IL-6 knockout mice had normal IgA production and could control *G. lamblia* infection but did so more slowly. Overall, the work from rodent models suggests that there is an early B-cell independent response to the parasite followed by an antibody-dependent phase, with secretory IgA being one of the most important components of the immune response. In humans its been suggested that failure to mount an IgA response is correlated with chronic giardiasis.

Work with rodent models also suggests that in addition to being important in clearing the infection, the host immune response may also lead to some of the pathology observed during infection. For example, one study found that adoptive transfer of either $CD8^+$ T-cells or whole unfractionated lymphocytes from *Giardia*-infected mice to naïve recipients resulted in diffuse shortening of microvilli and loss of brush border surface area in the recipients. In another study, *Giardia* infection in immunocompetent mice resulted in diffuse loss of the brush border microvillus surface whereas T-cell deficient mice infected with the same load of parasites had no change in microvillus surface area or structure.

Protection against both infection and disease is associated with breast feeding and it has been shown that both immune and nonimmune mechanisms are involved in this protection. Specific antibodies are transferred in milk from mothers previously infected with *Giardia* and are associated with a decreased risk of acquiring the infection and of developing severe giardiasis. Human breast milk also protects against *Giardia* infection because it contains a specific lipase that releases free fatty acids from milk triglycerides upon stimulation by bile salts (bile salt stimulated lipase), that kill *Giardia* trophozoites.

Pathogenesis

The mechanisms by which *Giardia* causes diarrhea and malabsorption have not been elucidated. There is no evidence that *Giardia* produces an enterotoxin or that it invades the intestinal epithelial cells. Electron microscopy shows that the ventral disk embeds the parasite into the epithelial microvillus layer and "footprints" of formerly adherent trophozoites are visible on the epithelial cell surface. However, even in a heavy infection, the surface area covered and possibly damaged by the adherent trophozoites cannot account for the symptoms. In humans, biopsy of the infected gut shows little abnormality. In a European study in which over 500 biopsy specimens from *Giardia*-infected patients were observed, slightly over 96% had normal looking mucosa and 3.7% had mild villous shortening with a small amount of neutrophil and lymphocyte infiltration. The lack of histologic abnormalities in the majority of symptomatic patients has also been observed in other, smaller studies. In one study in which patients with villous shortening and inflammatory infiltration were followed with serial biopsies, these abnormalities all resolved after the infection was eradicated. In the murine models of giardiasis, similar findings of villous atrophy and inflammatory infiltration of villous epithelium can be observed. However as with humans, the findings are subtle and the inflammatory changes mild.

In conclusion, the cause of diarrhea and malabsorption in Giardia infection is likely to be multifactorial, involving the host immune response to the pathogen as well as, yet to be identified, cytopathic substances that the parasite may secrete. Additionally, it has been suggested that *Giardia* may cause pathology by alteration of the bile content or endogenous flora of the small intestine which in turn could affect the absorptive function of gut. These hypotheses must now be formally tested before a more complete picture emerges.

Diagnosis

The traditional method of diagnosis is examination of stool for trophozoites or cysts (stool O&P). Both fresh and fixed stool specimens are usually examined. Cysts are normally found but motile trophozoites can be observed in a fresh specimen of loose stool (Fig. 27.3). Because the parasites are normally found in the small intestine and are shed intermittently, the sensitivity of one stool specimen is low, in the range of 50 - 70%. However, examination of three specimens, from three different days, increases the sensitivity to 85 - 90%; specificity is close to 100%. This assay remains the most widely used method to diagnose *Giardia* infection and is the gold standard to which other newer assays are usually compared. It is important to note that there can be a delay between the onset of symptoms and the excretion of cysts so that a negative stool sample in someone in whom giardiasis is suspected warrants reanalysis at a later time.

Recently, new assays have been developed based on detection of *Giardia* antigens. The direct fluorescent antibody test (DFA), uses a *Giardia*-specific antibody conjugated to a fluorophore to stain stool specimens. Because the parasites are labeled, much larger regions of the slide can be scanned more quickly and the likelihood of detecting the parasite is increased. On a single stool specimen the sensitivity is between 96 - 100%. Other antigen-detection tests detect soluble *Giardia*-specific proteins in the stool. There are two different types of soluble-antigen-detection

Cyst

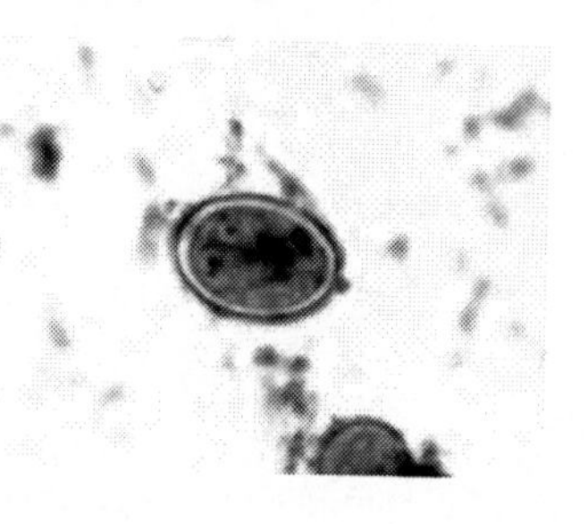

Trophozoite

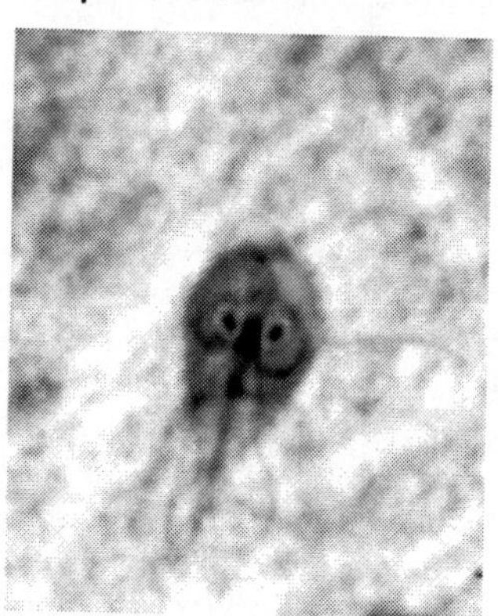

Figure 27.3. Light microscopy of *G. intestinalis* trophozoites and cyst in stool sample. Reproduced with permission from Apple Trees Productions, LLC, New York and generously supplied by Dr. Dickson Despommier.

27

tests, the enzyme immunoassays (EIA) and immunochromatographic dipsticks (ICT) which work on the same overall principal, i.e., immobilization of antibodies specific for *Giardia* proteins on a plate (EIA) or a piece of chromatography paper (ICT) that then capture *Giardia* proteins in the stool. ICTs are relatively new and have a lower sensitivity than EIAs (80% versus 94 to 98%). Importantly, these can be positive even after a person stops shedding intact organisms and can give false-positive results in someone who has been treated and cured of *Giardia* infection. Specificity for all of the antigen-based detection assays is 90 to 100%. Although these tests are faster and do not require a highly trained technician, in contrast to the stool O&P which does, stool for O&P can be more economical when screening for more than one intestinal parasite. However antigen tests can be very useful during outbreaks or in institutional settings when screening large numbers of people.

Both stool O&P and the antigen detection tests give false-negative results when there are very few organisms in the stool. In these cases, the physician may decide to directly sample the contents of the duodenum using either the string test (Entero-Test, HDC Corporation, San Jose, CA) or by endoscopy accompanied by duodenal fluid sampling and biopsy. The mucus adherent to the string and the duodenal aspirate fluid are examined for organisms, primarily trophozoites, directly and after staining. In cases where biopsy is performed, it is processed, stained and then examined by a pathologist for trophozoites. These more invasive tests are not more sensitive for detecting parasites but can be used when the more widely-used diagnostic methods fail.

Assays which detect *Giardia*-specific nucleic acid, using polymerase chain reaction (PCR), are being developed but are currently experimental and not in clinical use.

Treatment

Treatment of giardiasis is summarized in Table 27.1. The nitroimidazole, metronidazole is the most commonly used drug to treat *Giardia* infection although it has never been FDA approved for this use. In adults, 250 mg tid for 5-7 days has been shown to be effective in 80 to 95% of cases. Shorter courses of higher doses lead to better compliance but are less effective and have a worse side effect profile. Metronidazole is mutagenic to bacteria and carcinogenic in laboratory rodents; for this reason most physicians prefer not to use it in children although no mutagenic effects have been observed in humans.

Tinidazole is another nitroimidazole that until recently was not available in the United States but was commonly used in Europe and in parts of the world where the prevalence of *Giardia* infection is very high. It has a longer half-life than metronidazole and is 80 to 100% effective as a single 2 gm dose. It is likely that given its recent availability in the United States and FDA approval for use in giardiasis, combined with its efficacy as a single dose, tinidazole may replace metronidazole as the most frequently prescribed drug for the treatment of giardiasis.

Nitazoxanide, discovered in the 1980s by Jean Francois Rossignol at the Pasteur Institute, is a new drug in the anti-*Giardia* arsenal and was recently approved by the FDA to treat giardiasis in children aged 1-11 and in adults. It is a thiazolide that interferes with pyruvate-ferredoxin oxidoreductase-dependent electron transport

Table 27.1. Treatment of Giardia infection in adults and children

Drug	Adult Dose	Dosing in Children*	Comments
Metronidazole	250 mg tid x 5 -10 days	5 mg/kg tid x 5 -10 days	
Tinidazole	2 gm single dose	50 mg/kg single dose	Only recently available in the United States
Nitazoxanide	500 mg bid x 3 days	1-3 year-old: 100 mg bid x 3 days 4-11 yrs: 200 mg bid x 3 days	A recent addition to the anti-*Giardia* arsenal
Albendazole	400 mg qd x 5d	15 mg/kg/d x 5-7 d	
Paromomycin	500 mg tid x 5-10 days	10 mg/kg tid x 5-10 days	Used primarily in pregnant women
Quinacrine	100 mg tid x 5 - 7 days	2 mg/kg tid x 5 - 10 days	Used primarily in refractory cases in combination with a nitroimidazole

*Tinidazole, Nitazoxanide and Albendazole available in liquid form.

which is required for anaerobic respiration. It is not mutagenic and so is more appropriate for use in children. Efficacy is over 85% and it is well-tolerated with a good side effect profile. Nitazoxanide's efficacy extends well beyond the treatment of giardiasis, and it has been found to be effective in treating *Cryptosporidium*, *Entamoeba*, *Cyclospora*, *Isospora* as well as many intestinal helminthes.

Albendazole, a benzimidazole commonly used to treat intestinal helminth infections, is also effective against *Giardia*. Its cure rates are similar to metronidazole when 400 mg/d is given for 5 to 7 days. Interestingly, the other benzimidazole, mebendazole, is not as effective in the treatment of *Giardia* infection and so is generally not used for this purpose.

27

Pregnant women with *Giardia* infection can be difficult to manage because none of the aforementioned drugs has been shown to be safe for the fetus. The nitroimidazoles (metronidazole and tinidazole) rapidly enter the fetal circulation after absorption from the mother's GI tract. Although large retrospective studies have not demonstrated any adverse effects on the fetus in the 2nd and 3rd trimester, in one large study of pregnant women, a small increase in fetal malformations was associated with first trimester exposure to metronidazole. Albendazole, and the other benzimidazole, mebendazole, are teratogenic in animals at high doses and are also generally not used in pregnant women. The use of these drugs in mass treatment campaigns to eradicate filariasis or treat intestinal helminths has led to their inadvertent use in early pregnancy, and no increased risk of fetal malformations has been observed. However, given that the doses used were much lower than that needed to eradicate *Giardia* and that the numbers of pregnant women in these studies was small, it is not advisable to use these drugs to treat giardiasis in pregnant women. Lastly, there are no safety data available regarding the use of the recently discovered nitazoxanide in pregnant women. So how to treat the pregnant woman with giardiasis? If her symptoms are not severe and pose no threat to herself or the fetus, she can be given supportive treatment to maintain her fluid and nutritional status during the first trimester and if possible, the entire pregnancy. If treatment becomes necessary, paromomycin, an aminoglycoside that is not absorbed in significant quantities from the GI tract is usually tried first. It is not as effective as metronidazole, with efficacies ranging from 55 to 88% and is usually administered for 10 days. If necessary, metronidazole can be given during the 2nd and 3rd trimesters using the dosing schedule outlined in Table 27.1 rather than a high-dose short-course regimen.

Although most people respond to treatment with one of the aforementioned drugs, there is a small subset of patients who do not. In a series of six patients with refractory infections, the majority of which were immunodeficient, the combination of a nitroimidazole, either metronidazole or tinidazole, and quinacrine (100 mg tid), administered for 2 to 3 weeks, resolved the infection. Quinacrine is an old antimalarial drug that is no longer used to treat malaria, is no longer produced in the United States, and has a high incidence of side effects. However, in combination with a nitroimidazole, it appears to be effective in eradicating *Giardia* in refractory infections. It can be ordered from an independent compounding pharmacy, Panorama Pharmacy, Panorama City, CA (http://www.panoramapharmacy.com/).

Since a large proportion of infected people are asymptomatic, the question arises as to whether they should be treated. It is generally accepted that if they pose a risk to others and are unlikely to rapidly become re-infected, they should be treated. This would mean that asymptomatic children in daycare centers who are excreting cysts should be treated, as should asymptomatic carriers in institutional settings. In developing countries the likelihood that an infected, asymptomatic child will become re-infected after treatment is high and for this reason, treatment is not always recommended. However, in studies demonstrating that *Giardia* infection in preschool children with borderline nutritional status is associated with reduced growth, it was found that eradicating the infection allows for catch-up growth. For this reason, *Giardia* infection should be treated in children in areas where there is a high prevalence of giardiasis and malnutrition.

Prevention

Providing safe drinking water is critical if *Giardia* transmission is to be controlled. The traditional methods employed to insure water safety are protection of the watershed, flocculation and sedimentation (using compounds such as alum to bridge contaminating organisms into clumps that can then be removed by sedimentation), chemical disinfection usually with chlorine, and filtration. As the population increases and expands into areas once considered remote and pristine, our watersheds are becoming more difficult to protect from human and animal waste. A 1991 study found that 81% of the raw surface water entering 66 different water purification plants was contaminated with *Giardia* cysts. Short of assuring a pristine watershed, filtration is the most important component of the standard water purification process in the prevention of waterborne outbreaks of *Giardia*. Numbers of cysts can be decreased by flocculation and sedimentation but not to acceptable levels. Although *Giardia* cysts are sensitive to chlorine, their inactivation requires high levels of chlorine and long exposures that are not routinely implemented in water treatment plants. Proper filtration of surface water, or ground water under the direct influence of surface water, is required and can decrease cyst number by 2 to 3 orders of magnitude. However, filtration to remove all cysts on a large scale is not feasible so that drinking water continues to pose some risk. In New York City, the Department of Environmental Protection began, in 1992, to monitor the level of *Giardia* and *Cryptosporidium* contamination in the city's source water. These data can be viewed at: http://www.ci.nyc.ny.us/html/dep/html/pathogen.html.

Backpackers and hikers drinking untreated water and travelers to areas where *Giardia* is prevalent and safe drinking water is not available can perform one of the following to insure the safety of their water: Boil for 1 minute or at high altitude for 3 minutes; filter through a 2 micron pore size filter; treat with iodine (12.5 ml/liter for 30 minutes) or halazone (5 tablets/liter for 30 minutes).

Conclusion

Giardia is an important cause of diarrhea worldwide. Most symptomatic patients recover with treatment although disease is more prolonged and can be difficult to treat in certain groups of immunocompromised patients. The greatest impact of giardiasis is likely on children of marginal nutritional status where the malabsorption associated with infection can cause growth stunting. As our watersheds become increasingly contaminated with human waste, it is likely that the incidence of the disease will increase. The recent availability of tinidazole in the United States combined with the discovery of nitazoxanide should make the treatment of giardiasis simpler and more straightforward.

Suggested Reading

1. Appelbee AJ, Thompson ARC, Olson ME. Giardia and Cryptosporidium in mammalian wildlife - current status and future needs. Trends Parasitol 2005; 21:370-5.
2. Gardner TB, Hill DR. Treatment of Giardiasis. Clin Microbiol Rev 2001; 14:114-28.
3. Nash TE, Ohl CA, Thomas E et al. Treatment of patients with refractory Giardiasis. Clin Infect Dis 2001; 33:22-8.
4. Fox LM, Saravolatz LD. Nitazoxanide: A new Thiazolide antiparasitic agent. Clin Infect Dis 2005; 40:1173-80.
5. Farthing MJG, Mata L, Urrutia JJ, Kronmal RA. Natural history of Giardia infection of infants and children in rural Guatemala and its impact on physical growth. Am J Clin Nutr 1986; 43:395-405.
6. Berkman DS, Lescano AG, Gilman RG et al. Effects of stunting, diarrhoeal disease and parasitic infection during infancy on cognition in late childhood: A follow-up study. Lancet 2002; 359:564-71.
7. Lopez CE, Dykes AC, Juranek DD et al. Waterborne giardiasis: A communitywide outbreak of disease and a high rate of asymptomatic infection. Am J Epidemiol 1980; 112:495-507.
8. Adam RD. Biology of Giardia lamblia. Clin Microbiol Rev 2001; 14:447-475.
9. Petri WA. Treatment of Giardiasis. Curr Treat Options Gastroenterol 2005; 8:13-7.
10. Hill DR. Giardia lamblia. In: Mandell GL, Bennett JE, Dolin R, eds. Principles and Practics of Infectious Diseases, 6th Edition. Philadelphia: Churchill Livingstone, 2005:3198-203.

CHAPTER 28

Amebiasis

Daniel J. Eichinger

Introduction

The causative agent of intestinal amebiasis is the single-celled protozoan parasite *Entamoeba histolytica*. This parasite is endemic in most tropical and subtropical areas of the world, where it causes millions of cases of dysentery and liver abscess each year. Infected persons display a wide range of disease severity, reflecting the contributions of the patient's immune and nutritional status, the infective dose and the pathogenic potential of the infecting organism. It is only one of several Entamoeba organisms that can reside in the human GI tract, which in the past has made diagnosis difficult. It is uniquely adapted to the conditions found in the lumen of the large intestine and as a result is sensitive to drugs that target anaerobic organisms.

Life Cycle and Structure

There are only two stages to the life cycle of *E. histolytica*, the infectious cyst stage and the multiplying, disease-causing trophozoite stage. In the majority of cases infection results from the ingestion of fecally-contaminated water or food that contains *E. histolytica* cysts. Much less often the cyst or the trophozoite forms can be transmitted as a result of oral or oral/anal sexual practices. Cysts are 8-10 μm in diameter and when they are released from infected persons they are stable for days to weeks in low-temperature aqueous conditions. Re-ingested cysts are resistant to the low pH of the stomach and appear to be triggered to exit the cyst capsule (excyst) by components of bile and bicarbonate that are encountered in the small intestine. These metacystic amoebae are carried to the colon where the 10-30 μm diameter trophozoite form adheres to the mucus layer overlying the colonic epithelium. This adherence is mediated by lectins found on the surface of the trophozoite that have specificity for ligands expressing appropriately spaced terminal N-acetylgalactosamine and galactose sugar residues, as are found on the colonic form of mucin. The trophozoite stage of the parasite feeds on host ingesta, components of mucus and the resident bacteria and multiplies by asexual binary fission. In response to conditions and stimuli that are not yet completely defined, trophozoites stop multiplying and revert to the cyst form (encyst). These cysts are released in the tens of millions per gram of feces to allow for completion of the life cycle upon infection of another person. Most cysts are released from asymptomatic carriers of the parasite, suggesting that conditions within the intestine that are conducive to cyst formation are not significantly abnormal. Recent reports have also revealed an ability of products (short-chain fatty acids) normally produced

Medical Parasitology, edited by Abhay R. Satoskar, Gary L. Simon, Peter J. Hotez and Moriya Tsuji.

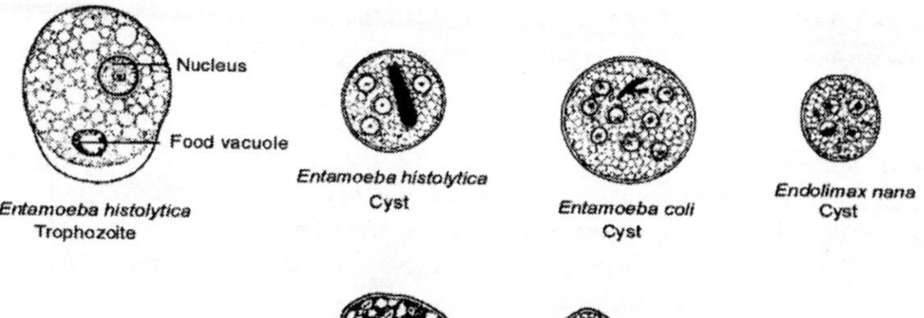

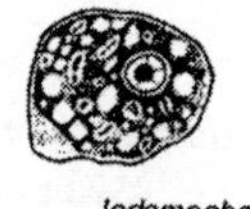

Figure 28.1. Schematic depictions of the morphology of the *E. histolytica* trophozoite and cyst, as compared to other amoebae found in the human intestine. Reproduced from Nappi AJ, Vass E, eds. Parasites of Medical Importance. Austin: Landes Bioscience, 2002:20.

by the enteric bacteria to regulate the encystment process. Mixtures of cysts and trophozoites are more commonly released by symptomatic (dysenteric) persons, but the trophozoite form is generally not considered infectious due to its sensitivity to hypoosmotic conditions outside the body and the low pH of the stomach.

The trophozoite form contains a single nucleus and many internal vesicles (Fig. 28.1). The nucleus has a thin continuous rim of heterochromatin and a centrally located nucleolus (karyosome). In fresh wet mounts the trophozoite displays ameboid motion with rapid extension of ectoplasmic pseudopods. The cyst form contains a refractile chitin-containing cyst wall, four nuclei and, in many cases, crystalline, bullet-shaped aggregations of ribosomes called chromatoid bodies. These features of the trophozoite nucleus and the cyst distinguish *E. histolytica* from other species of nonpathogenic amoebae that can also reside in the human colon, such as *E. coli*, *E. nana* or *Iodamoeba butschlii*.

Epidemiology

Most parasite transmission follows the oral-fecal route in which cysts are ingested. More rarely cysts are transmitted directly between persons. The worldwide distribution of *E. histolytica* therefore primarily reflects the incomplete separation of human feces from water and food sources typical of densely populated areas of developing countries although infrequent outbreaks in developed countries have occurred. The previous worldwide estimates of up to 500 million *Entamoeba*-infected persons now need to be qualified with the recent understanding that approximately 90% of those infections are with the morphologically indistinguishable but genetically distinct organism *E. dispar*, which does not cause disease. An estimated 40-100,000 persons die each year from infection with *E. histolytica*, however, and many more are either asymptomatically infected or present with varying degrees of dysentery and extraintestinal disease. Experimental establishment of infection with *E. coli*, a nonpathogenic parasite of humans, was shown to be possible with as few as 1-10 cysts, with a prepatent period (time following ingestion to the detection of cysts in stool) ranging from 6 to 23 days, an average of 10 days.

In closely examined areas where *E. histolytica* is endemic, most infected adults are asymptomatic. Depending on the age of the person, the time of asymptomatic infection can be considerable, with a half-life of parasite persistence of about 13 months in adults in one area of Vietnam where the 10% of adults are infected. The time of parasite persistence in children appears to be shorter, which may reflect the higher incidence of diarrhea episodes from multiple unrelated causes in that age group that serves to physically remove the parasite. The long (13 month) half life of asymptomatic parasite infection in adults suggests that up to 5% of individuals can retain the parasite for up to 5 years. Asymptomatic persons can therefore develop intestinal or extraintestinal disease (liver abscess) months to years after visiting areas where parasite transmission is known or suspected to occur. Most cases of diagnosed *E. histolytica* infection in the US occur in immigrants, particularly those from Mexico, central and South America and Asia. Seropositivity, reflecting the tissue-invasive capacity of *E. histolytica*, is approximately 8% in the entire population of Mexico.

In contrast to disease restricted to the lumen of the intestine, which occurs nearly equally in males and females, for unknown reasons extraintestinal liver abscess shows a 3-20 fold higher incidence in males vs females and in individuals that abuse alcohol.

Clinical Presentation

Presentation of amebiasis can take several forms, depending on the severity of the disease within the intestine and the involvement of extraintestinal sites.

Intestinal Disease

The greatest numbers of infected individuals have parasites restricted to the lumen of the intestine and are asymptomatic. 90-96% of these would normally clear the parasite spontaneously, but the potential for disease development in the remaining 4-10% requires that even asymptomatic individuals be treated if *E. histolytica* is definitively identified. Those persons presenting with colitis typically have a history of several weeks of gradually increasing abdominal cramping, tenderness, weight loss and a range of bowel function alterations, ranging from frequent mucoid stools to watery and bloody diarrhea, often with periods of dysentery alternating with constipation. As these symptoms are also typical of those elicited by a variety of bacterial pathogens, differential diagnosis should include other infectious as well as noninfectious causes of colitis. Stools are nearly always heme-positive due to the invasive nature of the parasite, and rectal release of blood without diarrhea is not uncommon in children. Fever is present in less than half of intestinal amebiasis patients. However, fulminant necrotizing colitis, developing in less than 1% of patients, has a high mortality rate (40%) and presents with fever, bloody mucoid diarrhea, leukocytosis and peritoneal tenderness. Intestinal perforation occurs in the majority of patients with fulminant colitis, and both it and toxic megacolon, associated with corticosteroid use, require surgical intervention. Another infrequent intestinal manifestation, ameboma, mimics colonic carcinoma in its radiologic presentation as an annular deposition of granulomatous tissue that locally narrows the lumen of the colon.

Liver Abscess

The most common site of extraintestinal infection (up to 9% of amebiasis cases) is the liver, resulting from hematogenous spread (via portal circulation primarily to the right lobe) of trophozoites that have eroded through the colonic mucosa. Within the liver trophozoites multiply while degrading liver tissue in a spherically expanding abscess that becomes filled with necrotic liver cells. Hepatic amebiasis can present months to years after an individual has traveled to or resided in an endemic area, during which time the patient may be completely asymptomatic and negative for stool parasites. Only 20% of patients with liver abscess have a prior history of clinical dysentery. Not all liver abscesses will progress to a symptomatic stage and some can self-resolve subclinically. Once they do become apparent, however, liver abscess symptoms develop relatively rapidly, over a course of 10 days to several weeks and can include right upper quadrant pain, fever, point tenderness of the liver, anorexia and weight loss. Abscesses located just below the diaphragm can lead to pleural pain or referred right shoulder pain. Liver alkaline phosphatase levels are normal and alanine aminotransferase levels are elevated in acute liver abscess, which may, however, reverse over time. Males are ten times more likely to present with liver abscess than females and middle-age and young adults more than children. Careful history elicitation is important because of the possible length of time ensuing between presentation with liver abscess and past residence or visitation in an endemic geographic region. Serum antibodies to amoeba antigens are nearly always found in such patients, but if residence in an endemic area was prolonged, such antibodies may be the result of prior infection.

Pathogenesis

The species name, histolytica, refers to the impressive ability of this organism to degrade a variety of host tissues. The parasite expresses a large number of factors, including lytic peptides (amebapores), cysteine proteinases and phospholipases that are presumably designed to aid in the ingestion and digestion of bacteria and other food materials. These products are considered virulence factors as they can also lyse colonic epithelial, liver and immune cells that come into contact with the trophozoite via its galNAc-specific lectin. *Entamoeba* trophozoites are actively phagocytic and the presence of ingested red blood cells is diagnostic of *E. histolytica* organisms that are found in suitably stained biopsy specimens (Fig. 28.2). As a result of the variability of these activities, the range of pathological findings in the colitic colon includes increased height of the mucosa, isolated ulcerations of the mucosal layer, invasion and erosion of the submucosa (Fig. 28.3), areas of necrosis with loss of large patches of mucosa and complete perforation of the colon. Invasion of the mucosal layer is often stopped by underlying muscularis layers, forcing the parasite to erode tissues laterally in a manner that generates ulcers with a flask-shaped cross sectional appearance.

Liver abscesses are well circumscribed with an outer wall of connective tissue surrounded by normal liver cells, an underlying layer of trophozoites and dying hepatocytes and a central area filled with dead hepatocytes and immune cells. A

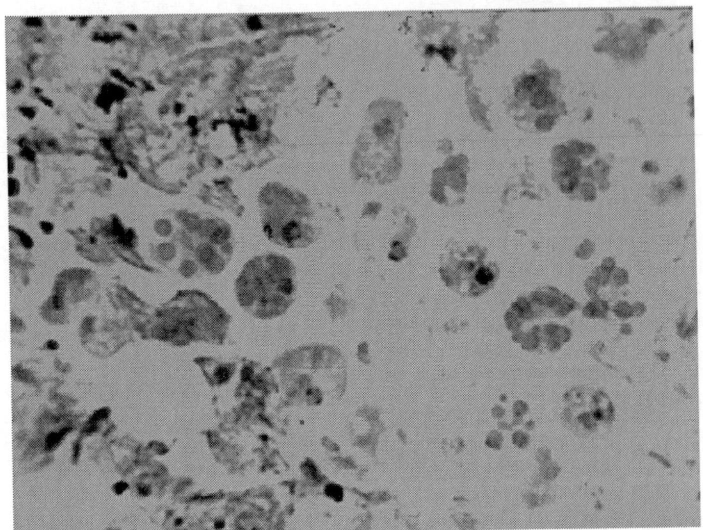

Figure 28.2. Biopsy specimen containing *E. histolytica* trophozoites with ingested red blood cells. Original image courtesy of Dr. William Petri, University of Virginia.

complication of liver abscess that results from sufficient expansion of the abscess to make contact the liver capsule is the rupture of the abscess into surrounding anatomic spaces, which occurs in up to 20% of abscess patients. Rupture of liver abscess through the diaphragm can yield pleuropulmonary amebiasis that presents with cough, chest pain and respiratory distress. Significant leakage of liver abscess material into the lung can yield cough producing brown sputum. Erosions of liver abscesses into the peritoneal and pericardial spaces are less frequent but can be of greater clinical significance even though the liver abscess contents are sterile. Much less common but with highest mortality is the dissemination of tropohozoites from liver abscesses via general circulation to the brain.

Diagnosis

28 Blood cell parameters of amebiasis patients are not grossly abnormal, but elevated total white cell counts are found in >75%. Unlike infection with invasive helminth parasites, there is typically no eosinophilia with amebiasis. Chemistry changes are usually limited to increased alkaline phosphatase levels in the majority of liver abscess patients.

Microscopic identification of trophozoites or cysts in stool samples or biopsy specimens is the most definitive method of diagnosis. However, cyst passage is known to be inconsistent in asymptomatic carriers and to require the examination of multiple samples. Visualization of the organism is therefore being supplanted by ELISA assays that detect stool antigen rather than whole cells and, importantly, are capable of distinguishing between pathogenic *E. histolytica* and nonpathogenic *E. dispar*. With high sensitivity (80%), specificity (99%) and

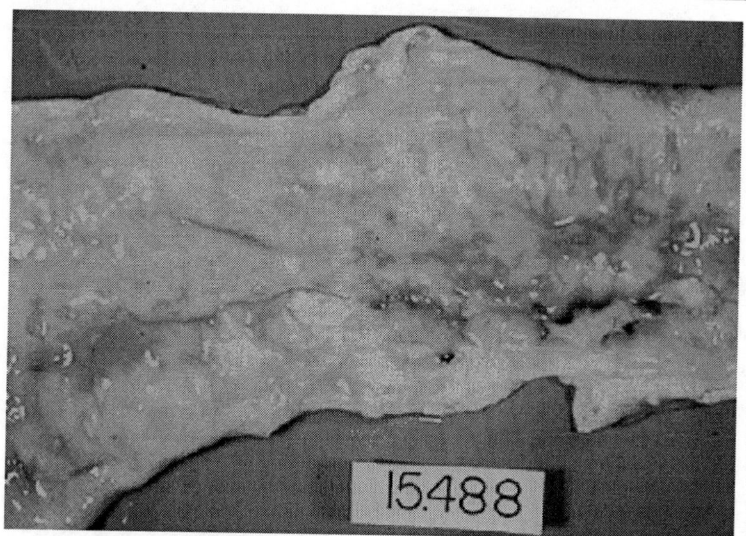

Figure 28.3. Isolated amoeba-induced lesions, as well as areas of sloughed mucosa resulting from fusion of adjacent lesions, are evident in this pathological colon specimen.

commercial availability, these tests will make diagnosis of intestinal amebiasis more reliable. Invasive disease presenting as colitis or liver abscess can additionally be diagnosed serologically, as antibodies are present in >90% of such patients and detectable by ELISA and various agglutination and electrophoresis methods. Antibody titers increase with length of infection, so patients presenting with acute suspected disease may be negative initially but positive within 2 weeks. Asymptomatic patients can also develop positive antibody titers, allowing for determination of the infecting organism when stool samples are positive for cysts and consideration of treatment.

Amebic liver abscesses are readily detected radiographically with ultrasound, CT scan or MRI methods, which, when combined with serology, can distinguish the amebic abscess from other space-occupying lesions such as hepatoma, pyogenic or hydatid abscesses. (Fig. 28.4). Nearly all amebic liver abscesses completely resolve, but a small number will leave residual radiographic lesions that do not require further treatment. Amebomas can be visualized by barium contrast radiography, taking into account the risk of perforation if colitis is also present.

Treatment

Since most asymptomatic persons diagnosed with amebiasis are infected with *E. dispar* and do not require treatment, it is important to establish that *E. histolytica* is present to justify treatment. In contrast, asymptomatic infections with *E. histolytica* do require treatment because of the possibility of eventual disease development. Paromomycin (Humatin) and diloxanide furoanate (Furamide) are the two drugs recommended for treatment of these patients. Recent analysis of

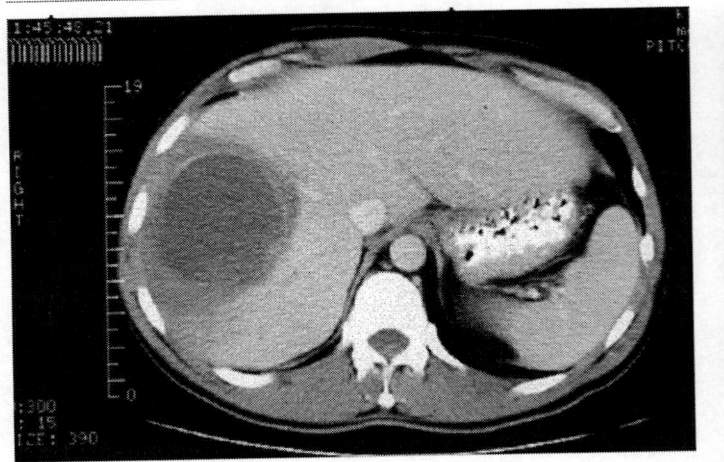

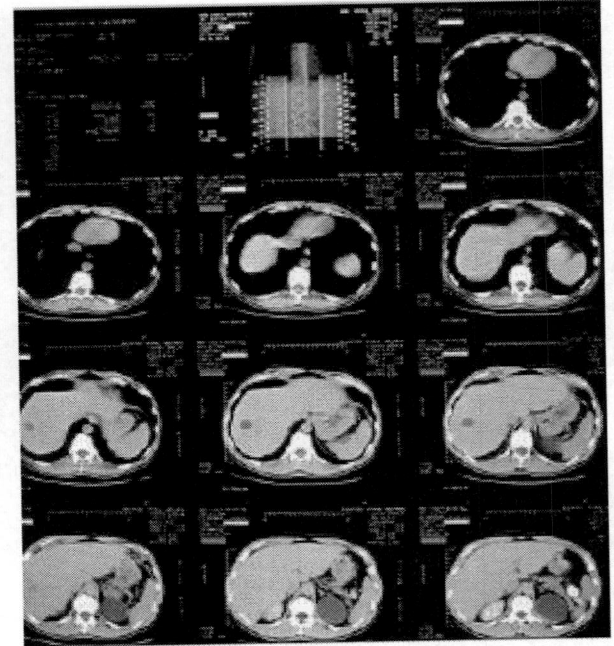

Figure 28.4. CT scan images of a patient with a right lobe amoebic liver abscess, upon admission (left), and six weeks following percutaneous drainage and a course of metronidazole (right).

the relative effectiveness of these two drugs against *E. histolytica* as compared to *E. dispar* showed that paromomycin (25-35 mg/kg/d in three doses for 7 days) is 85% effective against *E. histolytica* and an equal course of diloxanide furoate is 51% effective. Successful treatment is documented with follow-up stool (microscopic or ELISA) examination 2-4 weeks later. Treatment failure rates can be as high as 15%. An alternative drug is iodoquinol (Yodoxin), which, however, requires a longer treatment course (20 days) and has iodine-related toxicities.

As compared with luminal asymptomatic infection, the drug of choice for suspected or defined invasive (colitis) and extraintestinal (hepatic abscess) disease is metronidazole (Flagyl). Paroral metronidazole is completely absorbed and recent studies have shown that the previously recommended 750 mg tid 7-10 day course for adults can be reduced to one daily 2.4 g dose for 3 days with equal efficacy. Because it is so readily absorbed, treatment with metronidazole must be combined (either concurrently or sequentially) with a lumenally-retained agent to eliminate potential parasites within the intestine whether or not they are detected by stool examination. Metronidazole appears not to have adverse affects on fetal development during the first trimester. Percutaneous drainage of liver abscess is warranted only if medical therapy appears not to be working, abscess rupture appears to be imminent, or left lobe abscesses threaten to rupture into the pericardium.

Suggested Reading

1. Petri WA, Singh U. Diagnosis and management of amebiasis. Clin Infect Dis 1999; 29:1117-25.
2. Stanley SL. Amoebiasis. Lancet 2003; 361:1025-34.
3. Haque R, Huston CD, Hughes M et al. Amebiasis. New Engl J Med 2003; 348:1565-73.
4. Blessman J, Tannich E. Treatment of asymptomatic intestinal Entamoeba histolytica infection. New Engl J Med 2002; 347:1384.
5. Blessman J, Ali IKM, Nu PAT et al. Longitudinal study of intestinal Entamoeba histolytica infections in asymptomatic adult carriers. J Clin Microbiol 2003; 41:4745-50.

Chapter 29

Cryptosporidiosis

Gerasimos J. Zaharatos

Introduction

Although the first human cases of *Cryptosporidium* were described in 1976, the contribution of this protozoan parasite to gastrointestinal disease was not fully appreciated until the 1980s when scores of cases were described among patients with acquired immunodeficiency syndrome (AIDS). The disease gained greater notoriety after a massive outbreak of waterborne cryptosporidial infection in Milwaukee, Wisconsin in 1993. Watery diarrhea and malabsorption are the usual sequelae of symptomatic infection. In addition to *Cryptosporidium* causing chronic diarrhea and extraintestinal disease in immunocompromised individuals, the parasite is an important source of self-limited diarrhea in immunocompetent individuals and a major contributor to persistent diarrhea in children throughout the developing world. Routine fecal specimen testing, for enteric pathogens including ova and parasites, will fail to detect this organism and thus self-limited cases may remain undiagnosed. Limited therapeutic options for persistent and chronic disease present an additional challenge to clinicians and treatment of any underlying immunodeficiency is paramount.

The Parasite

The genus *Cryptosporidium* consists of at least 10 species. This group of organisms resides within the subphylum *Apicomplexa*, along with other protozoan parasites such as *Plasmodium* species. It is most closely related to coccidian parasites including other intestinal pathogens such as *Cyclospora* and *Isospora* species. *Cryptosporidium* species have been detected in the gastrointestinal tract of a number of mammalian and vertebrate species. Its presence in ruminants has been most widely described. *C. parvum*, a species commonly found in bovine hosts, was formerly the species most often associated with human disease. However, genotypic and phenotypic differences among isolates eventually led to the recognition of two separate species, *C. hominis* (formerly "human" genotype or *C. parvum* genotype 1) and *C. parvum* ("bovine" genotype or *C. parvum* genotype 2). Humans are most commonly infected by *C. hominis* or *C. parvum* and on occasion by species normally present in other animal hosts. It appears that *C. hominis* only infects humans. Mixed infections have been rarely described in immunocompromised patients. As the genetic diversity of various host-adapted species is better appreciated, it is likely that the present nomenclature will evolve and the relative importance of zoonotic transmission will undergo further re-evaluation.

Medical Parasitology, edited by Abhay R. Satoskar, Gary L. Simon, Peter J. Hotez and Moriya Tsuji.

Lifecycle, Pathogenesis and the Host Response

The lifecycle of *Cryptosporidium* can be completed entirely within a single host (Fig. 29.1A). Subsequent to oocyst ingestion and activation in the upper GI tract, the organisms excyst to release sporozoites. These sporozoites bind intestinal epithelial cells and via induction of actin polymerization, provoke their own engulfment to eventually reside in a parasitophorous vacuole within the microvillus layer. In this sequestered environment, the parasites undergo asexual reproduction (termed merogony) and ultimately produce merozoites that are released intraluminally. These forms in turn bind and are again engulfed by epithelial cells and thus perpetuate the cycle. Alternatively the engulfed merozoites may undergo sexual differentiation and ultimately the fertilization of macrogamonts by microgametes will yield new oocysts. These new oocysts may either be shed into the environment or excyst within the same host.

The organism can be found throughout the gastrointestinal tract; however it appears to have an affinity for epithelial cells in the jejunum, ileum and proximal colon. Cholangiocytes are also susceptible to infection, and apoptosis of these epithelial cells likely contributes to biliary tract disease. The respiratory tract also appears to be a site of infection in immunocompromised individuals. Epithelial cell death, by both apoptotic and necrotic mechanisms, has been noted in involved regions. There is evidence that infected epithelial cells can induce apoptosis in neighboring uninfected cells. Epithelial cell infection usually culminates in dysregulation of cell signaling pathways including upregulation of proinflammatory cascades as well as cyclooxygenase-2, prostaglandins and neuropeptide production. These perturbations result in epithelial barrier dysfunction, augmented intestinal permeability, dysregulation of electrolyte absorption and secretion and fluid malabsorption. Accordingly, symptomatic infection usually manifests as watery diarrhea.

Epidemiology

Cryptosporidium has a wide geographic distribution, though infection is more prevalent in regions of the world with poor sanitary conditions. Infection is more common during warm rainy months. The reported prevalence of infection varies widely and is influenced by geographic region, age, immune status, local outbreaks and the range of sensitivities and specificities offered by different diagnostic modalities. In general, exposure rates based on seroprevalence studies suggest that in North America at least 30% of adults have been previously exposed to *Cryptosporidium* species. However, seroprevalence rates are as high as 90% in the developing world. In moist environments, *Cryptosporidium* oocysts may remain infectious for 6 months. The infectious dose can be as low as 10 oocysts, though considerable variability exists among isolates and a much higher infectious dose is often required in previously exposed seropositive individuals. Oocysts have been detected in apparently pure surface water sources, though protected spring water sources are less likely to be contaminated. Untreated or raw waste water is substantially contaminated with oocysts. Moreover municipal wastewater treatment centers, runoff from animal agriculture and various wildlife populations all contribute to a remarkable release of oocysts into the aquatic environment. Oocysts are highly resistant to chlorination and can bypass certain filtration methods. Accordingly, sources of treated potable water

Figure 29.1, viewed on following page. A) Life cycle of *Cryptosporidium* species. From the CDC Public Health Image Library (PHIL #3386). Image credit: Alexander J da Silva and Melanie Moser (CDC). B) Modified Acid Fast Stain. Examination of fecal specimens may reveal round pink or red oocysts of 4-6 microns in diameter on a blue or blue-green background (represented here as black oocysts on a grey background). Sporozoites may be visualized inside individual oocysts. The assessment of three separate specimens and use of an oocyst concentration technique prior to staining may increase diagnostic yield. From the CDC Public Health Image Library (PHIL# 7829). Image credit: Melanie Moser (CDC/ DPDx).

can contain significant numbers of oocysts. Oocysts can survive for a period of time in seawater, and indeed shellfish in coastal areas have been found to be contaminated with infectious oocysts. Most cases among immunocompetent hosts have been associated with waterborne outbreaks and have involved either contaminated drinking water or recreational water sources such as swimming pools and lakes. Disease is also well-described in returning travelers, persons with animal contact (e.g., farmers) and amongst daycare personnel working with young children. Direct person-to-person transmission via the fecal-oral route is also common in a number of settings including during sexual activity. Health care workers should be cognizant of the potential for nosocomial transmission. Foodborne transmission, though relatively infrequent, has been reported from a number of sources including inadequately pasteurized beverages and raw fruits and vegetables.

Clinical Manifestations and the Host Response

Nonbloody diarrhea is the most common clinical presentation of cryptosporidiosis; however clinical findings may vary widely and are dependent on the affected host population being considered. The severity and duration of diarrhea may be quite variable. The incubation period is usually 7 to 10 days, though it can range from several days to weeks.

The Developed World: Immunocompetent Hosts

Among immunocompetent adults, the most common presentation is watery, occasionally mucoid, diarrhea. The severity and frequency of diarrhea can be variable ranging from small volume intermittent stools to continuous and voluminous unformed or watery stools. Diarrhea is usually self-limited and persists
29 for up to 14 days, though it can persist in normal hosts for a more prolonged period. Diarrhea may be accompanied by abdominal cramping, fever, malaise, nausea and occasionally vomiting. Concurrent respiratory symptoms have also been reported. Despite an initial resolution of symptoms, a considerable proportion of infected individuals eventually have recurrent disease within days to weeks. Some individuals with *C. hominis* infection report development of extraintestinal symptoms late in their course including recurrent headaches, fatigue, dizziness, ocular pain and arthralgias. Accumulating evidence suggests that mild or asymptomatic infection may also be common. Previously exposed seropositive individuals appear to be more resistant to reinfection and when reinfected have milder forms of disease.

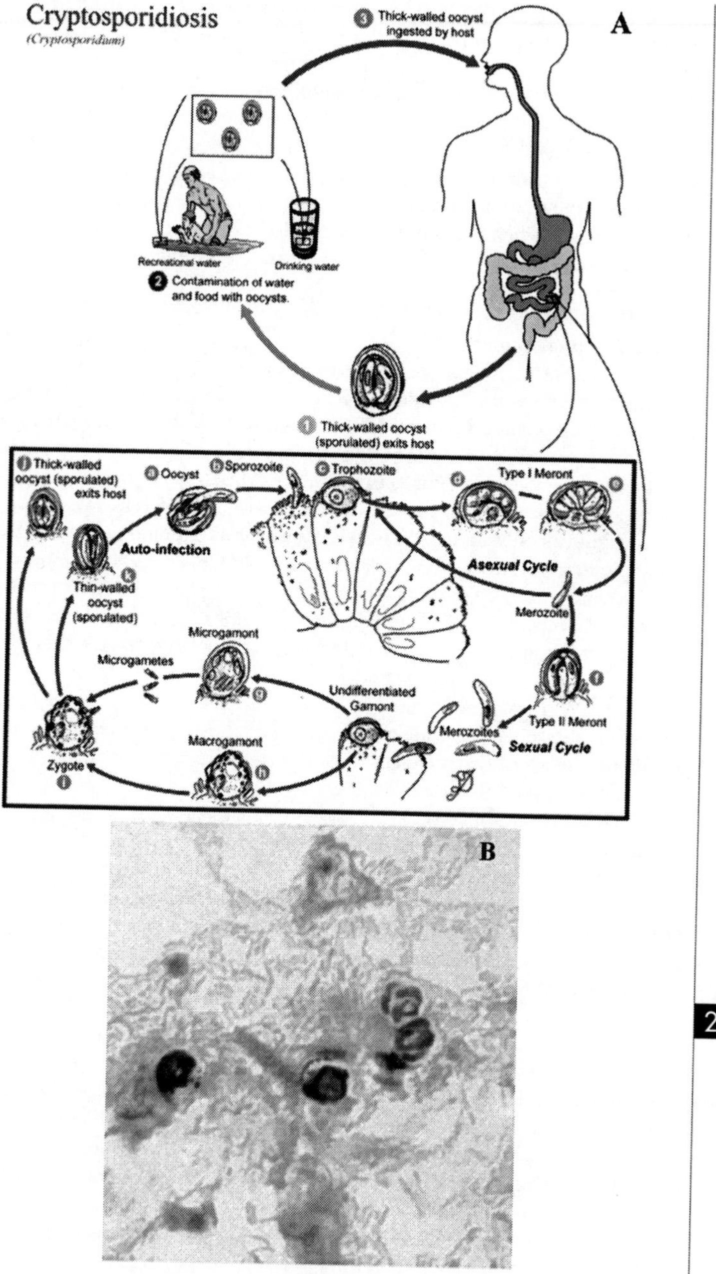

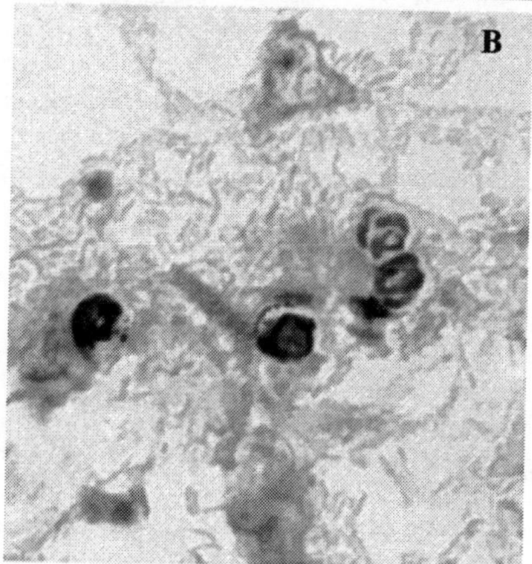

Figure 29.1. Please see figure legend on previous page.

The Developing World: Childhood Diarrhea and Malnutrition

In the developing world, diarrhea amongst children is the most common clinical presentation of cryptosporidiosis. Little distinguishes *Cryptosporidium*-associated diarrhea from other infectious causes of childhood diarrhea except for its propensity to cause persistent diarrhea beyond 2 weeks duration. Persistent diarrhea is associated with greater morbidity and mortality. Relapsing diarrhea, significant weight loss and growth rate reduction are common sequelae of infection in this population. The relationship between cryptosporidiosis and malnutrition is complex. It is unclear whether malnourished children have an increased susceptibility to infection and predilection for more severe and persistent disease.

The Immunocompromised Host: AIDS and Other Immunodeficiencies

In the patient with HIV, the course of cryptosporidiosis often correlates with the immune status of the individual. Patients with CD4 counts above 200 cells/μL are likely to have a clinical course similar to immunocompetent hosts. Patients with AIDS and progressively declining CD4 counts are more likely to present with foul smelling bulky stools in the context of chronic diarrhea and weight loss. Severely immunocompromised individuals with CD4 counts less than 50 cells/microL develop a more fulminant cholera-like disease with watery and voluminous diarrhea. Biliary and respiratory tract disease is more likely to manifest in severely immunocompromised persons with CD4 counts less than 50 cells/μL. Biliary tract involvement may result in biliary strictures, papillary stenosis, pancreatitis, acalculous cholecystitis, or sclerosing cholangitis. These may manifest with right upper quadrant pain, nausea, vomiting and low grade fever. Although oocysts have been detected in respiratory secretions of immunocompromised patients, a causal link between *Cryptosporidium* and pulmonary disease is usually difficult to establish given the occurrence of coexistent opportunistic pathogens in this population. More severe or persistent cryptosporidiosis has also been described in other immunocompromised settings, including organ transplantation, immunosuppressive therapy, chemotherapy, primary immunodeficiencies, hematologic malignancies, cytokine deficiencies and a variety of other conditions associated with cell mediated immune dysfunction. CD4 T-cells and the induction of certain cytokines, particularly interferon-gamma, play a critical role in controlling infection. The role of other cytokines in both pathogenesis and effective immunity or parasitologic clearance continues to be investigated. The role of antibodies in immune protection remains contentious and immunoglobulin deficits or B-cell dysfunction do not preclude contemporaneous deficits in T-cell or dendritic cell function. Moreover, patients with AIDS can mount apparently adequate humoral responses yet remain susceptible to severe or persistent cryptosporidiosis. Interestingly, fulminant disease has been noted in patients with severe combined immunodeficiency syndrome, immunoglobulin deficiencies and subset of hyper IgM syndromes. The latter syndromes are a heterogeneous and rare group of disorders which manifest high levels of IgM but markedly reduced levels of other Ig isotypes. The X-linked form is associated with inadequate levels of CD40L expression, defective immunoglobulin class switching and ineffective generation of antigen specific responses. Despite immunoglobulin replacement therapy, these patients remain susceptible to certain opportunistic infections including cryptosporidiosis.

Diagnosis

Routine stool examination for ova and parasites will generally fail to detect the organism and the clinician should specify that *Cryptosporidium* species is in the differential diagnosis. The sample should be preserved in 10% buffered formalin, though some laboratories use a concentration method with formalin-ethyl acetate fixation that may augment the sensitivity of detection. Examination of unfixed specimens should be discouraged to reduce the risk to laboratory workers. A number of diagnostic modalities with varying sensitivities and specificities are now available. Traditionally, the diagnosis of cryptosporidiosis has been made on the basis of microscopic identification of round oocysts of 4-6 μm in diameter (Fig. 29.1B). Standard staining techniques include modified acid fast staining in which oocysts appear pink or red on a blue or blue-green background, allowing clear differentiation from morphologically similar yeasts. Sporozoites may be seen in individual oocysts and their visualization may assist in the diagnosis. White blood cells are usually not seen. At least three separate specimens should be examined to improve the diagnostic yield. The yield may be enhanced if the stool specimen is unformed and if the laboratory uses a concentration technique prior to staining. Fluorescent auramine stains are potentially useful but provide little specificity by themselves. Malachite green, Giemsa and hematoxylin and eosin staining techniques have been used to detect the organism with varying success and are inferior to modified acid fast staining. Immunofluorescence assays are now commonly used and may be a log more sensitive than acid-fast stains.

Antigen-detection assays in the form of ELISA or immunochromatographic kits are now more widely used. A number of commercially available kits now have excellent sensitivity and specificity relative to the modified acid fast technique. The clinician should nevertheless review the diagnostic reliability of such kits before excluding the diagnosis. PCR remains investigational, is more labor intensive and is not widely used though significantly more sensitive than any microscopic technique. Validated in-house assays can be more sensitive than antigen-detection assays and may have the capability to distinguish between different species. Although serological testing has been instrumental in determining the prevalence of cryptosporidiosis, it is of limited usefulness given the persistence of antibody titers after remote infection in otherwise healthy individuals.

In the immunocompromised patient with biliary disease or pancreatitis, the clinician should maintain a high index of suspicion for a cryptosporidiosis. One should not exclude the diagnosis based on negative stool specimen testing; as such investigations may be negative in this setting. Without frank obstruction, hyperbilirubinemia may be unimpressive or absent though elevated alkaline phosphatase and glutamyl transpeptidase, as well as elevated amylase and lipase in the setting of pancreatitis, may point to the diagnosis of biliary tract infection. Ultrasonographic abnormalities may include a thickened gallbladder wall as well as dilated and irregular intrahepatic and extrahepatic bile ducts. A suspicion of *Cryptosporidium*-related cholangiopathy should prompt an endoscopic assessment of biliary involvement with tissue biopsy for histology as well as examination of bile for oocysts.

Management

Supportive care of the patient with cryptosporidiosis includes fluid and electrolyte replacement, adequate nutritional intake and a lactose-free diet. Fluid absorption may be improved by supplementation with glutamine. Antimotility agents like loperamide are a mainstay of symptomatic treatment, though opiates and octreotide have been used as adjuncts in more severe disease. In the setting of obstruction secondary to papillary stenosis, endoscopic sphincterotomy with or without stent placement may offer a symptomatic benefit.

The efficacy of antiparasitic therapy has historically been quite disappointing, particularly amongst immunocompromised or malnourished patients. Three antiparasitic agents are presently considered as having some in vivo activity against *Cryptosporidium:* paromomycin, azithromycin and nitazoxanide. Amongst immunocompetent individuals, nitazoxamide (500 mg po BID for 3 days) has more recently been shown to speed the resolution of diarrhea and oocyst shedding. Nitazoxanide is also available as a suspension for pediatric use and is approved for use in the United States for children older than 12 months of age at a range of age-adjusted doses. The clinician should be cognizant of the significant likelihood of relapsing disease, even in the immunocompetent host.

In patients with immune dysfunction, efforts to improve immune function are paramount. In persons with AIDS, diarrhea may improve dramatically and resolve on effective antiretroviral therapy as a result of immune reconstitution. Interestingly, some investigators have reported that protease inhibitors have some antiparasitic activity. Nevertheless, treatment of the patient with advanced AIDS may be challenging in that antiretroviral medications may be poorly absorbed in the setting of active cryptosporidiosis. For this reason, some authorities suggest initially controlling disease with antiparasitic drugs prior to instituting a new antiretroviral regimen. Antiparasitic therapy can speed resolution of diarrhea and oocyst shedding, however a transient response to therapy is to be expected in patients with advanced AIDS if immune reconstitution does not eventually occur.

Paromomycin has been shown to improve diarrhea to some extent in patients with AIDS and in combination with azithromycin has been effective in diminishing symptoms and oocyst shedding in some patients. However, this regimen does not usually result in resolution of disease in patients with AIDS. Numerous isolated case reports have suggested a response to various combinations of a number of alternative agents. Such regimens have contributed transciently, if at all, to improvement of symptoms and should be considered as adjuncts of limited effectiveness. These agents include spiramycin, clarithromycin, atovaquone, letrazuril and hyperimmune bovine colostrum. Interestingly, rifaximin has been shown to improve diarrhea to some extent in patients with HIV. A recent small case series has also described its success in treating patients with advanced AIDS, but such reports remain preliminary.

Promising data has been garnered from compassionate use programs for nitazoxamide in patients with AIDS, suggesting that most recipients have a sustained clinical response to this agent. However, short courses of nitazoxanide appear ineffective in children with AIDS not receiving ongoing effective antiretroviral therapy. Judging the effectiveness of the drug in HIV-infected and other immunocompromised hosts will require further formal investigation. Higher doses and/or a more extended duration of this therapy (500 to 1000 mg po BID for 14 days in adults) should be considered for such populations, and immune reconstitution remains fundamental to any sustained response. Therapy may need to be prolonged until sufficient immune reconstitution has been achieved. In hosts with intractable immunodeficiency, parasitologic cure is very unlikely irrespective of the regimen chosen and persistent fulminant diarrhea portends a dismal prognosis.

Suggested Reading

1. Bushen OY, Lima AA, Guerrant RL. Cryptosporidiosis. In: Guerrant RL, Walker DH, Weller PF: Tropical infectious diseases. Principle, pathogens and practice. Philadelphia, PA: Elsevier Churchill Livingstone 2006; 1003-14.
2. Huang DB, White AC. An updated review on Cryptosporidium and Giardia. Gastroenterol Clin N Am 2006; 35: 291-314.
3. White AC. Cryptosporidiosis (Cryptosporidium hominis, Cryptosporidium parvum and other species). In: Mandell, Bennett & Dolin: Principles and Practice of Infectious Diseases. 6th Edition. Philadelphia: Elsevier Churchill Livingstone, 2005:3215-28.
4. Xiao L, Fayer R, Ryan U et al. Cryptosporidium taxonomy: recent advances and implications for public health. Clin Microbiol Rev 2004; 17:72-97.
5. Hunter PR, Nichols G. Epidemiology and clinical features of Cryptosporidium infection in immunocompromised patients. Clin Microbiol Rev 2002; 15:145-54.
6. Centers for disease control and prevention and the national center for infectious diseases, division of parasitic diseases. Parasitic disease information: Cryptosporidium Infection/Cryptosporidiosis. http://www.cdc.gov/ncidod/dpd/parasites/cryptosporidiosis/default.htm

CHAPTER 30

Trichomoniasis

Raymond M. Johnson

Introduction

Trichomonas vaginalis is a flagellated single cell eukaryote with a relatively simple lifecycle. *T. vaginalis* exists only in the trophozoite form and divides by simple binary fission in its human host. There are no *T. vaginalis* cysts, portions of the parasite lifecycle that occur outside its human host, or known animal or environmental reservoirs. A closely related relative *Tritrichomonas foetus* causes commercially important reproductive tract and fetal infections in cattle.

T. vaginalis carries the distinction of being the only truly sexually transmitted parasitic infection in humans.[1] It is very successful as a pathogen causing roughly the same number of STDs as *Chlamydia trachomatis*, the most prevalent sexually transmitted bacterial pathogen. In the U.S. there are an estimated 3-5 million new cases of 'trich' every year with an infected pool of approximately 20 million individuals. Worldwide the prevalence of *T. vaginalis* varies from 2% to greater than 50% depending on region, country, gender and demographics of the population specifically evaluated.

T. vaginalis is highly adapted to the human urogenital tract and is never found in stool specimens. The unique adaptation of *T. vaginalis* to the urogenital tract allows it to be easily identified in urogenital tract clinical specimens without concern about other parasite species. *T. vaginalis* thrives in the microaerophilic environment of the vaginal mucosa. To live in the low oxygen tension it utilizes an organelle called a hydrogenosome to generate ATP. *T. vaginalis* lack mitochondria that generate ATP for oxygen-dependent eukaryotes. Instead the hydrogenosome generates ATP utilizing a pathway similar to mitochondria except that the final electron acceptor is hydrogen rather than oxygen, generating hydrogen gas as a byproduct of metabolism. The hydrogenosome is also the Achille's heel of *T. vaginalis* as it metabolizes the 5-nitroimidazole antibiotics metronidazole and tinidazole into toxic anion radicals that kill the parasite.

Clinical Disease

Trichomonas vaginalis typically comes to medical attention through one of four basic scenarios:

1. Women seeking evaluation for a vaginal discharge.
2. Women incidentally found to be infected with *T. vaginalis* during prenatal care visits or examination of routine Pap smears.
3. Male partners of women diagnosed with *T. vaginalis via* (1) or (2).
4. Men with nongonococcal urethritis (NGU) that does not respond to usual NGU therapy.

Medical Parasitology, edited by Abhay R. Satoskar, Gary L. Simon, Peter J. Hotez and Moriya Tsuji.

T. vaginalis is only found in the lower urinary and reproductive tracts of men and women. In women it causes a superficial infection of the vagina and urethra, occasionally ascending further to cause cystitis (bladder infections). In males the infection is largely confined to the urethra, but can ascend into the prostate and epididymis. *T. vaginalis* is not known to disseminate from the urogenital tract to cause deep seated infections in other parts of the body. Neonates acquiring *Trichomonas vaginalis* during transit through the birth canal rarely develop pneumonia and a single case of a trauma related *T. vaginalis* perinephric abscess has been reported.

Consistent with its tropism, *T. vaginalis* causes lower reproductive tract symptoms in women including vaginal discharge, vulvar irritation and dysparunia. Colonization of the urinary tract is associated with dysuria, urinary frequency, post voiding discomfort and lower abdominal pain. The vaginal discharge caused by *T. vaginalis* tends to be copious and can be 'frothy'. While women seeking medical attention for the above conditions are subjectively and objectively ill, many women carry *T. vaginalis* asymptomatically for protracted periods of time. 40-50% of women infected with *T. vaginalis* do not report a vaginal discharge.[2] A recent study in adolescent women over approximately 2 years had an incident rate of infection of 23%, with roughly a third of the infected adolescent women remaining asymptomatic.[3]

The majority of infected men are asymptomatic, however *T. vaginalis* can cause a symptomatic nongonococcal urethritis (NGU).[4] The urethral discharge associated with the infection tends to be minimal compared with other NGU etiologies. Because diagnostic testing for *T. vaginalis* in men is suboptimal, it is common practice in many STD clinics to empirically treat for *T. vaginalis* in men that have failed standard nongonococcal urethritis therapy and observe for resolution of symptoms. *T. vaginalis* rarely ascends into the prostate or epididymis to cause symptomatic infections.

The natural history of *T. vaginalis* infections is poorly understood. It is estimated that only 20% of women and 40% of men spontaneously clear their infections. The long duration of infection combined with a high rate of asymptomatic carriage likely account for the success of *T. vaginalis* as a sexually transmitted parasite. The relative insensitivity of currently utilized laboratory tests for diagnosing *T. vaginalis* allows it to escape detection and likely further facilitates its prevalence throughout the world.

Diagnosis

The approach to diagnosing *T. vaginalis* infection differs between men and women. A vaginal discharge may be caused by multiple etiologies and concurrent infections with two or more pathogens are common. The physical exam should include inspection of the cervix for presence of mucopurulent cervicitis associated with chlamydia and gonorrhea infections and the so-called 'strawberry cervix', a rare physical finding associated with *T. vaginalis* infection. The exam should also include evaluation for cervical motion tenderness, the physical exam finding that correlates with pelvic inflammatory disease.

The nature of the vaginal discharge itself is not particularly informative, though experienced STD clinic practitioners develop some ability to differentiate

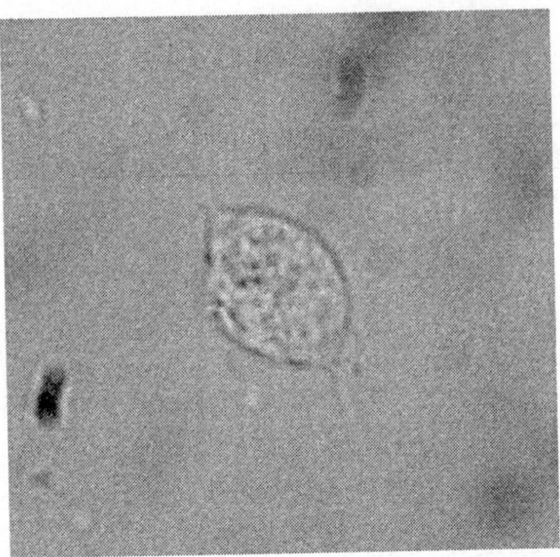

Figure 30.1. *Trichomonas vaginalis* visualized on normal saline wet prep microscopy. Image courtesy of Barbara Van Der Pol, Indiana University School of Medicine, Indianapolis, IN.

trichomonas infections from bacterial vaginosis and other infectious etiologies. An abundant frothy vaginal discharge is highly suspicious for *T. vaginalis* infection, but is not specific enough to establish a diagnosis and defer further diagnostic evaluation. Vaginal fluid specimens collected with swabs should be tested for pH and normal saline and KOH slides prepared. Typical cases of vaginal *T. vaginalis* infections have pH values greater than 4.5, but a normal vaginal pH does not rule out a *T. vaginalis* infection. This finding also does not distinguish *T. vaginalis* from bacterial vaginosis that is also associated with elevated pH. A positive KOH 'whiff' test for amines (fishy smell) implies bacterial vaginosis but does not exclude *T. vaginalis* co-infection. The normal saline wet prep slide on microscopy may show clue cells (vaginal epithelial cells coated with bacteria) consistent with bacterial vaginosis, or motile trichomonads diagnostic for *T. vaginalis* infection, or both. Visualization of motile trichomonads on microscopic examination is somewhat dramatic and is diagnostic of a 'trich' infection (Fig. 30.1). Unfortunately examination of the normal saline slide is only about 60% sensitive for making the diagnosis of a *T. vaginalis* infection. However, a positive *T. vaginalis* wet prep is all that is needed
30 to make the diagnosis in the majority of patients and therefore is an important part of the initial medical evaluation in women with vaginal discharges.

A sizeable minority of women with *T. vaginalis* infections will have a negative saline wet prep. There is no consensus among STD care providers on how the subsequent evaluation should proceed. For patients diagnosed with bacterial vaginosis (BV), systemic metronidazole given to treat BV would eradicate a coincident *T. vaginalis* infection while topical metronidazole gel therapy is not effective 'trich' therapy. Also, missing the diagnosis of a co-incident *T. vaginalis* infection means that the patient's male partner will go untreated and the patient will likely be reinfected immediately after coming off antibiotic therapy.

In medical settings where the test is available, the InPouch™ *Trichomonas vaginalis* culture system (Biomed) is probably the best and most affordable second screening test. It does however have several drawbacks. Samples must be inoculated into the culture medium within 30 minutes as the *T. vaginalis* must be viable for the culture to give a positive result. Also, because it is a culture system, there is an incubation period of 1 to 5 days before the test turns positive. In addition a microbiology technician must be available to evaluate the culture system microscopically for identification of motile *T. vaginalis* forms to finalize the test results. The culture technique is currently considered the gold standard for making a *T. vaginalis* diagnosis and is the most sensitive of any diagnostic test currently available for diagnosing *T. vaginalis* in women. PCR tests for *T. vaginalis* in vaginal fluid specimens are not commercially/clinically available and most of those in development do not appear to be more sensitive than *Trichomonas vaginalis* culture. Because a vaginal discharge represents more than a simple inconvenience for most women, a more rapid point of care diagnostic test would be desirable.

There are two nonculture clinical tests available for diagnosing *T. vaginalis* vaginal infections. A DNA probe test (Affirm™ VPII, Becton Dickinson) is capable of identifying *Trichomonas vaginalis*, *Gardnerella vaginalis* and *Candida species* in vaginal fluid specimens. The DNA probe test has a sensitivity of roughly 90% compared with culture. It has the advantage of potentially identifying polymicrobial infections and has more forgiving specimen handling requirements. However, it requires specialized equipment specific to the assay and has potential delays in diagnosis depending on individual laboratory protocols for processing specimens and reporting results. A simple point of care test based on *Trichomonas vaginalis* antigen (OSOM® Trichomonas Rapid Test, Genzyme Diagnositics) is commercially available and FDA approved in the US. The antigen-based test is about 80% sensitive compared with culture. Once vaginal swabs have been obtained, sample processing and test completion take about 10 minutes. In some clinical settings where culture is unavailable or immediate results are desired, examination of a saline wet mount combined with a point of care antigen test for negative wet preps may be a practical approach for diagnosing *T. vaginalis*.

In men, microscopic examination of urethral swabs or urine sediment are insensitive tests for diagnosing 'trich' infections and are not routinely performed. Culture is the test of choice for making a definitive diagnosis with a sensitivity of roughly 60-70%. To get maximum sensitivity both urethral swab and urine sediment cultures should be obtained. Semen may be the most sensitive clinical sample for culturing *Trichomonas vaginalis*, though it is not routinely used. In men, the development of urine sediment PCR assay for *T. vaginalis* holds the promise of being a more sensitive and convenient test. Currently urine sediment PCR testing is only available on an experimental basis or in clinical labs with 'home brew' PCR assays. Because of the difficulty inherent in diagnosing *T. vaginalis* in men with currently available tests, many men are treated empirically for *T. vaginalis* because a female partner has been diagnosed with *T. vaginalis* or because they have an NGU that did not respond to standard therapy. While there are no definite statistics available, asymptomatic infected men likely represent a major reservoir for ongoing transmission of *T. vaginalis* infections.

Treatment

Trichomonas vaginalis is susceptible to the 5-nitroimidazole antibiotics metronidazole and tinidazole. The standard therapies for both men and women are:

1. Metronidazole 2 g orally in a single dose, or
2. Metronidazole 500 mg orally twice a day for 7 days, or
3. Tinidazole 2 g orally in a single dose
4. A single 2 g dose of metronidazole is recommended for pregnant women.

There are no major advantages of tinidazole over metronidazole, with the latter being less costly than the former. Tinidazole, commonly used outside the U.S. and recently approved in the U.S., has a longer half-life and may be better tolerated. Patients and their partners are generally treated with a single 2 g dose of metronidazole or tinidazole. Patients should abstain from sexual activity until asymptomatic or, more specifically, for at least one week after completing antibiotic therapy. The seven day course of metronidazole therapy is generally reserved for initial treatment failures. Multiple treatment failures reflect either untreated ongoing sexual partners or 5-nitroimidazole resistance. 5-nitroimidazole resistance is estimated to be 2-5%.[5] Suspicious cases should be referred to an infectious diseases specialist or gynecologist with expertise and in the U.S. arrangements should be made to send cultures to the Centers for Disease Control and Prevention (CDC; Consultation is available via tel: 770-488-4115 or website: http://www.cdc.gov/std/). Therapy with metronidazole or tinidazole is generally well tolerated, but patients should be warned to avoid drinking alcohol for at least 48 hours after their last dose of either metronidazole or tinidazole to avoid severe 'hangovers' that result from antibiotic inhibition of alcohol metabolism. In addition, patients should be warned about a metallic taste in the back of their mouths especially those on a 7 day course of therapy. For a very small minority of patients nausea/vomiting is a significant side effect of the medications.

The possibility the *Trichomonas vaginalis* infections contribute to the spread of HIV combined with its high prevalence has altered "trich's" status from that of a mere 'nuisance' to that of an important sexually transmitted microbial pathogen.[6]

References

1. Schwebke JR, Burgess D. Trichomoniasis. Clin Microbiol Rev 2004; 17:794-803, table of contents.
2. Fouts AC, Kraus SJ. Trichomonas vaginalis: reevaluation of its clinical presentation and laboratory diagnosis. J Infect Dis 1980; 141:137-43.
3. Van Der Pol B, Williams JA, Orr DP et al. Prevalence, incidence, natural history and response to treatment of Trichomonas vaginalis infection among adolescent women. J Infect Dis 2005; 192:2039-44.
4. Krieger JN. Trichomoniasis in men: old issues and new data. Sex Transm Dis 1995; 22:83-96.
5. Cudmore SL, Delgaty KL, Hayward-McClelland SF et al. Treatment of infections caused by metronidazole-resistant Trichomonas vaginalis. Clin Microbiol Rev 2004; 17:783-93, table of contents.
6. Cohen MS, Hoffman IF, Royce RA et al. Reduction of concentration of HIV-1 in semen after treatment of urethritis: implications for prevention of sexual transmission of HIV-1. AIDSCAP Malawi Research Group. Lancet 1997; 349:1868-73.

CHAPTER 31

Pneumocystis Pneumonia

Allen B. Clarkson, Jr. and Salim Merali

Introduction

Pneumocystis is an opportunistic fungal pathogen causing *Pneumocystis* pneumonia (PcP) in mammals with compromised immunity. It was most often classified as a protozoan until, beginning in the late 1980s, genetic and other data showed conclusively that *Pneumocystis* is a fungus, albeit an unusual one. Because it appears similar in all mammals and produces similar pneumonias, all *Pneumocystis* was once classified as the single species, *P. carinii*. When genetic analyses in recent years showed wide divergence amongst *Pneumocystis* infecting different animals, the need for differentiation became clear. Initially the species *P. carinii* was divided into several *form specialis*, but that clumsy nomenclature has been superseded by a widely, but not yet universally, accepted classification with multiple species within the genus *Pneumocystis* with *P. jiroveci* for the species infecting humans, *P. carinii* for one of two species specific for rats and other names for *Pneumocystis* species infecting other mammals. Identification of genetically distinct strains within *P. jiroveci* has proven helpful for epidemiological studies. For practical reasons, the established acronym PCP (or PcP) remains in use, but now refers to *Pneumocystis* pneumonia rather than *P. carinii* pneumonia. Figure 31.1 shows *Pneumocystis* in the lung at the electron microscope level and Figure 31.2 at the light microscope level.

Populations at Risk

Pneumocystis was described in animal lungs in 1912, but was not known to be a human pathogen until 1951 when it was associated with interstitial plasma-cell pneumonia in malnourished institutionalized children. Subsequently, PCP was recognized as important complication for children being treated for acute lymphocytic leukemia and the introduction of cotrimoxazole to treat PCP comorbidity in 1975 greatly improved outcomes. PCP remains a threat for patients given drugs that suppress immunity. Without prophylaxis, the rate of PCP after organ transplant ranges from 5-25% and for cancer treatment from 1-25%. Drugs used for rheumatic disease can also increase susceptibility. However, advanced HIV disease accounts for the greatest number of cases. The rate of PCP associated with HIV infection dropped when prophylaxis was widely adopted in the early 1990s and dropped further in the mid 1990s with the introduction of HAART to suppress viral load. But for untreated or nonresponsive advanced HIV disease, the rate remains above 50%. Although the risk of HIV-associated PCP can be nearly eliminated by good compliance with a correctly chosen and tolerated chemoprophylactic regimen, PCP remains the most common opportunistic infection associated with AIDS.

Medical Parasitology, edited by Abhay R. Satoskar, Gary L. Simon, Peter J. Hotez and Moriya Tsuji.

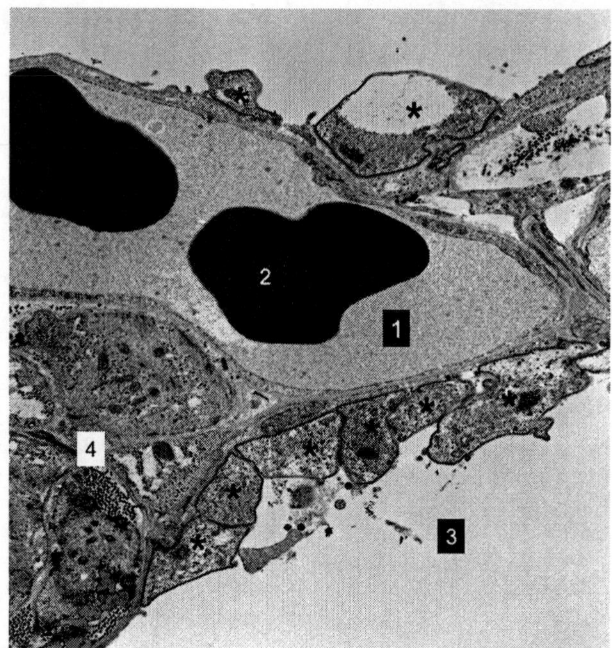

Figure 31.1. Transmission electron micrograph showing *Pneumocystis* trophozoites attached to Type 1 pneumocytes: These organisms reduce gas exchange, although the degree of involvement seen here is light. As the disease progresses and alveoli of lobules become completely filled with organisms, gas exchange becomes totally blocked. However, progression is slow with areas of high involvement becoming consolidated and fibrous tissue replacing portions of lung parenchyma. *) *Pneumocystis* organisms; 1) capillary lumen; 2) erythrocyte; 3) alveolar air space; 4) area of consolidation with cross-sectioned collagen fibers seen as small dots.

Recent anecdotal information suggests the overall rate has risen with the shift in HIV infection patterns to populations with a lower probability of seeking and less access to timely medical care. PCP was thought to be rare in the huge population of HIV-infected people in sub-Saharan Africa, but recent studies show a high prevalence there as well—especially in children.

Life Cycle and Transmission

Because *Pneumocystis* cannot yet be cultured well and genetic manipulation is not yet possible, it is difficult to study and much remains unknown. Two distinct life cycle forms are known: cyst and trophozoite. The mature cyst has a thick cell wall and eight intracystic bodies. Good evidence indicates that intracystic bodies are the product of meiosis, but details are lacking. Upon cyst rupture, intracystic bodies are released and become trophozoites which either grow and divide by binary fission or develop into precysts then cysts. Mature cysts have a characteristic indented-sphere

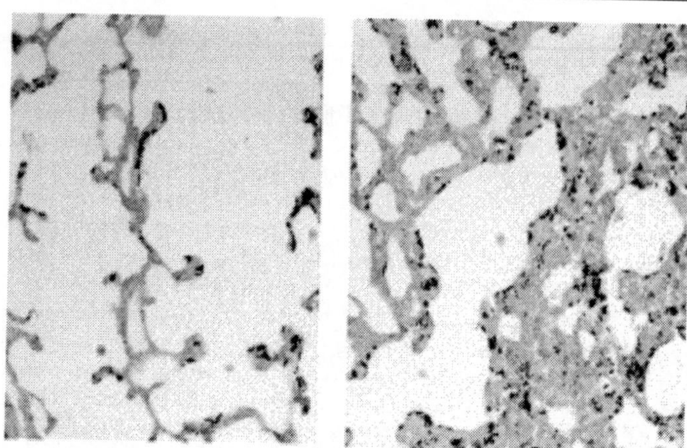

Figure 31.2. Lung sections silver-stained and counterstained with eosin: The dark bodies are cysts, but the far more numerous trophozoites are not selectively stained. The left panel shows a region with moderate pathology; septa are not yet markedly thickened, although gas exchange will have already been compromised by trophozoites attached to Type 1 pneumocytes. The right panel shows a region with greater involvement; septa are thickened and some regions show consolidation. As the disease progresses, gas exchange becomes more compromised eventually to the point of asphyxiation.

shape and a narrow size range, but trophozoites are highly variable in both size and shape. Transmission from animal to animal by air has been directly demonstrated and air sampling filters placed in hospital rooms of patients with PCP trap particulates containing more copies of *Pneumocystis*-specific nucleic acid sequences than filters from control areas clinical. The thick presumably environmentally resistant cyst wall suggests a role in transmission, but animal studies show both forms can transmit infection when instilled directly into lungs. Speculation exists regarding a cryptic life cycle stage outside mammalian hosts, but evidence is lacking.

A long-standing question has been whether PCP develops upon immunosuppression due to activation of nearly universal colonization or as the result of a new inoculation. If the former is true, patients with PCP present no risk to other immunosuppressed persons. However, accumulating data support the second hypothesis. These data include reports of clusters of PCP cases, analyses showing related groups of patients to have the same genetic strain and examinations of *Pneumocystis* taken from patients during multiple episodes of PCP that showed different genetic strains to be involved in each episode. The risk of those with active PCP present to other immunosuppressed persons is unclear, but several recent reports suggest it is significant.

Diagnosis

Presenting symptoms of PCP can include general fatigue, dry cough, night sweats, low grade fever, tachypnea and history of progressive dyspnea most noticeable on exertion. Auscultation is characteristically uninformative, but fine dry rales may be noted. Presentation can include pneumothorax, but this is rare. For those with advanced HIV disease, symptoms may progress slowly so the patient isn't aware of the developing pneumonia and attributes breathlessness upon exertion as general fatigue. Clinical data associated with PCP include chest X-ray showing a bilateral fine smooth "ground glass" opacity, positive gallium scan, hypoxia indicated by arterial oxygen saturation <85%, abnormal diffusing capacity and elevated serum lactate dehydrogenase. Although all these symptoms and data can all result from other causes and thus are not diagnostic, the presence of any requires PCP be considered, especially for persons with impaired immune function. Furthermore, since PCP may be the first indication of developing HIV disease, particularly for those unaware of their infection, it should be considered even in the absence of known immunosuppression. A presumptive diagnosis of PCP is sometimes given based on knowledge of immunosuppression in combination with a set of symptoms and clinical data; response to specific therapy is considered confirmation. However, the particular set of symptoms and laboratory data varies among physicians and hospitals and there is no accepted standard. Although presumptive diagnoses by experienced clinicians can be timely, accurate, cost effective and beneficial to patients, most authorities recommend the effort to obtain a definitive diagnosis by demonstrating presence of *Pneumocystis* using histologic, cytologic, or nucleic acid-based methods. For some years following discovery of HIV infection, diagnosis based on open lung biopsy was common, but this has been replaced by less invasive procedures. Bronchial alveolar lavage (BAL) is the current gold standard and is sometimes combined with transbronchial biopsy, depending on hospital policy. Figure 31.3 shows sediment from BAL fluid stained to reveal cysts. Examination of sputum induced by inhalation of an aerosol of hypertonic saline is also very effective. In some hospitals, diagnoses have been made using oral washings. Success of these progressively less invasive diagnostic procedures has depended upon progressive improvements in both detection methods and laboratory experience. Because *P. jiroveci* cysts have a thick fungal-type, glycan-containing cell wall, they are revealed by fungal cell wall stains such as methenamime silver, cresyl echt violet and toluidine blue as well as by selective uptake of fluorescent dyes such as calcofluor white. These are not specific for *Pneumocystis* so differentiation from other fungi depends on the characteristic size (~5 μM) and shape of cysts (individual indented spheres with a smooth surface). As a consequence, identification of individual cysts in sections is not as easy as in preparations such as smears that preserve the shape of the intact organism. However, sections reveal alveoli filled with "foamy exudate" characteristic of PCP and can be useful for diagnosis. Although trophozoite numbers exceed cysts by an order of magnitude, they can be less useful for diagnosis. They are not stained by fungal cell wall reagents. They do take up hematological stains such as Wrights and Giemsa, but host material and other pathogens are also stained by these. Trophozoites are highly variable in both size (0.5-8 μm) and shape (from round to almost stellate and from a smooth to a highly irregular surface) so that reliable identification requires training, skill and experience. Because intracystic bodies within cysts stain similarly to trophozoites, they are sometimes seen as a cluster of eight intracystic bodies. IFA kits licensed for PCP diagnosis are commercially available; these label both trophozoites and cysts selectively and strongly.

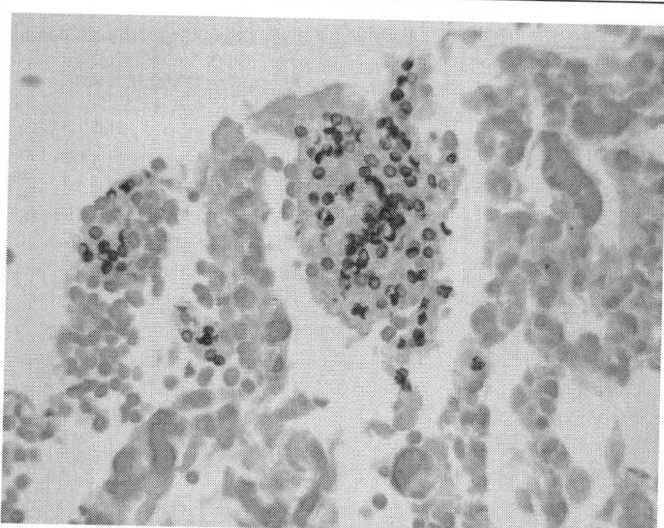

Figure 31.3. BAL sample silver-stained and counterstained with Fast Green: Due to their shape and dark gray staining, cysts are readily recognizable. When present in adequate numbers, the diagnosis can be made with confidence.

When used with BAL samples, they often reveal distinctive fluorescent clusters of trophozoites and cysts enmeshed in exudate that is also labeled. Although IFA provides high sensitivity and specificity, this diagnostic tool may not be available due to drawbacks of reagent cost, fluorescence microscope requirement and the time, skill and equipment needed for processing the specimens. PCR-based nucleic acid detection offers exquisite sensitivity and specificity, but suffers from lack of FDA-approved kits as well as the general problems of PCR-based diagnoses: costs of special reagents, special equipment, technical training, inclusion of extensive positive and negative controls and need for rigid procedures to prevent false positives due to contamination. Furthermore, the extreme sensitivity of PCR presents a problem in that colonization with small numbers of *P. jiroveci* can be detected even if they are not causing disease. A well-designed and fully developed quantitative-PCR protocol with a clear "initial target copy number" threshold could resolve many of these issues and encouraging clinical studies have been done; however, even if fully developed, general disadvantages of PCR diagnosis will remain. Reported success rates for the various diagnostic methods vary from study to study, partially due to the experience and expertise of physicians and laboratories, but trends are clear. Open lung or transbronchial biopsy specimens stained for cysts provide diagnostic sensitivity and specificity of 98+ and 100%, respectively, but both procedures are invasive and present risks to the patient. Sensitivities and specificities for BAL fluid samples stained for cysts are reported as high as 95 and 100%, respectively; IFA can improve sensitivity to 98% while retaining selectivity. Reports of efficacy using induced sputum samples have a much broader spread with sensitivity ranging from 30 to 90%; specificity remains high.

Finally, because those persons vulnerable to PCP can and often do have multiple respiratory infections, the presence of other pathogens and their possible contribution to disease must be considered. This is true even if all symptoms and data are consistent with PCP, *P. jiroveci* is positively identified in patient specimens and the disease responds to treatment.

Course of Disease and Pathology

PCP in infants from age 1 to 7 months of age, particularly from 2 to 4 months, develops more rapidly and becomes more severe than in older children or adults. Onset in infants can be insidious with the first signs being tachypnea and slight perioral cyanosis; coughing and fever may not be significant. The disease may progress over several weeks with increased tachypnea, dyspnea with sternal contraction and cyanosis. However, progression can be very rapid with death preceding diagnosis. There is no information on the effectiveness of maternally transferred immunity, but animal studies showing antibody protection suggests a role for maternal antibodies. Since antibodies specific for *Pneumocystis* commonly develop in early childhood without any history of PCP, the severity of PCP in immunosuppressed infants may be due to a combination of fading maternal protection and primary exposure. Interestingly, recent information indicates that even in otherwise healthy infants primary infections may not be completely silent. When nasopharyngeal samples from 105 infants (2 or 3 months old with respiratory infection symptoms) were tested for the presence of *P. jiroveci*, 48% were positive; the frequency was markedly lower for younger or older infants with similar symptoms. Examination of lungs from sudden infant death syndrome victims revealed a high proportion infected with *P. jiroveci*. However, the question remains open as to whether PCP contributed to death, whether the presence *P. jiroveci* reflects another underlying pathological condition, or whether this finding is irrelevant because *P. jiroveci* colonization is common in infants of that age and causes no significant pathology without immunosuppression. Progression of symptoms in older children and adults is slower, often over several months. Progression in HIV-infected children and adults is still slower than in those immunosuppressed from other causes. The best evidence suggests that, despite immunosuppression, disease is more the result of residual immune function than any direct damage by *P. jiroveci*. This may explain why, despite numbers of organisms being greater for PCP associated with HIV, symptoms develop more slowly and the probability of successful treatment is better than other causes of immunosuppression. Regardless the cause of immunosuppression, disease progression always involves increasing numbers of alveoli becoming filled with organisms and exudate which leads to impairment of air exchange and ultimately asphyxiation. While mechanical ventilation can be helpful in maintaining blood gases, that requirement is associated with poor prognosis. Postmortem examination reveals heavy, firm lungs that retain shape after removal from the thoracic cavity. H&E stained sections from biopsy and postmortem specimen show alveoli filled with an eosinophilic foamy or honeycomb exudate. When treatment is successful, there may be residual fibrosis and pulmonary deficit. While rare, extrapulmonary pneumocystosis does occur and infections have been reported at sites such brain, spleen, middle ear and peritoneum.

Treatment

The antifolate combination of trimethoprim (15 mg/kg/d) and sulfamethoxazole (75 mg/kg/d in 3-4 doses) (cotrimoxazole, TMP-SMZ) for 14-21 d is widely considered first line treatment for PCP. Advantages include efficacy equal or better than for all other treatments, favorable cost and availability and generally good tolerance. Drawbacks include gathering genetic evidence suggesting development of resistance as well as more frequent and more severe adverse drug reactions (ADRs) in patients immunosuppressed due to HIV infection (40-60% or higher depending on the study vs 8-10% for patients without HIV). However, clinical resistance has not been demonstrated to date and ADRs are not always of such severity to necessitate a change in therapy. Second line drugs include clindamycin with primaquine, other antifolate combinations such as trimethoprim or pyrimethamine combined with dapsone, pentamidine either by aerosol or i.v. and atovaquone as a single agent or combined with azithromycin. All treatments for PCP have associated ADRs and some interact negatively with drugs used to suppress HIV replication. While ADRs are frequently mild, they may be severe and changes in therapy are more often due to ADRs than lack of response. For TMP-SMZ and dapsone, the ADR spectra are similar, yet it is not uncommon for a patient to be intolerant of one and do well with the other. The most frequent ADRs for these drugs are fever, rash, nausea and vomiting, elevation of plasma liver enzymes and leukopenia. For TMP-SMZ the most severe ADR is debilitating Stevens-Johnson syndrome-like skin reaction that sometimes leads to extensive skin loss. TMP-SMZ and dapsone can also cause hemolytic anemia, especially for patients with G-6-PD deficiency. Occasionally other severe systemic reactions involve liver, kidney, bone marrow and heart. Clindamycin plus primaquine can cause rash, liver enzyme elevation, leukopenia, or anemia. Pentamidine can result in nephrotoxicity, hypoglycemia, pancreatitis and arrhythmias. Pentamidine delivery by aerosol has a lower ADR rate, but such delivery may not provide adequate dosage to upper portions of the lung and cannot treat rare extrapulmonary infection sites. Atovaquone when used as a single agent can cause rash, nausea, diarrhea or headache, but the rate of ADRs is lower than for other drugs; efficacy, however, is lower as well. Combining azithromycin with atovaquone improves efficacy, but the rate of ADRs also rises. Atovaquone can slow metabolism of zidovudine thus may significantly increase ADRs for that drug. Documentation of differences in efficacies or ADR rates for approved anti-*Pneumocystis* regimens is made difficult by the wide variation in clinical responses with any one drug, differences in study populations and differences in study designs. Furthermore, actual differences may be small. Should a change in treatment be necessary due to poor response or ADRs, any of the approved drugs can be substituted. However, one large study did find clindamycin/primaquine and atovaquone to be the most successful salvage therapies, regardless of the primary treatment. Trimetrexate with leucovorin to reduce toxicity has been used successfully for salvage, but suffers from a higher rate of relapse.

The evidence for spreading resistance to TMP-SMZ is emergence of mutations in *P. jiroveci* similar to those of other microbes that are known to confer resistance to the sulfamethoxazole component of TMP-SMZ. These mutations are found with higher frequency in populations where TMP-SMZ has been used extensively to treat and/or prevent PCP as well as in individual patients who have used TMP-SMZ for prophylaxis. The same relationship exists for emergence of mutations associated with resistance to atovaquone. However, direct evidence of clinical resistance is lacking and obtaining such evidence is problematic because response to treatment is variable and in vitro testing is not possible because the organism cannot be cultured from patient samples.

Response to treatment is often slow with improvement sometimes seen only after a week or more; lung function improvement often precedes changes detectable by radiology. Slow response can be problematic if there is any doubt of the diagnosis and, even if the diagnosis is confirmed cytologically, the possibility of one or more comorbidities arises. Several approaches to a response-to-treatment assay have been suggested and promising data have been collected. These include a plasma assay for glycan shed from cyst walls, a sputum assay using PCR to detect changes in *P. jiroveci* DNA or RNA content and a plasma assay for changes in host metabolic intermediates influenced by *P. jiroveci*. None, however, are available for clinical use. Controlled trials needed to determine optimal duration of therapy are lacking, but clinical experience has led to treatment usually lasting 14 to 21 days, even if response is relatively rapid. Adjuvant corticosteroid treatment for severe PCP was evaluated by controlled trials and the consensus recommendation of an expert panel in 1990 was they be used for treatment of severe PCP (arterial O_2 partial pressure <70 mm Hg or alveolar-arterial gradient >35 mm Hg, both on room air). Meta-analysis of data from randomized trials published through 2004 confirmed and quantified this benefit. The advantage of steroid treatment can be demonstrated by radiology and lung performance studies showing that anti-*Pneumocystis* drugs used alone cause an initial increase in lung opacity and a decline in lung function parameters. Animal model data show treatment exacerbates inflammation leading to a reduction in gas exchange efficiency. For moderate PCP the case is not as strong and there is no indication for steroids in treatment of mild disease.

Treatment Outcomes

Factors associated with poor outcomes include age, poor oxygenation on admission, elevated serum LDH, low hemoglobin, low serum albumin, presence of pulmonary co-pathogens, neutrophils in BAL fluid, delay in needed ICU admission after start of anti-PCP treatment, high APACHE II score and pneumothorax either at presentation or after initiation of mechanical ventilation. PCP is a serious disease but it is often now described as a "treatable disease", reflecting the perception that outcomes are generally better now than early in the HIV epidemic. Any improvements must be attributed to clinical skills derived from experience since, other than the introduction of adjuvant steroid treatment, the list of approved drugs has not changed for over a decade and efficacies from oldest to newest drugs are very similar. A sobering view is provided by two recent papers, one based on data from a major UK and the other from a major US medical center. Both studies included all patient records with laboratory-confirmed PCP diagnoses. The UK data covered the period from 1985 -2006 and included 494 patients; the US data from 1990 -2001 and 488 patients. Both reported reductions in PCP mortality after the introduction of HAART in 1996. UK data showed a 16.9% mortality rate for the period 1990 -1996 and 9.7% for the period 1996 -2006; however, the 1985 -1989 rate was 10.1% so real improvement is certain. US data showed a 47% mortality rate for the period 1990 -1995 and 37% for the period 1996 -2001; the higher US mortality likely reflects the "inner city" patient population of the study. Since these data are from major centers where there is considerable experience and expertise in treating PCP, broader mortality rates may be even higher.

Prophylaxis

The overall incidence of PCP associated with HIV infection dropped with wide acceptance of prophylaxis guidelines published in 1989 and dropped further upon introduction of HAART in the mid-1990s. Currently, discontinuation of prophylaxis is now recommended when HAART allows restoration of CD4 cells to >200 cell μL^{-1}, but risk of PCP is not a threshold phenomenon: an increase in CD4 count from 100 to 200 cells μL^{-1} is associated with a 10x reduction in risk of PCP, but there is a further 10X reduction in risk with an increase from 200 to 500 CD4 cells μL^{-1}. Drugs used for prophylaxis include all the drugs used for treatment. TMP-SMZ is the mainstay and the advantages/disadvantages are the same as for treatment; however, due to time factors, desensitization using a slowly increasing dosage plays a bigger role in prophylaxis. Aerosolized pentamidine has advantages of a low ADR rate and good protection but has drawbacks: upper lung regions are poorly treated and this can allow intense local involvement without symptoms so pneumothorax may be a result. Extrapulmonary sites are also not protected, but infection beyond the lungs is rare. Difficulties and costs associated with aerosol administration are also disadvantages. Dapsone used as a single prophylactic agent has efficacy and ADR profiles similar to aerosolized pentamidine, but without the upper lung and extrapulmonary site limitations. However, one of the few studies of prophylaxis failure reported the vast majority of probable failures to be associated with dapsone.

Suggested Reading

Biology of *Pneumocystis*

1. Wakefield AE. Pneumocystis carinii. British Medical Bulletin 2002; 61:175-88.
2. Stringer JR, Beard CB, Miller RF et al. A new name (Pneumocystis jiroveci) for pneumocystis from humans. Emerg Infect Dis 2002; 8:891-6.
3. Huang L, Morris A, Limper AH et al. ATS pneumocystis workshop participants. An official ATS workshop summary: recent advances and future directions in pneumocystis pneumonia (PCP). Proc Am Thorac Soc 2006; 3:655-64.

Course of Disease and Prognosis

1. Miller RF, Allen E, Copas A et al. Improved survival for HIV infected severe Pneumocystis jirovecii pneumonia independent of highly active antiretroviral. Thorax 2006; 61:716-72.
2. Festic E, Gajic O, Limper AL et al. Acute respiratory failure due to pneumocystis pneumonia in patients without human immunodeficiency virus infection: outcome and associated features. Chest 2005; 128:573-9.
3. Fujii T, Nakamura T, Iwamoto A. Pneumocystis pneumonia in patients with HIV infection: clinical manifestations, laboratory findings and radiological features. J Infect Chemother 2007; 13:1-7.
4. Walzer, PD, Evans HE, Copas AJ et al. Early predictors of mortality from Pneumocystis jirovecii pneumonia in HIV-infected patients: 1985-2006. Clin Infect Dis 2008; 46(4):625-33.
5. Tellez I, Barragán M, Franco-Paredes C et al. Pneumocystis jiroveci pneumonia in patients with AIDS in the inner city: a persistent and deadly opportunistic infection. Am J Med Sci 2008; 335(3):192-7.

Adjuvant Steroids for Treating Severe PCP and Insight into Beneficial Effect

1. Briel M, Boscacci R, Furrer H et al. Adjunctive corticosteroids for Pneumocystis jiroveci pneumonia in patients with HIV infection: a meta-analysis of randomised controlled trials. BMC Infect Dis 2005; 5:101-8.
2. Gigliotti F, Wright TW. Immunopathogenesis of Pneumocystis carinii pneumonia. Expert Rev Mol Med 2005; 7:1-16.

Prophylaxis

1. Rodriguez M, Fishman JA. Prevention of infection due to Pneumocystis spp. in human immunodeficiency virus-negative immunocompromised patients. Clin Microbiol Rev 2004; 17:770-82.
2. Podlekareva D, Mocroft A, Dragsted UB et al. EuroSIDA study group. Factors associated with the development of opportunistic infections in HIV-1-infected adults with high CD4+ cell counts: a EuroSIDA study. J Infect Dis 2006; 194:633-41.
3. Green H, Hay P, Dunn DT et al. STOPIT Investigators. A prospective multicentre study of discontinuing prophylaxis for opportunistic infections after effective antiretroviral therapy. HIV Med 2004; 5:278-83.

Current Recommendations for Treatment and Prophylaxis

1. Morbidity and Mortality Weekly Report (MMWR) http://www.cdc.gov/mmwr/
2. The Medical Letter on Drugs and Therapeutics http://www.medicalletter.org/ (fee-based access).

CHAPTER 32

Malaria

Moriya Tsuji and Kevin C. Kain

Introduction

Malaria parasites are members of the *Apicomplexa*, characterized by the presence of a unique organelle called an apical complex. There are four malaria species that infect and cause disease in humans, *Plasmodium falciparum*, *P. vivax*, *P. malariae* and *P. ovale*, although cases of human infection with *Plasmodium knowlesi*, a monkey malaria, have very recently been reported in Malaysia. Each *Plasmodium* sp. is associated with a specific cyclic fever, caused by the synchronous release of parasites from the erythrocytes in which they had been developing and multiplying. The different species may be distinguished by their morphological characteristics on Giemsa-stained thin blood smears. They are also distinct in their clinical course, including incubation period, pathophysiology and associated morbidity and mortality.

Life Cycle

Malaria infection may be acquired congenitally from mother to baby across the placenta, from platelet or blood transfusions and from the use of shared needles; however it is most frequently initiated with the bite of an infected, female *Anopheles* mosquito, which injects the sporozoite stage of the parasite with its bite. A typical life cycle of *Plasmodium* is shown in Figure 32.1. Usually 20-30 sporozoites are transmitted to the host by a single mosquito bite and some of the sporozoites rapidly reach the liver of the host via the blood circulation and thereby invade hepatocytes. Once inside the hepatocyte, the parasite undergoes asexual division and develops into liver schizonts within the infected hepatocytes over a period of approximately 1-2 weeks, depending on the species of *Plasmodium*. Because the initial stage of development within the liver occurs outside the bloodstream, this hepatic stage is generally referred to as the exo-erythrocytic stage. At the end of the hepatic stage of development, a single sporozoite can develop into a schizont that contains thousands of daughter parasites that fill the hepatocyte. Infected hepatocytes burst and release numerous merozoites into the bloodstream. *P. falciparum* can complete this liver stage within 7 days and each of its sporozoites produces about 40,000 daughter parasites. For *P. vivax*, these values are 6-8 days and 10,000 merozoites; for *P. malariae*, 12-16 days and 2000 merozoites; and for *P. ovale*, 9 days and 15,000 merozoites.

The next stage of development, called the erythrocytic or blood stage, is initiated when exo-erythrocytic merozoites from the liver invade red blood cells (RBCs). Merozoites of *P. falciparum* can infect RBCs of all ages, whereas those of *P. vivax* and *P. ovale* infect reticulocytes and those of *P. malariae* invade only older RBCs.

Medical Parasitology, edited by Abhay R. Satoskar, Gary L. Simon, Peter J. Hotez and Moriya Tsuji.

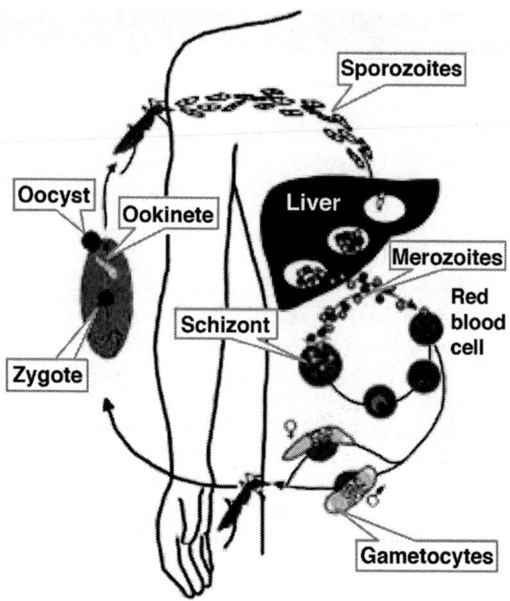

Figure 32.1. Life cycle of malaria parasite. Modified from an original, kindly provided by Drs. Chris Janse and Andy Waters at the Leiden University Medical Centre, The Netherlands.

Shortly after merozoites are released from hepatocytes, they invade RBCs and over a period of 2 or 3 days, develop asexually. The stages of asexual development include the ring (early trophozoite), trophozoite and schizont stages. The diagnosis of malaria can be made upon the identification of the parasites within erythrocytes on Giemsa-stained blood smears (see Table 32.1). The distinct appearance of these stages of development on Giemsa-stained thin blood smears, allows one to determine the specific species infecting the host.

The ring stage derives its name from its signet ring-like appearance, with a blue-stained nucleus and a pink-stained ring of cytoplasm. The trophozoite is the feeding stage of the parasite and contains a single nucleus with pigment granules, called hemozoin (a product of hemoglobin digestion), located within the cytoplasm of the parasite. The schizont stage is initiated by the division of the trophozoite nucleus. Further nuclear division leads to enlargement of the parasite. Each individual nucleus then becomes surrounded by parasite cytoplasm to form
32 a merozoite.

At maturation, the schizont bursts and releases merozoites into the blood circulation. Most of the released merozoites re-invade a new erythrocyte, thereby repeating their asexual life cycle (blood stage cycle). In some instances, however, invasion of an erythrocyte by a merozoite initiates sexual development instead of asexual development. Thus, merozoites may develop into male gametocytes (microgametocytes) or female gametocytes (macrogametocytes). These gametocytes

Table 32.1. Differential features of infected RBCs among various plasmodial species

	P. falciparum	*P. vivax*	*P. malariae*	*P. ovale*
Characteristics of infected red blood cell (RBC)	RBC not enlarged	Schüffner's dots in cytoplasm of RBC; RBC enlarged	RBC not enlarged	Schüffner's dots in cytoplasm of RBC; Compact trophozoite; RBC enlarged
Ring forms	Smaller rings; Multiple rings per cell; Double nuclei; Appliqué forms (Fig. 32.2)	Large rings	Medium sized rings	Large rings
Trophozoites	Seldom seen in peripheral blood	Parasite "active"; amoeboid shape with pseudopodia	"Band" forms	Compact
Mature schizonts	Seldom seen in peripheral blood	Schüffner's dots in cytoplasm of RBC; RBC enlarged	"Rosette" forms	Schüffner's dots; fewer merozoites per cell; RBC elongated
Gametocytes	"Crescent" shaped	Round; within enlarged RBC	Round; within non-enlarged RBC	Round; within enlarged RBC

can develop further only when they are taken up by an appropriate species of *Anopheles* mosquito during a blood meal. They subsequently mate within the gut of the mosquito, the definitive host. The parasites eventually become sporozoites, which reach the salivary gland of the mosquito. With the next bite, the infected mosquito releases sporozoites into the host, thereby completing the life cycle.

Pathology

Malaria is often classified as uncomplicated or complicated/severe. Uncomplicated malaria can be caused by all four species and is characterized by periodic fever and chills, mild anemia and splenomegaly. Uncomplicated malaria is rarely fatal unless it is left untreated and it progresses to severe disease. Severe or complicated malaria is almost exclusively caused by *P. falciparum* infections (although occasionally by *P. vivax* and other species) and is associated with higher parasite burdens and vital organ dysfunction including CNS (coma, seizures etc.) and pulmonary compromise (pulmonary edema, ARDS, respiratory distress etc.), acute renal failure, severe anemia and metabolic acidosis. Anemia arises in part from the destruction of erythrocytes when merozoites burst out of the infected RBC and RBC production is further compromised by bone marrow suppression or dyserythropoeisis. In falciparum malaria, anemia can be dramatic and life threatening. The rise in temperature is also correlated with the rupture of schizonts with release of pyrogens together with merozoites from the bursting infected RBCs. The pathogenesis of general malaise, myalgia and headache is ill-defined. The classic periodicity of the fever (*P. vivax/ovale* = every 2nd day; *P. malariae* = every 3rd day), based on synchronous infections, is often not observed particularly early in the course of infection. In the early phase of infection, the growth of the parasites is not synchronous, RBC rupture is more random and consequently fever can be erratic. In addition, some infections may be due to two or more broods of parasites, with the periodicity of one being independent of that of the others. This is more often seen in the case of severe falciparum malaria.

Most malaria deaths are associated with *P. falciparum* infections. RBCs infected with the maturing forms of this parasite express parasite proteins called PfEMP-1 associated with morphological structures ("knobs") that permit them to stick to endothelial cells lining the blood vessels and result in sequestration of these infected RBCs within the vascular bed of vital organs. When this occurs in the brain, the resulting cerebral malaria may lead to coma and death. Renal, pulmonary and GI complications may also be seen. Congenital malaria and infection of the placenta may result in stillbirth, low birth weight infants, or perinatal mortality.

After the initiation of blood stage infection by the parasite, the repeated infection of erythrocytes by merozoites results in exponential growth. As a result, the parasitized RBCs accumulate in the capillaries and sinusoids of blood vessels,
 causing general congestion in the peripheral blood circulation. The congestion causes organomegaly, notably splenomegaly and possibly hepatomegaly and contributes to anemia, leukopenia and thrombocytopenia. In vivax malaria, these processes occur rather acutely and the affected organs, particularly the spleen, become susceptible to rupture following trauma. In severe falciparum malaria, the kidneys may show punctate hemorrhages and tubular necrosis. Severe hemolysis and damage in the renal tubules results in hemoglobinuria or in its most severe

form "blackwater fever". In fact, the latter condition has been often associated with massive intravascular hemolysis in the context of prior treatment with quinine or treatment with primaquine in those with glucose-6 phosphate dehydrogenase deficiency (G6PD). Chronic *P. malariae* infection can be associated with nephrotic syndrome, a condition in which the kidney shows histological hypertrophy, caused by the deposition of immune complexes.

Epidemiology

Malaria is the most important parasitic disease in the world accounting for over 500 million clinical infections and 1 million deaths every year. Ecologic change, economic and political instability, combined with escalating malaria drug resistance, has led to a worldwide resurgence of this parasitic disease. However, malaria is not just a problem in the developing world. The combination of burgeoning international travel and increasing drug resistance has resulted in a growing number of travelers at risk of contracting malaria. It is estimated that as many as 30,000 travelers from industrialized countries contract malaria each year. However, this incidence is likely to be an underestimate because of the prevalence of underreporting. The majority of *P. falciparum* cases imported into North America and Europe are acquired in Africa (85%) and travel to the African continent is currently on the rise.

P. ovale infection can be distinguished from *P. vivax* infection in part by its epidemiology, i.e., the distribution of *P. ovale* is limited to tropical Africa and to discrete areas of the Western Pacific. Most West Africans are negative for the Duffy blood-type, which is shown to be associated with receptor sites for *P. vivax* merozoites on the RBC. Therefore, many West Africans are not susceptible to infection with *P. vivax*. Falciparum malaria is generally confined to tropical and subtropical regions, particularly in sub-Saharan Africa, the Amazon region of South America, rural forested areas of Southeast Asia and urban and rural areas of the Indian subcontinent. Individuals with sickle-cell trait (AS) are more resistant to severe falciparum malaria than normal homozygotes (AA). The SS individuals are also protected, but their sickle-cell disease leads to an early death. *P. malariae* has a wide, but spotty distribution throughout the world.

Clinical Manifestations

After being bitten by a malaria-infected *Anopheles* mosquito, the first symptoms appear after an incubation period ranging from 7 to 30 days. A shorter incubation period is most frequently observed with *P. falciparum,* whereas the incubation period for *P. malariae* can be quite lengthy. Typical symptoms include fever, chills, sweats, rigors, headache, nausea and vomiting, body aches and general malaise. These symptoms may be seen in all types of malaria and the malaria paroxysm is typically accompanied by sudden shaking chills. This may last 10 to 15 minutes or longer. During this stage, the patient complains of feeling extremely cold, despite a steady elevation of body temperature. Chills may be followed by severe frontal headache and myalgia (muscular pain) in the limbs and back. This stage lasts 2-6 hours in *P. vivax* and *P. ovale* infections, 6 hours or more in *P. malariae* infection and considerably longer in falciparum malaria. Finally, the patient starts to sweat profusely for several hours and usually begins to feel better until the onset of the next paroxysm. Fever occurs on alternate days with *P. vivax* and *P. ovale* and every 3 days with *P. malariae*. With falciparum malaria, fever may be asynchronous, recurring every 36 to 48 hours. *P. falciparum* is a deadly parasite, causing death as quickly as 36 hours from the onset of symptoms in non-immune individuals. *P. vivax* is a relatively benign parasite that elicits alternate day fever without causing mortality. *P. ovale* also produces alternate day fever and is clinically similar to vivax malaria.

Relapsing and recrudescent disease must also be differentiated. Relapsing disease implies the reappearance of parasitemia in sporozoite-induced infection, following adequate antiblood stage therapy. In the case of *P. vivax* and *P. ovale*, the development of exo-erythocytic forms allows the parasite to remain dormant within the hepatocyte. These dormant parasites are called hypnozoites. Accordingly, despite eradication of parasites from the peripheral circulation with conventional antimalarial drugs, a fresh wave of exo-erythrocytic merozoites can emerge from the hepatocytes and reinitiate the infection. The hypnozoites can remain quiescent in the liver for more than five years. In order to achieve radical cure, therefore, it is necessary to destroy not only the blood circulating parasites, but also the hypnozoites. *P. falciparum* and *P. malariae* do not develop hypnozoites and do not cause relapsing disease. Recrudescence is the recurrence of symptoms of malaria after a subclinical or asymptomatic level of parasitemia for a certain period of time. This recrudescence likely occurs in cases where the blood stages of malaria are maintained at very low levels after inadequate drug treatment. Such parasites may become drug resistant. All malaria species can cause recrudescence. In the case of *P. falciparum*, the parasites can recrudesce after one or two days, whereas *P. malariae* can do so for up to 30 years.

Diagnosis

For over a century, malaria diagnosis has relied on the microscopic detection of *Plasmodium* sp. on Giemsa-stained blood smears as no other reliable and relatively rapid method for the detection of infection and quantification of parasite burden has been available. A detailed study of malaria parasites can be accomplished with Giemsa stained thin blood smears (Fig. 32.2), whereas thick smears are more sensitive, because greater volume of blood can be examined. As shown in Table 32.1, there are a number of differential features of infected RBCs among the various species. It should be noted that a diagnosis often cannot be made on the basis of a single slide. In practice, several slides must be read systematically before a negative report may be given. It is noteworthy that *P. ovale*, the least common of the malaria species, resembles *P. vivax* in morphology and in biology.

Although microscopic detection of parasites has been the reference standard, its reliability highly depends on the technical expertise of the microscopist. The ability to maintain the required level of expertise in malaria diagnostics may be problematic especially in peripheral medical centers in countries where the disease is not endemic. Consequently recent efforts have focused on developing sensitive and specific nonmicroscopic malaria-diagnostic devices including those based on PCR or the detection of malaria antigen in whole blood. Many first

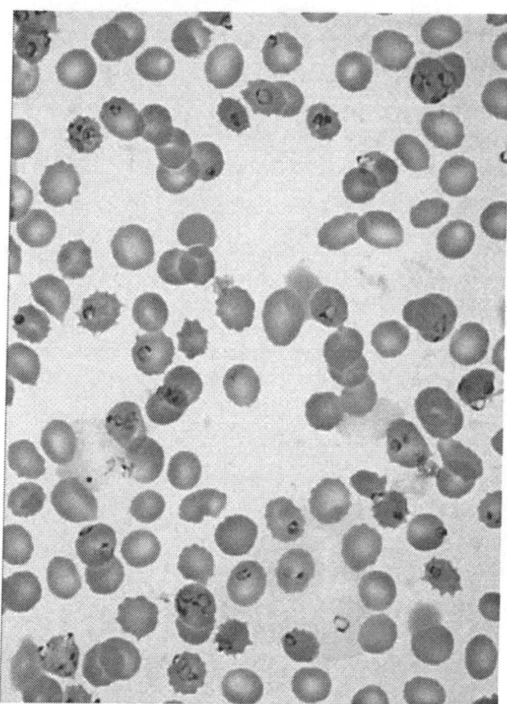

Figure 32.2. Thin blood smear showing *P. falciparum*-infected erythrocytes.

generation rapid diagnostic products relied on the detection of the histidine-rich protein II (HRP II) antigen of *P. falciparum* and therefore could not detect other *Plasmodium* species. The next generation of rapid diagnostic devices based on antigen capture with immunochromatographic (ICT) strip technology, utilize monoclonal antibodies to HRP II for the detection of *P. falciparum* as well as aldolase, a pan- *Plasmodium* antigen, thus facilitating identification of nonfalciparum infections. One such assay has now been licensed and is available for sale in the United States. Within the last decade, PCR-based diagnostic methods for malaria have been developed and surpass microscopic methods with respect to sensitivity and specificity. Currently reported amplification-based methods for malaria diagnosis, particularly nested PCR-based methods, are sensitive and specific but are also labour intensive with turnaround times that are generally too long for routine clinical application. Moreover these are open systems that require considerable pre- and post-sample handling and therefore special efforts need to be employed in order to prevent false positive assays. Real-time quantitative PCR technology has recently been developed and overcomes these limitations, offering a simple, time-effective and quantitative diagnostic option, if an appropriate laboratory setting is readily available.

Table 32.2. Summary of CDC guidelines for treatment of malaria in the United States

Clinical Diagnosis/ Plasmodial Species	Region Infection Acquired	Recommended Drug
Uncomplicated malaria/*P. falciparum* or Species unidentified (If subsequently diagnosed as *P. vivax* or *P. ovale*, then treatment with primaquine is required provide G6PD levels are normal)	Chloroquine-sensitive (Central America west of Panama Canal; Haiti; the Dominican Republic; and most of the Middle East)	Chloroquine phosphate
	Chloroquine-resistant or unknown resistance (All malarious regions except those specified as chloroquine-sensitive listed above)	A. Quinine sulfate plus Doxycycline, Tetracycline or Clindamycin B. Atovaquone-proguanil C. Mefloquine
Uncomplicated malaria/*P. malariae*	All regions	Chloroquine phosphate
Uncomplicated malaria/*P. vivax or P. ovale*	All regions	Chloroquine phosphate followed by Primaquine phosphate
Uncomplicated malaria/*P. vivax*	Chloroquine-resistant (Papua New Guinea and Indonesia)	A. Quinine sulfate plus Doxycycline or Tetracycline, followed by Primaquine phosphate B. Mefloquine followed by Primaquine phosphate
Uncomplicated malaria: alternatives for pregnant women	Chloroquine-sensitive	Chloroquine phosphate
	Chloroquine-resistant *P. falciparum* (regions except Chloroquine-sensitive regions listed above)	Quinine sulfate plus Clindamycin
	Chloroquine-resistant *P. vivax* (Papua New Guinea and Indonesia)	Quinine sulfate
Severe falciparum malaria	All regions	A. Parenteral quinidine or quinine plus Doxycycline, Tetracycline or Clindamycin B. Artesunate* followed by Atovaquone, Doxycycline or Mefloquine

*Investigational new drug available from CDC.

Treatment

Table 32.2 summarizes the CDC guidelines for treatment of malaria in the United States.

Key tips to stepwise approaches for malaria:

i. Fever in a returned traveler or immigrant from a malaria endemic area is considered malaria until proven otherwise.
ii. Malaria is a medical emergency and needs STAT diagnosis and treatment.
iii. Treat all falciparum malaria as drug-resistant unless you know without question that the infection was acquired in an area of chloroquine sensitivity.
iv. Decide if the case is falciparum or not and what the parasitemia is (percentage of RBCs infected with parasites on a blood smear).
v. Is it complicated malaria (i.e., any evidence of organ dysfunction etc.) or not.
vi. If complicated then you need to use parenteral quinine/quinidine including a loading dose (parenteral drug should also be used in any case for any species in which the patient can not tolerate oral therapy); alternatively, parenteral Artesunate is available through the CDC as an investigational new drug.
vii. If falciparum and the parasitemia is high (>10%) consider exchange transfusion and seek expert consultation.

Prevention and Control

There are four principles—adapted from the WHO's ABCD of malaria protection—of which all travelers to malarious areas should be informed:

A. Be **A**ware of the risk, the symptoms and understand that malaria is a *serious* infection.
B. Avoid mosquito **B**ites.
C. Take **C**hemoprophylaxis when appropriate.
D. Seek immediate **D**iagnosis and treatment if they develop fever during or after travel.

Protection against malaria can be summarized into the following four principles:

Assessing Individual Risk

Estimating a traveler's risk is based on a detailed travel itinerary and the specific risk behaviors of the traveler. The risk of acquiring malaria varies according to the geographic area visited (e.g., Southeast Asia versus Africa), the travel destination within different geographic areas (urban versus rural area), type of accommodations (well-screened or air conditioned versus camping), duration of stay (1 week business travel versus 3-month overland trek), season of travel (low versus high malaria transmission season) and elevation of destination (malaria transmission is rare above 2000 meters). In addition to the location, travelers can influence their own risk by how well they comply with preventive measures such as treated bed nets and chemoprophylactic drugs and the efficacy of these measures. Additional information can be obtained from good sources of updated malaria information and country-specific risk are available on line from the WHO and Centers for Disease Control and Prevention (CDC).

Preventing Mosquito Bites (Personal Protection Measures)

All travelers to malaria-endemic areas need to be instructed in how best to avoid bites from *Anopheles* mosquitoes that transmit malaria. Any measure that reduces exposure to the evening and night-time feeding female *Anopheles* mosquito will reduce the risk of acquiring malaria. Insecticide-impregnated bed nets (permethrin or similarly treated) are safe for children and pregnant women and are an effective prevention strategy that is underused by travelers. Insect repellents that contain DEET (N,N-diethyl-meta-toluamide) can also help you to avoid contracting malaria. Insect repellents should be applied sparingly, only to exposed skin or clothing, and from dusk to dawn—the time malaria mosquitoes bite the most. The use of repellents should be minimized in pregnant and nursing women.

Use of Chemoprophylactic Drugs where Appropriate

The use of antimalarial drugs and their potential adverse effects must be weighed against the risk of acquiring malaria (as described previously). The following questions should be addressed before prescribing any antimalarial:

a. Will the traveler be exposed to malaria?
b. Will the traveler be in a drug-resistant *P. falciparum* zone?
c. Will the traveler have prompt access to medical care (including blood smears prepared with sterile equipment and then properly interpreted) if symptoms of malaria were to occur?
d. Are there any contraindications to the use of a particular antimalarial drug?

If the traveler will be in chloroquine-sensitive areas, the drug of choice is chloroquine phosphate at 5 mg base/kg/wk beginning 1-2 weeks before travel until 4 weeks after leaving the malarious area. For the adult traveler heading to chloroquine-resistant areas, the primary choice of the prophylaxis is atovaquone/proquantil at 1 adult tab/d beginning 1-2 days before travel until 1 week after leaving, or mefloquine at 250 mg once/wk beginning 1-2 weeks before travel until 4 weeks after leaving. It is important to note that a number of travelers to low-risk areas, such as urban areas and tourist resorts of Southeast Asia, continue to be inappropriately prescribed antimalarial drugs that result in unnecessary adverse events but offer little protection. Improved traveler adherence with antimalarial drugs is more likely when travel medicine practitioners make a concerted effort to identify and carefully counsel the high-risk traveler and avoid unnecessary drugs in the low-risk individual.

Seeking Early Diagnosis and Treatment if Fever Develops during or after Travel

Travelers should be informed that although personal protection measures and antimalarials can markedly decrease the risk of contracting malaria, these interventions do not guarantee complete protection. Symptoms resulting from malaria may occur as early as 1 week after first exposure and as late as several years after leaving a malaria zone whether or not chemoprophylaxis has been used. Most travelers who acquire falciparum malaria will develop symptoms within 2 months of exposure. Falciparum malaria can be effectively treated early in its course, but delays in therapy may result in a serious and even fatal outcome. The most important factors that determine outcome are early diagnosis and appropriate therapy. Travelers and health care providers alike must consider and urgently rule out malaria in any febrile illness that occurs during or after travel to a malaria-endemic area.

In addition to travelers from endemic areas, an important source of malaria infection includes immigrants from endemic areas. Also, it is noteworthy that a person who has never travelled to endemic areas could get malaria by the bites of malaria-infected mosquitoes carried in airplanes from endemic areas. This is called "Airport (Baggage) malaria."

Both adequate mosquito control and an effective malaria vaccine remain unavailable, though promising work is proceeding towards development of these pivotal preventive strategies.

Suggested Reading

1. Centers for Disease Control and Prevention: Health information for international travel, Washington DC, 2005-2006, Government Printing Office.
2. Katz, Despommier, Gwadz. Parasitic Diseases, Second Edition. Springer-Verlag 1988.
3. Fiona E, Lovegrove, Kevin C. Kain. Malaria prevention. In: Jong E, ed. The Travel and Tropical Medicine Manual. 4th Edition. Philadelphia: WB Saunders, 2007: Chapter 5.
4. Neva FA, Brown HW. Basic Clinical Parasitology, 6th Edition. Norwalk: Appleton & Lange, 1994.
5. World Health Organization: International travel and health 2006. Geneva Switzerland, 2006.
6. Nappi AJ, Vass E, eds. Parasites of Medical Importance. Austin: Landes Bioscience, 2002:29.

Section V

Arthropods

CHAPTER 33

Clinically Relevant Arthropods

Sam R. Telford III

Introduction

Arthropods are animals with segmented bodies covered by a chitin exoskeleton. Familiar arthropods include flies, fleas, bees, spiders, crabs and shrimp. Medically important arthropods are typically thought of as vectors of infectious agents or as infestations causing direct injury. Arthropods may also actively or passively defend themselves against crushing or swatting by biting, stinging, piercing, or secreting noxious chemicals; or by means of urticarial hairs. Arthropods may also be medically important due to a patient's fear of insects, delusional parasitosis, or allergy due to dust mites. The diversity of ways in which arthropods may affect human health reflects the great diversity of these animals. More than 80% of all known animal species are arthropods. Given their presence in all environments, pathology may also sometimes be erroneously attributed to their presence.

Arthropods may transmit infectious agents including viruses, bacteria, protozoa and helminths. Such agents may require an arthropod for perpetuation (biological transmission) or may simply contaminate an arthropod (mechanical transmission). Leishmania parasites have a complex developmental cycle within certain phlebotomine sandflies and could not move from animal to animal without the sandfly. Houseflies may have salmonella contaminating their outer surfaces or mouthparts, which may be transferred to animals by the act of landing and crawling; but, salmonella does not need the housefly because it is an enteric commensal of many kinds of animals. Few agents can survive long enough to be transmitted by mechanical means: HIV, for example, has never been epidemiologically linked with mosquitoes or other arthropods even though it might be taken up during the act of bloodfeeding, probably because the virus is labile outside of host tissues and enough infected lymphocytes may never contaminate the small surface area of mosquito mouthparts.

There are five major groups of vectors, the diptera (flies and mosquitoes); hemiptera (kissing bugs); siphonaptera (fleas); anoplura (lice); and acarines (ticks and mites). In addition, several other groups of arthropods may be medically important. The general life history strategies and clinical significance for each group are briefly reviewed; specific vector-pathogen relationships are discussed in detail in chapters focusing on the respective agents.

Medical Parasitology, edited by Abhay R. Satoskar, Gary L. Simon, Peter J. Hotez and Moriya Tsuji.

Hexapoda (Insects)

Diptera

The dipteran vectors are winged insects that include mosquitoes, sandflies, blackflies, gnats, horse/deerflies and tsetse flies. These range in size from minute (ceratopogonid midges less than 2 mm in length) to large (horseflies more than 2 cm in length). Unlike other winged insects, dipterans have only one pair of wings. Those that take bloodmeals as adult females may serve as vectors or pests. Bloodmeals are used as nutrient to produce eggs; once those eggs are laid, another bloodmeal may be taken and more eggs produced. Thus, unless a mosquito (as an example of a dipteran) inherits infection (transovarial or vertical transmission), the first bloodmeal infects it and the second allows the agent to be transmitted; under favorable environmental circumstances, a mosquito may survive for several weeks and take more than two bloodmeals. Eggs, larvae and pupae would never be associated with illness; the only medical relevance is that these stages are often targeted when trying to control infestations.

Host-seeking cues for the bloodfeeding diptera include body heat, carbon dioxide (exhaled by the host), mechanical vibrations and lactic acid or other skin associated compounds. Repellants such as DEET (diethyltoluamide) work by disrupting the airborne gradients of chemical cues that allow diptera to home in to their hosts. Once a host has been identified, feeding is initiated and completed within minutes. Sandflies, blackflies and mosquitoes have a diverse salivary armamentarium of pharmacologically active substances that promote finding and removing blood. People may develop immediate and delayed-type hypersensitivity to dipteran bites, manifesting as erythema or welts accompanied by itch. Itch may be severe enough to cause scratching and secondary infection. The transient nature of infestation by bloodfeeding diptera means that few specimens are submitted to clinical laboratories. Other than analyses related to confirming the diagnosis of a vector borne infection, dipterans would come to the attention of physicians mainly for issues related to hypersensitivity, or for myiasis.

The muscoid diptera include the muscids (houseflies, stable flies); the Calliphoridae (blowflies); and the Sarcophagidae (the flesh flies). Egg deposition and larval development occurs in characteristic materials, viz., fecal material for houseflies, decaying plant material for stable flies, live flesh for blowflies and carrion for the flesh flies. Houseflies have received much attention for their potential health burden because of their association with poor hygiene. A large fleshy structure at the apex of the proboscis provides a surface for contamination, as does the hairy body and legs of the fly. In addition, flies may regurgitate while feeding and the vomitus may contain organisms that were acquired in a previous landing. Houseflies commonly feed on human excrement and virtually every possible enteric pathogen (those causing amebiasis, cholera, typhoid, hepatitis A, poliomyelitis; even roundworms and *Helicobacter pylori*) has been detected within or upon them. With few exceptions, such findings are epidemiologically irrelevant inasmuch as all of the agents perpetuate in their absence. It is likely that individual cases of enteric disease may derive from fly contamination of food, but whether the risk of such an event merits worry by patients or their health care providers remains unclear.

Myiasis is the infestation of human or animal tissue by fly larvae, deposited as eggs or first stage larvae; the larvae develop by feeding on the surrounding tissue, emerge as third stage larvae and pupate in the environment. A variety of clinical presentations are evident depending on the site where the larvae are present. Botflies cause furuncular lesions or migratory integumomyiasis (a serpiginous track may be produced in the skin). Wound myiasis comprises shallow or pocketlike initial lesions that become more deeply invasive. Maggots may invade the nose and accompanying structures, causing nasal or oral myiasis. Maggots may get into the ears, producing aural myiasis. Ophthalmomyiasis is due to external or internal infestation. Enteric, vaginal, or vesicomyiasis is due to invasion of the gut or genitalia. In all presentations, pathology may be due to tissue trauma or local destruction, but is more often associated with secondary bacterial infection. On the other hand, many maggots do not promote bacterial infection, but rather secrete bacteriolytic compounds and have been used as a surgical intervention to debride wounds. Thus, the development of secondary bacterial infection in myiasis may suggest the death of the maggot. In all cases, treatment is by removal of the maggots, laboriously picking them out using forceps. The maggots causing furuncular myiasis or migratory integumomyiasis will have their posterior end visible within the lesion, exposing the spiracular plates that cover their tracheolar breathing apparatus. Although much lore exists on how to best remove such maggots, including "luring" the maggots with bacon or pork fat, the simplest method is to cover the lesion with vaseline, thereby preventing the maggot from breathing. It will eventually move out enough in an attempt to get air so that it may be grasped with forceps.

Hemiptera

The only hemiptera that serve as vectors are the kissing or reduviid bugs, minute to large insects with compound eyes, antennae, sucking mouthparts, two pairs of wings (one delicate pair hidden under an outer pair of more robust ones) and a segmented abdomen. All are easily seen without magnification, adult bedbugs being slightly less than 1 cm in length and adult triatomines ranging in size from 1 cm to more than 5 cm. Reduviid bugs are cryptic, living within cracks of mud walls or other narrow, confined spaces. They serve as vectors of trypanosomes (*Trypanosoma cruzi*, the agent of Chagas disease) in Latin America. Transmission of *T. cruzi* to humans requires contamination of the site of the bite or mucosa by trypanosomes that are excreted in the bug feces. As with mosquitoes, salivary products from reduviids contain a variety of pharmacologically active compounds. Unlike those of mosquitoes, repeated reduviid feedings may cause a dangerous anaphylactoid reaction in residents of houses where the bugs are common. The resulting itching and irritation promote the entry of bug feces directly into the bite site or indirectly by contamination of mucosa. Although it is possible that a true reduviid bug may be presented by a patient for identification in clinical settings outside of Latin America, it is more likely that such specimens are related heteropterans such as assassin bugs, which are insect predators, or the plant-feeding stink bugs, chinch bugs, harlequin bugs, or squash bugs.

Bedbugs (*Cimex lectularius, C. hemipterus*) are hemipterans, but are not known to serve as vectors for any pathogen. These bugs with short broad heads, oval bodies and 4-jointed antennae undergo incomplete metamorphosis, with four nymphal stages, each taking a bloodmeal in order to develop. They are cryptic and require hiding places such as cracks in walls, mattress foundations, or rattan furniture. At night, bedbugs will emerge and infest sleeping people, taking 10-20 minutes for engorgement. Feeding is often interrupted by the movements of people during their sleep and so multiple bites may result from a single bedbug. Erythematous, itching bites, often several in number, are typically observed as a result; treatment is symptomatic. Bedbug infestations are currently increasing in prevalence, in all socioeconomic levels. Elimination of these pests can be difficult, particularly in multi-family dwellings. Mattresses and box springs must be decontaminated or destroyed; baseboards of walls must be sprayed.

Blattaria

Roaches are dorsoventrally flattened, smooth-bodied, winged insects that resemble the hemiptera but are closely related to termites and mole crickets, with long antennae, biting-type mouthparts and abdominal projections. The outer pair of wings is thick and leathery and the inner pair membranous. Roaches may fly, but usually scuttle about on long spiny legs. Roaches can live for months without food, but water seems critical. They are omnivorous, feeding on the finest of foods to the vilest of waste, usually at night. Secretions deposited by scent glands (including trail and aggregation pheromones) give rise to a characteristic disagreeable odor that confirms an infestation even when live roaches cannot be found. Common roaches range in size from the small German cockroach (*Blatella germanica*) about half an inch in length, to the American cockroach, nearly two inches. As with flies, the presence of roaches suggests poor environmental hygiene, but only rare instances of enteric disease might be associated with them. Dense infestations may produce large amounts of antigenic material (feces, molted cuticle, or parts from carcasses), which may be implicated in allergic or asthmatic reactions. Control of roach infestations can be difficult. Boric acid, deposited along walls, behind moldings and around other sites where they may hide, is effective in killing adults and nymphs by abrading cuticle between abdominal segments, rendering the roach prone to dessication. Removing standing water (wiping up and getting rid of clutter around sinks) can also reduce infestations by preventing access to water.

Siphonaptera

The fleas are bilaterally compressed, heavily chitinized insects with greatly modified hind legs for their characteristic jumping mode of locomotion; they lack wings. Fleas are generally small, no larger than 5 mm in length. The female flea requires blood for egg production. Most fleas are what are known as "nest parasites", particularly of rodents. Individual fleas may live as long as a year but usually only for a couple of months, laying eggs daily. Only a few species are of medical importance. These include the "human" flea (*Pulex irritans*); the dog and cat fleas (*Ctenocephalides canis* and *C. felis*); the main plague vector (*Xenopsylla cheopis*) and the sticktight flea (*Echidnophaga gallinacea*). Of these, *C. felis* are the most notorious pests, feeding voraciously and rapidly developing dense infestations. Chronic infestation of homes are largely due to the presence of a cat or dog (despite its name, this flea will feed on either animal), their bedding, wall-to-wall carpeting and relatively great humidity within the home. Cold climates rarely have self sustaining infestations because winter heating tends to dry out the carpets and molding and other places where the larval fleas tend to be hidden. Although bites sustained over several weeks will usually induce a typical delayed type hypersensitivity reaction, with an intensely itching red spot developing at the site of the bite (usually around the ankles), note that not all members of a household will react in the same manner.

An unusual flea, the chigoe or jigger (*Tunga penetrans*) will attach to a host and maintain a feeding site. Originally found in Latin America, the chigoe has been carried across to sub-Saharan Africa by humans and may be found anywhere there. This flea will often penetrate under toenails, or burrow into skin between toes, or in the soles of feet, an infestation known as tungiasis. When the flea dies, the lesion becomes secondarily infected, causing great irritation and pain. Tourists will often become infested by walking barefoot in shady spots around beaches.

Phthiraptera

Infestation by lice is called pediculosis. The lice are wingless, flat (dorsoventrally), elongate, small (0.4-10 mm) insects that are generally characterized by strong host specificity. Lice undergo incomplete (hemimetabolous) development, with nymphal forms resembling adult lice and are often found concurrently with the adults. Thus, size may appear to greatly vary within a single collection of specimens from one host. All lice are delicate and very sensitive to temperature and humidity requirements; all die within days without the host. The lice of greatest clinical importance are the head louse (*Pediculus humanus capitis*); body louse (*Pediculus humanus corporis*); and the pubic louse (*Phthiris pubis*). All of these sucking lice have prominent claws attached to each of their legs, morphologically adapted for grasping the hairs of their host. They feed at least daily and deposit 1-10 eggs from each bloodmeal, gluing one egg at a time onto the shafts of hairs (or in the case of the body louse, onto threads within clothing). The eggs, or nits, are almost cylindrical in shape and have an anterior operculum. Lice are transferred between hosts by close physical contact, or by sharing clothing in the case of body lice.

Body or pubic louse infestation may result in intense irritation for several days, with each bite generating a red papule. Chronically infested individuals may become desensitized, or may develop a nonspecific febrile illness with lymphadenopathy, edema and arthropathy (although such signs and symptoms should prompt a search for the agent of trench fever). Although body louse or pubic louse infestation may be considered evidence of poor hygiene or poor judgement, infestation by head lice should not be a stigma, occurring in the best of families. Nor should head louse infestation be considered to be a public health menace, or even a clinical problem. Very few infestations are dense enough to cause signs or symptoms and head lice are not vectors. Head or pubic louse infestations may be easily treated by shaving the head or pubes; body lice, by changing clothing. Several permethrin-based shampoos have great efficacy, although resistance has been reported.

33

Hymenoptera

Wasps, bees and ants all belong to the order Hymenoptera, which are minute to medium-sized insects with compound eyes, mandibles and two pairs of transparent wings (although in the ants, wings may be seen on males or females only at certain times of the year). These insects have a complex social behavior, with males, females and worker castes. Workers have a stinging organ, which is used for defending the colony or capturing prey. Most stings by hymenoptera cause localized reactions, sometimes with extreme pain and resulting in a transient induration with hyperemia. By virtue of their living in large colonies, bees and ants may swarm an intruder and dozens if not hundreds of stings may be sustained. Airway obstruction may result should multiple stings be received on the face or neck. Bees differ from wasps and ants in that their stinging apparatus is forcibly torn out during the act of stinging, thereby ensuring the death of individual bee as a result. Wasps and ants may sting multiple times; some fire ants may hang on by their mandibles and repeatedly insert their posteriorly located stinger. Bee venom has been well characterized and consists of a large amount of a polypeptide (melittin), phospholipase A2, histamine, hyaluronidase, mast cell discharging peptide and apamin. Histamine appears to be the cause of the acute pain of bee stings. The most important clinical manifestation of bee or wasp stings is anaphylaxis. Chest tightness, nausea, vertigo, cyanosis and urticaria may be seen even in individuals who apparently had never previously been exposed. Dozens of people die each year in the US due to bee sting anaphylaxis. Ants, on the other hand, rarely pose a risk for anaphylaxis but produce a reaction that may persist for a longer duration. An induration or wheal may be observed immediately after the sting and a papule may develop that itches or remains irritated for several days.

Lepidoptera

Caterpillars with urticating hairs are associated with rashes caused by contact with hairs (erucism, or erucic rash). A common shade tree pest, the brown tailed moth (*Nygmia phaeorrhoea*), in Europe and northeastern North America, liberates tiny barbed hairs when the caterpillar molts. These hairs are blown about by the wind and when skin is exposed, a severe dermatitis results; ingestion or inhalation can also cause significant irritation of the mucosa or bronchospasm. Contact with the eye may induce conjunctivitis. Dermatitis produced by urticating hairs typically comprises itchy, erythematous patches associated with small vesicles and edema. These lesions are only where the hairs have free access to the skin or where contact is made. Use of a masking tape-type lint roller can be very effective in removing urticating hairs, which may or may not be visible.

Beetles

Some beetles may induce vesication: direct contact with live or crushed beetles may expose a person to cantharidin. Bombardier beetles will spray a boiling hot jet of benzoquinone as a defense, causing burns and blistering. Blistering is also produced by crushing the staphylinid beetle *Paederus fusca* of Southeast Asia, which contain a toxic alkaloid, pederin. More commonly, carpet beetles (dermestids) which feed on wool rugs and other animal fur products are associated with a papulovesicular eruption. In particular, larval dermestids and their hairs or shed skins (exuviae) cause a contact dermatitis.

Arachnida

The acarines are a subclass within the Class Arachnida, which also contains the spiders and scorpions. Acarina comprise the mites and ticks, tiny to small arthropods with eight legs as nymphs and adults (as opposed to six for insects) and with fused main body segments as opposed to three discrete ones for insects.

Rickettsialpox and scrub typhus are currently the only known infections transmitted by mites. Mites are, however, important for their pest potential, causing itch, dermatitis and allergy. The most common of the ectoparasite-caused direct injury is scabies, caused by infestation with the human scabies mite (*Sarcoptes scabiei*). Infestation may occur anywhere in the world. Canine sarcoptic mange is commonly associated with scabies infestations in the owners of the dogs. In either animal or human scabies, transmission is by direct personal contact and infestations often cluster among groups of people, particularly families. There is little evidence that environments become contaminated; fomites have not been identified. The female scabies mite burrows beneath the stratum corneum, leaving behind eggs and highly antigenic feces within a track like trail. Nocturnal itching begins within a month of the first infestation, but may begin within a day in previously exposed individuals. Erythematous papules and vesicles first appear on the webs of fingers and spread to the arms, trunk and buttocks. Interestingly, in individuals who are immunocompromised, hundreds of female mites may be found, itching is minimal, but a hyperkeratosis is prominent. Such "crusted" or Norwegian scabies are highly infectious to other people.

Scabies infestations can be easily diagnosed by scraping a newly developed papule (not one that has been scratched) with a scalpel coated with mineral oil. The scrapings in oil may be transferred to a slide and examined at X100 or X400 brightfield for 300-400 μm mites or the smaller black fecal granules. Scabies may be treated by 5% permethrin cream or 1% permethrin rinse (Nix, same as is used for head lice). The method of choice until recently was topical lindane (Kwell), but FDA now suggests the use of permethrin first and lindane only if that fails. Lindane can be neurotoxic to children and small adults.

Demodex folliculorum, the follicle mite, infests virtually everybody. This elongate 500 μm long mite may be found within follicles on the face, the ear and breast and are often brought to the attention of a physician because a tweezed eyelash may have a few mites at the base of the hair shaft. Although they appear to largely be nonpathogenic, the same species causes demodectic mange in dogs and cattle. Pityriasis folliculorum, with small pustules appearing on the forehead, has been attributed to them.

Dust mite allergies (one of many causes of asthma, rhinitis and atopic dermatitis) are due to inhalation of feces excreted by *Dermatophagoides farinae* or related pyroglyphid mites, which are human commensals that feed on flakes of skin shed from a person. The mites themselves do not infest a person, but remain in the environment (usually within bedding or carpets) to feed and develop. About a half gram of skin may be shed from a person each day; one female mite lays one egg a day, for about 2 months; thus, large accumulations of mites may readily develop. The 300-400 μm long mites may be presented by patients as suspects for other nonspecific lesions or sets of signs and symptoms because they may be found in virtually all houses and can be detected if dust is allowed to settle on standing water; the mites float and will move and can be seen at X20. Humidity less than 60% will greatly reduce dust mite infestations, as will periodic vacuuming and washing of bedding and carpets.

Ticks are prolific vectors, with more recognized transmitted agents than any arthropod other than mosquitoes. Hard ticks are so named because of the hardened dorsal shield or scutum. In female hard ticks, the scutum is on the anterior third of the body with the remainder consisting of pleated, leathery cuticle that allows for tremendous expansion during bloodfeeding. In male hard ticks, which may or may not feed at all, the scutum extends the length of the body. In contrast, soft ticks have no scutum; their entire body is leathery. The "head", or capitulum, consists of the holdfast (hypostome), chelicerae (which are homologs of insect mandibles) and the palps, which cover the mouthparts (hypostome and chelicerae). Chelicerae act as cutting organs, the two sides sliding past each other, with the cutting teeth at the end gaining a purchase into a host's skin. The hypostome is thereby inserted and allows anchoring of the entire tick due to recurved, backward facing teeth or denticles. Many hard ticks also secrete a cement around the hypostome. Often, when removing an attached tick from a host, a large piece of skin comes with the hypostome, mostly cement and surrounding epidermis. Thus, by virtue of the cement and denticles, tick hypostomes rarely emerge intact when a tick is removed. "Leaving the head in" is not critical and most times the remnant will be walled off as a foreign body, or will work itself out, perhaps by the act of scratching. Treatment, therefore, should simply be disinfection of the site of the bite and certainly not excavation of the epidermis looking for the "head". Soft ticks are transient feeders and will only rarely be found attached.

Hard ticks require several days to complete their bloodmeal; the number of days depends on the species and stage of the tick. (In contrast, soft ticks are more like mosquitoes in their feeding, spending tens of minutes to no more than a few hours feeding, usually as their host is sleeping.) North American deer ticks (*Ixodes dammini*) will feed 3 days as a larva, 4 days as a nymph and 7 days as the female. During the first 70% of the feeding process, very little blood or lymph appears to be present within ticks, which remain dorsoventrally flat. Hemoglobin is excreted from the anus, lipids are retained and water from the blood is recycled back into the host as saliva. In the last day, usually in the last 3 or 4 hours of the bloodmeal, the tick takes what has been termed "the big sip", removing a large volume of whole blood, then detaching and dropping from the host. Because they must remain attached for days, hard ticks have evolved means of temporarily disabling a host's local inflammatory response, which might inhibit its feeding. Hard tick saliva is an extremely complex mixture of anticoagulant, anti-inflammatory and antihemostatic agents that act mainly at the site of the feeding lesion. Hosts that have never been exposed to ticks will not realize that a tick is attached. Indeed, most cases of Lyme disease or spotted fever never knew that they had been "bitten". The most dangerous tick is not necessarily the one that a patient finds, removes and aborts the transmission process, but the one that he or she never knew was there and which was able to complete its feeding. Because most tick-borne pathogens require at least 24 hours to attain infectivity, prompt removal of an attached tick will usually abort infection.

An unusual toxicosis due to tick bite is tick paralysis. The presence of certain feeding ticks (Australian *Ixodes holocyclus*, Western American *Dermacentor andersoni*) induces an acute ascending paralysis. Children are the usual victims of tick paralysis, with ticks attached at the nape of the neck. The illness is characterized by fatigue, irritability, distal paresthesias, leg weakness with reduced tendon reflexes, ataxia and lethargy. Unless the tick is removed, quadriplegia and respiratory failure may result; the case fatality rate without treatment can be 10%. Removal of the tick induces a miraculous recovery within 48 hours.

Scorpions

Scorpions are arachnids with a characteristic crablike appearance. The segmented body ends in a segmented tail, which terminates in a prominent stinging apparatus. The four pairs of legs includes well defined pincers on the first pair of legs. Scorpions may range in size from 2-10 cm. Scorpions are problems mainly in the warmer climates. The bulbous end of the tail contains muscles that force venom through the stinger. Humans are stung by walking barefoot at night; by not shaking their shoes out in the morning in an endemic area; lifting rocks or logs; or in bedding that is on the floor. The stings cause local pain (probably due to the great biogenic amine content of the venom), edema, discoloration and hypesthesia. Systemic signs can include shock, salivation, confusion or anxiety, nausea, tachycardia and tetany. Venom characteristics differ depending on the genus of scorpion: some stimulate parasympathetic nerves and can lead to secondary stimulation of catecholamines resulting in sympathetic stimulation, which in turn may contribute to respiratory failure. Others affect the central nervous system, are hemolytic, or cause local necrosis. Treatment is usually symptomatic.

Spiders

Even though all spiders have stout chelicerae and can bite and all spiders have a venom with which to immobilize their prey, most are too small to be noticed by humans even if they were to be bitten. The clinical manifestations and complications of envenomenation may differ between the four main kinds of medically important spiders and the syndromes caused by each have been given names that reflect the identity of the spider. With spider bites in general there is local pain and erythema at the site of the bite and this may be accompanied by fever, chills, nausea and joint pains. In loxoscelism (recluse spiders), the site of the bite will ulcerate and become necrotic. Skin may slough and there may be destruction of the adjacent tissues. Hemolysis, thrombocytopenia and renal failure may ensue. Necrotic dermal lesions are often classified as loxoscelism even if the appropriate spiders are not known to be present in the area. In addition, severe reactions to tick bite, or even the erythema migrans of Lyme disease may be confused with recluse spider bites. With latrodectism (black widow spiders), phoneutrism and funnel-web neurotoxicity, the venoms have a strong neurotoxic action. Muscle rigidity and cramping (similar to acute abdomen) is seen with latrodectism; complications include EKG abnormalities and hypotension. With phoneutrism, visual disturbances, vertigo and prostration may occur; complications include hypotension and respiratory paralysis. Funnel web spiders induce autonomic nervous system excitation, with muscular twitching, salivation and lachrymation,

nausea and vomiting and diarrhea; fatal respiratory arrest may result from apnea or laryngospasm. Funnel web spider bites require prompt first aid and the same recommendations could be used for latrodectism or phoneutrism. A compression bandage should be applied over the site of the bite and the affected limb immobilized by splinting with a compression bandage if bitten on an extremity (standard procedures for snakebite). This may help prevent the venom from moving from the local lymphatics. The patient should seek medical attention as soon as possible for antivenom treatment. Otherwise, treatment is symptomatic with analgesics and antipyretics.

Myriapods

The centipedes and millipedes are elongate, vermiform arthropods with dozens of segments, each of which bears a pair of legs. "Centipedes" would suggest having 100 segments (and pairs of legs) or fewer; "millipedes", more than 100 and up to 1000. The diversity of both millipedes and centipedes is greatest in the tropics and virtually all those of medical importance are found in warm climates. Millipedes may squirt a noxious, corrosive fluid from pores on their segments. Such fluid may contain benzoquinone, aldehydes and hydrocyanic acid and cause an immediate burning sensation followed by erythema and edema, even progressing to blistering. Most millipedes also have a repugnant smell; both the corrosive fluid and smell tend to protect them from predation. People become exposed when they step on or sleep on millipedes, or provoke them (children playing with them are often victims). Treatment consists of washing the affected site as soon as possible to dilute and remove the corrosive fluids and symptomatic for the skin lesions and pain.

Centipedes have powerful biting mandibles and small fang-like structures situated between them and derived from the first pair of legs that may inject a venom. Centipede bites occur when people step on or sleep on them, or play with them. Envenomation is manifested by local pain and swelling, with proximal lymphadenopathy. Headache, nausea and anxiety are common. Skin lesions may ulcerate and become necrotic.

Suggested Reading

1. Centers for Disease Control and Prevention. Pictorial keys to arthropods, reptiles, birds and mammals of public health significance. [online] http://www.cdc.gov/nceh/ehs/Pictorial_Keys.htm
2. Eldridge BF, Edman JD. Medical Entomology: A Textbook on Public Health and Veterinary Problems Caused by Arthropods. New York: Kluwer Academic Publishers, 2004.
3. Harwood RF, James MT. Entomology in Human and Animal Health, 7th Edition. New York: Macmillan, 1979.
4. Herms WB. Medical Entomology, 3rd Edition. New York: Macmillan, 1939.
5. Horsfall WR. Mosquitoes: Their Bionomics and Relation to Disease. London: Constable, 1955.
6. Kettle DS. Medical and Veterinary Entomology. London: Croom Helm, 1984.
7. Marquardt WC, Black WC IV, Freier JE et al. Biology of Disease Vectors, 2nd Edition. Burlington: Elsevier Academic Press, 2005.
8. Peters W. A Colour Atlas of Arthropods in Clinical Medicine. London: Wolfe Publishing Ltd, 1992.
9. Smith KGV. Insects and Other Arthropods of Medical Importance. London: British Museum of Natural History, 1973.
10. Sonenshine DE. Biology of Ticks, vol 1. New York: Oxford Press, 1991.
11. Sonenshine DE. Biology of Ticks, vol 2. New York: Oxford Press, 1994.

Drugs for Parasitic Infections

Infection/Drug	Adult Dosage	Pediatric Dosage
***Acanthamoeba* keratitis—see footnote 1**		
Amebiasis (*Entamoeba histolytica*)		
Asymptomatic		
Iodoquinol[D,2]	650 mg PO tid x 20 d	30-40 mg/kg/d (max 2 g) PO in 3 doses x 20 d
Paromomycin[3]	25-35 mg/kg/d PO in 3 doses x 7 d	25-35 mg/kg/d PO in 3 doses x 7 d
Diloxanide furoate[4*]	500 mg PO tid x 10 d	20 mg/kg/d PO in 3 doses x 10 d
Mild to moderate intestinal disease		
Metronidazole[D,5]	500-750 mg PO tid x 7-10 d	35-50 mg/kg/d PO in 3 doses x 7-10 d
Tinidazole[6]	2 g once PO daily x 3 d	≥3 yrs: 50 mg/kg/d (max 2 g) PO in 1 dose x 3 d
either followed by		
Iodoquinol[2]	650 mg PO tid x 20 d	30-40 mg/kg/d (max 2 g) PO in 3 doses x 20 d
Paromomycin[3]	25-35 mg/kg/d PO in 3 doses x 7 d	25-35 mg/kg/ d PO in 3 doses x 7 d
Severe intestinal and extraintestinal disease		
Metronidazole[D]	750 mg PO tid x 7-10 d	35-50 mg/kg/d PO in 3 doses x 7-10 d
Tinidazole[6]	2 g once PO daily x 5 d	≥3 yrs: 50 mg/kg/d (max 2 g) PO in 1 dose x 3 d
either followed by		
Iodoquinol[2]	650 mg PO tid x 20 d	30-40 mg/kg/d (max 2 g) PO in 3 doses x 20 d
Paromomycin[3]	25-35 mg/kg/d PO in 3 doses x 7 d	25-35 mg/kg/d PO in 3 doses x 7 d
Amebic meningoencephalitis—primary and granulomatous		
Naegleria		
Amphotericin B[D,7,8]	1.5 mg/kg/d IV in 2 doses x 3 d, then 1 mg/kg/d x 6 d plus 1.5 mg/d intrathecally x 2 d, then 1 mg/d every other day x 8 d	1.5 mg/kg/d IV in 2 doses x 3 d, then 1 mg/kg/d x 6 d plus 1.5 mg/d intrathecally x 2 d, then 1 mg/d every other day x 8 d

Appendix information from: The Medical Letter, "Drugs for Parasitic Infections", vol. 5, Supplement 2007, last modified August 2008; with permission.

Medical Parasitology, edited by Abhay R. Satoskar, Gary L. Simon, Peter J. Hotez and Moriya Tsuji. ©2009 Landes Bioscience.

Appendix

Infection/Drug	Adult Dosage	Pediatric Dosage
Acanthamoeba—see footnote 9		
Balamuthis mandrillaris—see footnote 10		
Sappinia diploidea—see footnote 11		
***Ancylostoma caninum* (Eosinophilic enterocolitis)**		
Albendazole[D,7,12]	400 mg PO once	400 mg PO once
Mebendazole	100 mg PO bid x 3 d	100 mg PO bid x 3 d
Pyrantel pamoate[7,13*]	11 mg/kg (max 1 g) PO x 3 d	11 mg/kg (max 1 g) PO x 3 d
Endoscopic removal		
***Ancylostoma duodenale*—see Hookworm**		
Angiostrongyliasis (*Angiostongylus cantonensis, Angiostrongylus costaricensis*)[14]		
Anisakiasis (*Anisakis* spp.)—see footnote 15		
Surgical or endoscopic removal		
Ascariasis (*Ascaris lumbricoides*, roundworm)		
Albendazole[D,5,7,12]	400 mg PO once	400 mg PO once
Mebendazole	100 mg bid PO x 3 d or 500 mg once	100 mg PO bid x 3 d or 500 mg once
Ivermectin[7,16]	150-200 mcg/kg PO once	150-200 mcg/kg PO once
Babesiosis (*Babesia microti*)		
Clindamycin[D,7,17,18]	1.2 g bid IV or 600 mg tid PO x 7-10 d	20-40 mg/kg/d PO in 3 doses x 7-10 d
plus quinine[7,19]	650 mg PO tid x 7-10 d	30 mg/kg/d PO in 3 doses x 7-10 d
or Atovaquone[7,20]	750 mg PO bid x 7-10 d	20 mg/kg/d PO in 2 doses x 7-10 d
plus azithromycin[7]	600 mg PO daily x 7-10 d	12 mg/kg/d PO x 7-10 d
***Balamuthia mandrillaris*—see Amebic meningoencephalitis, primary**		
Balantidiasis (*Balantidium coli*)		
Tetracycline[D,7,21]	500 mg PO qid x 10 d	40 mg/kg/d (max 2 g) PO in 4 doses x 10 d
Metronidazole[A,7]	750 mg PO tid x 5 d	35-50 mg/kg/d PO in 3 doses x 5 d
Iodoquinol[A,2,7]	650 mg PO tid x 20 d	30-40 mg/kg/d (max 2 g) PO in 3 doses x 20 d
Baylisascariasis (*Baylisascaris procyonis*)—see footnote 22		
Blastocystis *hominis* infection—see footnote 23		
Capillariasis (*Capillaria philippinensis*)		
Mebendazole[D,7]	200 mg PO bid x 20 d	200 mg PO bid x 20 d
Albendazole[A,7,12]	400 mg PO daily x 10 d	400 mg PO daily x 10 d
Chagas' disease—see Trypanosomiasis		
***Clonorchis sinensis*—see Fluke infection**		
Cryptosporidiosis (*Cryptosporidium*)		
Non-HIV infected		
Nitazoxanide[D,5]	500 mg PO bid x 3 d	1-3 yrs: 100 mg PO bid x 3 d 4-11 yrs: 200 mg PO bid x 3 d >12 yrs: 500 mg PO q 12 h x 3 d
HIV infected—see footnote 24		

Infection/Drug	Adult Dosage	Pediatric Dosage
Cutaneous larva migrans (creeping eruption, dog and cat hookworm)		
Albendazole[D,7,12,25]	400 mg PO daily x 3 d	400 mg PO daily x 3 d
Ivermectin[7,16]	200 mcg/kg PO daily x 1-2 d	200 mcg/kg PO daily x 1-2 d
Cyclosporiasis (*Cyclospora cayetanensis*)		
Trimethoprim/sulfamethoxazole[D,7,26]	TMP 160 mg/SMX 800 mg (1 DS tab) PO bid x 7-10 d	TMP 5 mg/kg/SMX 25 mg/kg/d PO in 2 doses x 7-10d
Cysticercosis—see Tapeworm infection		
***Dientamoeba fragilis* infection[27]**		
Iodoquinol[D,2,7]	650 mg PO tid x 20 d	30-40 mg/kg/d (max 2 g) PO in 3 doses x 20 d
Paromomycin[3,7]	25-35 mg/kg/d PO in 3 doses x 7 d	25-35 mg/kg/d PO in 3 doses x 7 d
Tetracycline[7,21]	500 mg PO qid x 10 d	40 mg/kg/d (max 2 g) PO in 4 doses x 10 d
Metronidazole[7]	500-750 mg PO tid x 10 d	35-50 mg/kg/d PO in 3 doses x 10 d
***Diphyllobothrium latum*—see Tapeworm infection**		
***Dracunculus medinensis* (guinea worm) infection—see footnote 28**		
Echinococcus—see Tapeworm infection		
***Entamoeba histolytica*—see Amebiasis**		
***Enterobius vermicularis* (pinworm) infection**		
Mebendazole[D,29]	100 mg PO once; repeat in 2 wks	100 mg PO once; repeat in 2 wks
Pyrantel pamoate[13*]	11 mg/kg base PO once (max 1 g); repeat in 2 wks	11 mg/kg base PO once (max 1 g); repeat in 2 wks
Albendazole[7,12]	400 mg PO once; repeat in 2wks	400 mg PO once; repeat in 2wks
***Fasciola hepatica*—see Fluke infection**		
Filariasis[30]		
Wuchereria bancrofti, Brugia malayi, Brugia timori		
Diethylcarbamazine[D,31]*	6 mg/kg/d PO in 3 doses x 12 d[32,33]	6 mg/kg/d PO in 3 doses x 12 d[32,33]
Loa loa		
Diethylcarbamazine[34]*	6 mg/kg/d PO in 3 doses x 12 d[32,33]	6 mg/kg/d PO in 3 doses x 12 d[32,33]
***Mansonella ozzardi*—see footnote 35**		
Mansonella perstans		
Albendazole[D,7,12]	400 mg PO bid x 10 d	400 mg PO bid x 10 d
Mebendazole[7]	100 mg PO bid x 30 d	100 mg PO bid x 30 d
Mansonella streptocerca		
Diethylcarbamazine[D,36]*	6 mg/kg/d PO x 12 d[33]	6 mg/kg/d PO x 12 d[33]
Ivermectin[7,16]	150 mcg/kg PO once	150 mcg/kg PO once

Infection/Drug	Adult Dosage	Pediatric Dosage
Filariasis (continued)[30]		
Tropical Pulmonary Eosinophilia (TPE)[37]		
Diethylcarbamazine[D]*	6 mg/kg/d in 3 doses x 12-21 d[33]	6 mg/kg/d in 3 doses x 12-21 d[33]
Onchocerca volvulus **(River blindness)**		
Ivermectin[D,16,38]	150 mcg/kg PO once, repeated every 6-12 mos until asymptomatic	150 mcg/kg PO once, repeated every 6-12 mos until asymptomatic
Fluke—hermaphroditic, infection		
Clonorchis sinensis **(Chinese liver fluke)**		
Praziquantel[D,39]	75 mg/kg/d PO in 3 doses x 2 d	75 mg/kg/d PO in 3 doses x 2 d
Albendazole[7,12]	10 mg/kg/d PO x 7 d	10 mg/kg/d PO x 7 d
Fasciola hepatica **(sheep liver fluke)**		
Triclabendazole[D,40]*	10 mg/kg PO once or twice[41]	10 mg/kg PO once or twice[41]
Bithionol[A,]*	30-50 mg/kg on alternate days x 10-15 doses	30-50 mg/kg on alternate days x 10-15 doses
Nitazoxanide[A,5,7]	500 mg PO bid x 7 d	1-3 yrs: 100 mg PO q 12 h x 7 d 4-11 yrs: 200 mg PO q 12 h x 7 d >12 yrs: 500 mg PO q 12 h x 7 d
Fasciolopsis buski, Heterophyes heterophyes, Metagonimus yokogawai **(intestinal flukes)**		
Praziquantel[D,7,39]	75 mg/kg/d PO in 3 doses x 1 d	75 mg/kg/d PO in 3 doses x 1 d
Metorchis conjunctus **(North American liver fluke)**		
Praziquantel[D,7,39]	75 mg/kg/d PO in 3 doses x 1 d	75 mg/kg/d PO in 3 doses x 1 d
Nanophyetus salmincola		
Praziquantel[D,7,39]	60 mg/kg/d PO in 3 doses x 1 d	60 mg/kg/d PO in 3 doses x 1 d
Opisthorchis viverrini **(Southeast Asian liver fluke)**		
Praziquantel[D,39]	75 mg/kg/d PO in 3 doses x 2 d	75 mg/kg/d PO in 3 doses x 2 d
Paragonimus westermani **(lung fluke)**		
Praziquantel[D,7,39]	75 mg/kg/d PO in 3 doses x 2 d	75 mg/kg/d PO in 3 doses x 2 d
Bithionol[A,42]*	30-50 mg/kg on alternate days x 10-15 doses	30-50 mg/kg on alternate days x 10-15 doses
Giardiasis (*Giardia duodenalis*)		
Metronidazole[7]	250 mg PO tid x 5-7 d	15 mg/kg/d PO in 3 doses x 5-7 d
Tinidazole[6]	2 g PO once	50 mg/kg PO once (max 2 g)
Nitazoxanide[5]	500 mg PO bid x 3 d	1-3 yrs: 100 mg PO q 12 h x 3 d 4-11yrs: 200 mg PO q12h x 3d >12yrs: 500 mg PO q12h x 3d
Paromomycin[A,3,7,43,44]	25-35 mg/kg/d PO in 3 doses x 5-10 d	25-35 mg/kg/d PO in 3 doses x 5-10 d
Furazolidone[A,43]*	100 mg PO qid x 7-10 d	6 mg/kg/d PO in 4 doses x 7-10 d
Quinacrine[A,4,43,45]*	100 mg PO tid x 5 d	2 mg/kg/d PO in 3 doses x 5 d (max 300 mg/d)

Infection/Drug	Adult Dosage	Pediatric Dosage
Gnathostomiasis (*Gnathostoma spinigerum*)[46]		
Albendazole[T,7,12]	400 mg PO bid x 21 d	400 mg PO bid x 21 d
Ivermectin[7,16]	200 mcg/kg/d PO x 2 d	200 mcg/kg/d PO x 2 d
either		
± Surgical removal		
Gongylonemiasis (*Gongylonema* sp.)[47]		
Surgical removal[T]		
Albendazole[7,12]	400 mg/d PO x 3 d	400 mg/d PO x 3 d
Hookworm infection (*Ancylostoma duodenale, Necator americanus*)		
Albendazole[D,7,12]	400 mg PO once	400 mg PO once
Mebendazole	100 mg PO bid x 3 d or 500 mg once	100 mg PO bid x 3 d or 500 mg once
Pyrantel pamoate[7,13]*	11 mg/kg (max 1 g) PO x 3 d	11 mg/kg (max 1 g) PO x 3 d
Hydatid cyst—see Tapeworm infection		
***Hymenolepis nana*—see Tapeworm infection**		
Isosporiasis (*Isospora belli*)		
Trimethoprimsulfa-methoxazole[D,7,48]	TMP 160 mg/SMX 800 mg (1 DS tab) PO bid x 10 d	TMP 5 mg/kg/d/SMX 25 mg/kg/d PO in 2 doses x 10 d
Leishmania		
Visceral[49,50]		
Liposomal amphotericin B[D,51]	3 mg/kg/d IV d 1-5, 14 and 21[52]	3 mg/kg/d IV d 1-5, 14 and 21[52]
Sodium stibogluconate*	20 mg Sb/kg/d IV or IM x 28 d	20 mg Sb/kg/d IV or IM x 28 d
Miltefosine[53]*	2.5 mg/kg/d PO (max 150 mg/d) x 28 d	2.5 mg/kg/d PO (max 150 mg/d) x 28 d
Meglumine antimonate[A]*	20 mg Sb/kg/d IV or IM x 28 d	20 mg Sb/kg/d IV or IM x 28 d
Amphotericin B[A,7]	1 mg/kg IV daily x 15-20 d or every second day for up to 8 wks	1 mg/kg IV daily x 15-20 d or every second day for up to 8 wks
Paromomycin[7,13,54]*	15 mg/kg/d IM x 21 d	15 mg/kd/d IM x 21 d
Cutaneous[49,55]		
Sodium stibogluconate[D]*	20 mg Sb/kg/d IV or IM x 20 d	20 mg Sb/kg/d IV or IM x 20 d
Meglumine antimonate*	20 mg Sb/kg/d IV or IM x 20 d	20 mg Sb/kg/d IV or IM x 20 d
Miltefosine[53]*	2.5 mg/kg/d PO (max 150 mg/d) x 28 d	2.5 mg/kg/d PO (max 150 mg/d) x 28 d
Paromomycin[A,7,13,54,56]*	Topically 2 x/d x 10-20 d	Topically 2 x/d x 10-20 d
Pentamidine[7,56]	2-3 mg/kg IV or IM daily or every second day x 4-7 doses[57]	2-3 mg/kg IV or IM daily or every second day x 4-7 doses[57]

Appendix

Infection/Drug	Adult Dosage	Pediatric Dosage
Leishmania (continued)		
Mucosal[49,58]		
Sodium stibogluconate[D,*]	20 mg Sb/kg/d IV or IM x 28 d	20 mg Sb/kg/d IV or IM x 28 d
Meglumine antimonate*	20 mg Sb/kg/d IV or IM x 28 d	20 mg Sb/kg/d IV or IM x 28 d
Amphotericin B[7]	0.5-1 mg/kg IV daily or every second day for up to 8 wks	0.5-1 mg/kg IV daily or every second day for up to 8 wks
Miltefosine[53]*	2.5 mg/kg/d PO (max 150 mg/d) x 28d	2.5 mg/kg/d PO (max 150 mg/d) x 28d
Lice infestation (*Pediculus humanus, P. capitis, Phthirus pubis*)		
0.5% Malathion[D,60]	Topically	Topically
1% Permethrin[61]	Topically	Topically
Pyrethrins with piperonyl butoxide[A,61]	Topically	Topically
Ivermectin[A,7,16,62]	200 mcg/kg PO	>15 kg: 200 mcg/kg PO
Loa loa—see Filariasis		
Malaria, treatment of (*Plasmodium falciparum,*[63] *P. vivax,*[64] *P. ovale, and P. malariae*[65])		
Oral:[66]		
***P. falciparum* or unidentified species acquired in areas of chloroquine-resistant *P. falciparum*[63]**		
Atovaquone/ proguanil[D,68]	2 adult tabs bid[69] or 4 adult tabs once/d x 3 d	<5 kg: not indicated 5-8 kg: 2 peds tabs once/d x 3 d 9-10 kg: 3 peds tabs once/d x 3 d 11-20 kg: 1 adult tab once/d x 3 d 21-30 kg: 2 adult tabs once/d x 3 d 31-40 kg: 3 adult tabs once/d x 3 d >40 kg: 4 adult tabs once/d x 3 d
Quinine sulfate	650 mg q 8 h x 3 or 7 d[70]	30 mg/kg/d in 3 doses x 3 *or* 7 d[70]
plus		
doxycycline[7,21,71]	100 mg bid x 7 d	4 mg/kg/d in 2 doses x 7 d
or plus		
tetracycline[7,21]	250 mg qid x 7 d	6.25 mg/kg/d in 4 doses x 7 d
or plus		
clindamycin[7,21,72]	20 mg/kg/d in 3 doses x 7d[73]	20 mg/kg/d in 3 doses x 7 d
Mefloquine[A,67,74,75]	750 mg followed 12 hrs later by 500 mg	15 mg/kg followed 12 hrs later by 10 mg/kg
Artemether/lumefantrine[A,67,76,77]*	6 doses over 3d (4 tabs/dose at 0, 8, 24, 36, 48 and 60 hrs)	6 doses over 3d at same intervals as adults; <15 kg: 1 tab/dose 15-25 kg: 2 tabs/dose 25-35 kg: 3 tabs/dose >35 kg: 4 tabs/dose
Artesunate[A,76]*	4 mg/kg/d x 3 d	4 mg/kg/d x 3 d
plus see footnote 78		

Infection/Drug	Adult Dosage	Pediatric Dosage
Malaria, treatment of (continued)		
***P. vivax* acquired in areas of chloroquine-resistant *P. vivax*[64]**		
Mefloquine[D,67,74]	750 mg PO followed 12 hrs later by 500 mg	15 mg/kg PO followed 12 hrs later by 10 mg/kg
Atovaquone/ proguanil[68]	2 adult tabs bid[69] or 4 adult tabs once/d x 3 d	<5 kg: not indicated 5-8 kg: 2 peds tabs once/d x 3 d 9-10 kg: 3 peds tabs once/d x 3 d 11-20 kg: 1 adult tab once/d x 3 d 21-30 kg: 2 adult tabs once/d x 3 d 31-40 kg: 3 adult tabs once/d x 3 d >40 kg: 4 adult tabs once/d x 3 d
either followed by		
primaquine phosphate[79]	30 mg base/d PO x 14 d	0.6 mg/kg/d PO x 14 d
Chloroquine phosphate[A,67,80]	25 mg base/kg PO in 3 doses over 48 hrs[81]	25 mg base/kg PO in 3 doses over 48 hrs[81]
Quinine sulfate[A]	650 mg PO q 8 h x 3-7 d[70]	30 mg/kg/d PO in 3 doses x 3-7 d[70]
plus		
doxycycline[7,21,71]	100 mg PO bid x 7 d	4 mg/kg/d PO in 2 doses x 7 d
either followed by		
primaquine phosphate[79]	30 mg base/d PO x 14 d	0.6 mg/kg/d PO x 14 d
All *Plasmodium* species except chloroquine-resistant *P. falciparum*[63] and chloroquine-resistant *P. vivax*[64]		
Chloroquine phosphate[D,67,80]	1 g (600 mg base) PO, then 500 mg (300 mg base) 6 hrs later, then 500 mg (300 mg base) at 24 and 48 hrs[81]	10 mg base/kg (max 600 mg base) PO, then 5 mg base/kg 6 hrs later, then 5 mg base/kg at 24 and 48 hrs[81]
Parenteral:[66]		
All *Plasmodium* species (Chloroquine-sensitive and resistant)		
Quinidine gluconate[D,67,82,83]	10 mg/kg IV loading dose (max 600 mg) in normal saline over 1-2 hrs, followed by continuous infusion of 0.02 mg/kg/min until PO therapy can be started	10 mg/kg IV loading dose (max 600 mg) in normal saline over 1-2 hrs, followed by continuous infusion of 0.02 mg/kg/min until PO therapy can be started
Quinine dihydro- chloride[83]*	20 mg/kg IV loading dose in 5% dextrose over 4 hrs, followed by 10 mg/kg over 2-4 hrs q 8 h (max 1800 mg/d) until PO therapy can be started	20 mg/kg IV loading dose in 5% dextrose over 4 hrs, followed by 10 mg/kg over 2-4 hrs q 8 h (max 1800 mg/d) until PO therapy can be started
Artesunate[76]*	2.4 mg/kg/dose IV x 3d at 0, 12, 24, 48 and 72 hrs	2.4 mg/kg/dose IV x 3d at 0, 12, 24, 48 and 72 hrs
plus see footnote 78		

Appendix

Infection/Drug	Adult Dosage	Pediatric Dosage
Malaria, prevention of[84]		
All *Plasmodium* species in chloroquine-sensitive areas[63-65]		
Chloroquine phosphate[D,67,80,85,86]	500 mg (300 mg base) PO once/wk[87]	5 mg/kg base PO once/wk, up to adult dose of 300 mg base[87]
All *Plasmodium* species in chloroquine-resistant areas[63-65]		
Atovaquone/ proguanil[D,67,68]	1 adult tab/d[88]	5-8 kg: 1/2 peds tab/d[68,88] 9-10 kg: 3/4 peds tab/d[68,88] 11-20 kg: 1 peds tab/d[68,88] 21-30 kg: 2 peds tabs/d[68,88] 31-40 kg: 3 peds tabs/d[68,88] >40 kg: 1 adult tab/d[68,88]
Doxycycline[7,21,71]	100 mg PO daily[89]	2 mg/kg/d PO, up to 100 mg/d[89]
Mefloquine[74,75,90]	250 mg PO once/wk[91]	5-10 kg: 1/8 tab once/wk[91] 11-20 kg: 1/4 tab once/wk[91] 21-30 kg: 1/2 tab once/wk[91] 31-45 kg: 3/4 tab once/wk[91] >45 kg: 1 tab once/wk[91]
Primaquine phosphate[A,7,79]	30 mg base PO daily[93]	0.6 mg/kg base PO daily[93]
Malaria, prevention of relapses: *P. vivax* and *P. ovale*[67]		
Primaquine phosphate[D,79]	30 mg base/d PO x 14 d	0.6 mg base/kg/d PO x 14 d
Malaria, self-presumptive treatment[94]		
Atovaquone/ proguanil[D,7,68]	4 adult tabs once/d x 3 d[69]	<5 kg: not indicated 5-8 kg: 2 peds tabs once/d x 3 d 9-10 kg: 3 peds tabs once/d x 3 d 11-20 kg: 1 adult tab once/d x 3 d 21-30 kg: 2 adult tabs once/d x 3d 31-40 kg: 3 adult tabs once/d x 3 d >40 kg: 4 adult tabs once/d x 3 d[69]
Quinine sulfate	650 mg PO q 8 h x 3 or 7 d[70]	30 mg/kg/d PO in 3 doses x 3 or 7 d[70]
plus		
doxycycline[7,21,71]	100 mg PO bid x 7 d	4 mg/kg/d PO in 2 doses x 7 d
Artesunate[76]*	4 mg/kg/d PO x 3 d	4 mg/kg/d PO x 3 d
plus see footnote 78		
Microsporidiosis		
Ocular (*Encephalitozoon hellem, E. cuniculi, Vittaforma corneae* [*Nosema corneum*])		
Albendazole[D,7,12]	400 mg PO bid	
plus fumagillin[95]*		
Intestinal (*E. bieneusi, E. [Septata] intestinalis*)		
E. bieneusi		
Fumagillin[D,96]*	20 mg PO tid x 14 d	
E. intestinalis		
Albendazole[D,7,12]	400 mg PO bid x 21 d	
Disseminated (*E. hellem, E. cuniculi, E. intestinalis, Pleistophora sp., Trachipleistophora sp.* and *Brachiola vesicularum*)		
Albendazole[D,7,12,97]*	400 mg PO bid	

Infection/Drug	Adult Dosage	Pediatric Dosage
Mites—see Scabies		
***Moniliformis* infection**		
Pyrantel pamoate[D,7,13]*	11 mg/kg PO once, repeat twice, 2wks apart	11 mg/kg PO once, repeat twice, 2wks apart
***Naegleria* species—see Amebic meningoencephalitis, primary**		
***Necator americanus*, see Hookworm infection**		
***Oesophagostomum bifurcum*—see footnote 98**		
***Onchocerca volvulus*—see Filariasis**		
***Opisthorchis viverrini*—see Fluke infection**		
***Paragonimus westermani*—see Fluke infection**		
***Pediculus capitis, humanus, Phthirus pubis*—see Lice**		
Pinworm—see Enterobius		
Pneumocystis jiroveci (formerly *carinii*) pneumonia (PCP)[99]		
Trimethoprim/ sulfamethoxazole[D]	TMP 15 mg/SMX 75 mg/kg/d, PO or IV in 3 or 4 doses x 21 d	TMP 15 mg/SMX 75 mg/kg/d, PO or IV in 3 or 4 doses x 21 d
Primaquine[A,7,79]	30 mg base PO daily x 21 d	0.3 mg/kg base PO daily x 21 d
plus clindamycin[7,18]	600 mg IV q 6 h x 21 d, or 300-450 mg PO q 6 h x 21 d	15-25 mg/kg IV q 6 h x 21 d, or 10 mg/kg PO q 6 h x 21 d
Trimethoprim[A,7]	5 mg/kg PO tid x 21 d	5 mg/kg PO tid x 21 d
plus dapsone[7]	100 mg daily x 21 d	2 mg/kg/d PO x 21 d
Pentamidine[A]	3-4 mg/kg IV daily x 21 d	3-4 mg/kg IV daily x 21 d
Atovaquone[A]	750 mg PO bid x 21 d	1-3 mos: 30 mg/kg/d PO x 21 d 4-24 mos: 45 mg/kg/d PO x 21 d >24 mos: 30 mg/d PO x 21 d
Primary and secondary prophylaxis[100]		
Trimethoprim/ sulfamethoxazole[D]	1 tab (single or double strength) daily or 1 DS tab PO 3 d/wk	TMP 150 mg/SMX 750 mg/m²/d PO in 2 doses 3 d/wk
Dapsone[A,7]	50 mg PO bid or 100 mg PO daily	2 mg/kg/d (max 100 mg) PO or 4 mg/kg (max 200 mg) PO each wk
Dapsone[A,7]	50 mg PO daily or 200 mg PO each wk	
plus pyrimethamine[101]	50 mg PO or 75 mg PO each wk	
Pentamidine[A]	300 mg aerosol inhaled monthly via *Respirgard II* nebulizer	≥5 yrs: 300 mg inhaled monthly via *Respirgard II* nebulizer
Atovaquone[A,7,20]	1500 mg PO daily	1-3 mos: 30 mg/kg/d PO 4-24 mos: 45 mg/kg/d PO >24 mos: 30 mg/kg/d PO

Infection/Drug	Adult Dosage	Pediatric Dosage
River Blindness—see Filariasis		
Roundworm—see Ascariasis		
***Sappinia diploidea*—see Amebic meningoencephalitis, primary**		
Scabies (*Sarcoptes scabiei*)		
5% Permethrin[D]	Topically once[102]	Topically once[102]
Ivermectin[A,7,16,103,104]	200 mcg/kg PO once[102]	200 mcg/kg PO once[102]
10% Crotamiton[A]	Topically once/d x 2	Topically once/d PO x 2
Schistosomiasis (*Bilharziasis*)		
S. haematobium		
Praziquantel[D,39]	40 mg/kg/d PO in 2 doses x 1 d	40 mg/kg/d PO in 2 doses x 1 d
S. japonicum		
Praziquantel[D,39]	60 mg/kg/d PO in 3 doses x 1 d	60 mg/kg/d PO in 3 doses x 1 d
S. mansoni		
Praziquantel[D,39]	40 mg/kg/d PO in 2 doses x 1 d	40 mg/kg/d PO in 2 doses x 1 d
Oxamniquine[A,105]*	15 mg/kg PO once[106]	20 mg/kg/d PO in 2 doses x 1 d[106]
S. mekongi		
Praziquantel[D,39]	60 mg/kg/d PO in 3 doses x 1 d	60 mg/kg/d PO in 3 doses x 1 d
Sleeping sickness, see Trypanosomiasis		
Strongyloidiasis (*Strongyloides stercoralis*)		
Ivermectin[D,16,107]	200 mcg/kg/d PO x 2 d	200 mcg/kg/d PO x 2 d
Albendazole[A,7,12]	400 mg PO bid x 7 d	400 mg PO bid x 7 d
Tapeworm infection		
Adult (intestinal stage):		
***Diphyllobothrium latum* (fish), *Taenia saginata* (beef), *Taenia solium* (pork), *Dipylidium caninum* (dog)**		
Praziquantel[D,7,39]	5-10 mg/kg PO once	5-10 mg/kg PO once
Niclosamide[A,108]*	2 g PO once	50 mg/kg PO once
***Hymenolepis nana* (dwarf tapeworm)**		
Praziquantel[D,7,39]	25 mg/kg PO once	25 mg/kg PO once
Nitazoxanide[A,5,7]	500 mg PO once/d or bid x 3 d[109]	1-3 yrs: 100 mg PO bid x 3 d[109] 4-11 yrs: 200 mg PO bid x 3 d[109]
Larval (tissue stage):		
***Echinococcus granulosus* (hydatid cyst)**		
Albendazole[D,12,110]	400 mg PO bid x 1-6 mos	15 mg/kg/d (max 800 mg) x 1-6 mos
***Echinococcus multilocularis*—see footnote 111[T]**		
***Taenia solium* (Cysticercosis)**		
Treatment of choice—see footnote 112		
Albendazole[A,12]	400 mg PO bid x 8-30 d; can be repeated as necessary	15 mg/kg/d (max 800 mg) PO in 2 doses x 8-30 d; can be repeated as necessary
Praziquantel[A,7,39]	100 mg/kg/d PO in 3 doses x 1 d then 50 mg/kg/d in 3 doses x 29 d	100 mg/kg/d PO in 3 doses x 1 d then 50 mg/kg/d in 3 doses x 29 days

Infection/Drug	Adult Dosage	Pediatric Dosage
Toxocariasis—see Visceral larva migrans		
Toxoplasmosis (*Toxoplasma gondii*)		
Pyrimethamine[D,113,114]	25-100 mg/d PO x 3-4 wks	2 mg/kg/d PO x 2d, then 1 mg/kg/d (max 25 mg/d) x 4 wks[115]
plus		
sulfadiazine[116]	1-1.5 g PO qid x 3-4 wks	100-200 mg/kg/d PO x 3-4 wks
Trichinellosis (*Trichinella spiralis*)		
Steroids for severe symptoms[D]		
plus		
albendazole[7,12]	400 mg PO bid x 8-14 d	400 mg PO bid x 8-14 d
Mebendazole[A,7]	200-400 mg PO tid x 3 d, then 400-500 mg tid x 10 d	200-400 mg PO tid x 3 d, then 400-500 mg tid x 10 d
Trichomoniasis (*Trichomonas vaginalis*)		
Metronidazole[D,117]	2 g PO once or 500 mg bid x 7 d	15 mg/kg/d PO in 3 doses x 7 d
Tinidazole[6]	2 g PO once	50 mg/kg once (max 2 g)
Trichostrongylus infection		
Pyrantel pamoate[D,7,13]*	11 mg/kg base PO once (max 1 g)	11 mg/kg PO once (max 1 g)
Mebendazole[A,7]	100 mg PO bid x 3 d	100 mg PO bid x 3 d
Albendazole[A,7,12]	400 mg PO once	400 mg PO once
Trichuriasis (*Trichuris trichiura*, whipworm)		
Mebendazole[D]	100 mg PO bid x 3 d or 500 mg once	100 mg PO bid x 3 d or 500 mg once
Albendazole[A,7,12]	400 mg PO x 3 d	400 mg PO x 3 d
Ivermectin[A,7,16]	200 mcg/kg PO daily x 3 d	200 mcg/kg/d PO x 3 d
Trypanosomiasis		
T. cruzi **(American trypanosomiasis, Chagas' disease)**		
Nifurtimox[D]*	8-10 mg/kg/d PO in 3-4 doses x 90-120 d	1-10 yrs: 15-20 mg/kg/d PO in 4 doses x 90-120 d 11-16 yrs: 12.5-15 mg/kg/d in 4 doses x 90-120 d
Benznidazole[119]*	5-7 mg/kg/d PO in 2 doses x 30-90 d	≤12 yrs: 10 mg/kg/d PO in 2 doses x 30-90 d >12 yrs: 5-7 mg/kg/d in 2 doses x 30-90 d
T. brucei gambiense **(West African trypanosomiasis, sleeping sickness)**		
Hemolymphatic stage:		
Pentamidine[D,7,120]	4 mg/kg/d IM x 7 d	4 mg/kg/d IM x 7 d
Suramin[A]*	100-200 mg (test dose) IV, then 1 g IV on days 1,3,7,14 and 21	20 mg/kg on d 1,3,7,14 and 21

Appendix

Infection/Drug	Adult Dosage	Pediatric Dosage
Trypanosomiasis (continued)		
T. brucei gambiense **(continued)**		
Late disease with CNS involvement:		
Eflornithine[D,121]*	400 mg/kg/d IV in 4 doses x 14d	400 mg/kg/d IV in 4 doses x 14 d
Melarsoprol[122]	2.2 mg/kg/d IV x 10 d	2.2 mg/kg/d IV x 10 d
T. b. rhodesiense **(East African trypanosomiasis, sleeping sickness)**		
Hemolymphatic stage:		
Suramin[D]*	100-200 mg (test dose) IV, then 1 g IV on days 1,3,7,14 and 21	20 mg/kg on d 1,3,7,14 and 21
Late disease with CNS involvement		
Melarsoprol[D,122]	2-3.6 mg/kg/d IV x 3 d; after 7 d 3.6 mg/kg/d x 3 d; repeat again after 7d	2-3.6 mg/kg/d x 3 d; after 7 d 3.6 mg/kg/d x 3 d; repeat again after 7d
Visceral larva migrans[123] (*Toxocariasis*)		
Albendazole[D,7,12]	400 mg PO bid x 5 d	400 mg PO bid x 5 d
Mebendazole[7]	100-200 mg PO bid x 5 d	100-200 mg PO bid x 5 d
Whipworm—see Trichuriasis		
Wuchereria bancrofti**—see Filariasis**		

D Drug of choice
A Alternative drug choice
T Treatment of choice
* Availability problems. See following table.

1. Topical 0.02% chlorhexidine and polyhexamethylene biguanide (PHMB, 0.02%), either alone or in combination, have been used successfully in a large number of patients. Treatment with either chlorhexidine or PHMB is often combined with propamidine isethionate (*Brolene*) or hexamidine (*Desmodine*). None of these drugs is commercially available or approved for use in the US, but they can be obtained from compounding pharmacies (see footnote 2). Leiter's Park Avenue Pharmacy, San Jose, CA (800-292-6773; www.leiterrx.com) is a compounding pharmacy that specializes in ophthalmic drugs. Propamidine is available over the counter in the UK and Australia. Hexamidine is available in France. The combination of chlorhexidine, natamycin (pimaricin) and debridement also has been successful (K Kitagawa et al, Jpn J Ophthalmol 2003; 47:616). Debridement is most useful during the stage of corneal epithelial infection. Most cysts are resistant to neomycin; its use is no longer recommended. Azole antifungal drugs (ketoconazole, itraconazole) have been used as oral or topical adjuncts (FL Shuster and GS Visvesvara, Drug Resist Update 2004; 7:41). Use of corticosteroids is controversial (K Hammersmith, Curr Opinions Ophthal 2006; 17:327; ST Awwad et al, Eye Contact Lens 2007; 33:1).
2. Iodoquinol should be taken after meals.
3. Paromomycin should be taken with a meal.
4. Not available commercially. It may be obtained through compounding pharmacies such as Panorama Compounding Pharmacy, 6744 Balboa Blvd, Van Nuys, CA 91406 (800-247-9767) or Medical Center Pharmacy, New Haven, CT (203-688-6816). Other compounding pharmacies may be found through the National Association of Compounding Pharmacies (800-687-7850) or the Professional Compounding Centers of America (800-331-2498, www.pccarx.com).
5. Nitazoxanide may be effective against a variety of protozoan and helminth infections (DA Bobak, Curr Infect Dis Rep 2006; 8:91; E Diaz et al, Am J Trop Med Hyg2003; 68:384).

It was effective against mild to moderate amebiasis, 500 mg bid x 3 d, in a recent study (JF Rossignol et al, Trans R Soc Trop Med Hyg 2007 Oct; 101:1025 E pub 2007 July 20). It is FDA-approved only for treatment of diarrhea caused by Giardia or Cryptosporidium (Med Lett Drugs Ther 2003; 45:29). Nitazoxanide is available in 500-mg tablets and an oral suspension; it should be taken with food.

6. A nitroimidazole similar to metronidazole, tinidazole appears to be as effective as metronidazole and better tolerated (Med Lett Drugs Ther 2004; 46:70). It should be taken with food to minimize GI adverse effects. For children and patients unable to take tablets, a pharmacist can crush the tablets and mix them with cherry syrup (*Humco*, and others). The syrup suspension is good for 7 days at room temperature and must be shaken before use (HB Fung and TL Doan et al, Clin Ther 2005; 27:1859). Ornidazole, a similar drug, is also used outside the US.
7. Not FDA-approved for this indication.
8. Although a *Naegleria fowleri* infection was treated successfully in a 9-year-old girl with combination of amphotericin B and miconazole both intravenous and intrathecal, plus oral rifampin (JS Seidel et al NEJM 1982;306:346). Amphotericin B and miconazole appear to have a synergistic effect, but Medical Letter consultants believe the rifampin probably had no additional effect (GS Visvesvara et al, FEMS Immunol Med Microbiol 2007; 50:1). Parenteral miconazole is no longer available in the US. Azithromycin has been used successfully in combination therapy to treat *Balmuthia* infection, but was changed to clarithromycin because of toxicity concerns and for better penetration into the cerebrospinal fluid. In vitro, azithromycin is more active than clarithromycin against *Naegleria*, so may be a better choice combined with amphotericin B for treatment of *Naegleria* (TR Deetz et al, Clin Infect Dis 2003; 37:1304; FL Schuster and GS Visvesvara, Drug Resistance Updates 2004; 7:41). Combinations of amphotericin B, ornidazole and rifampin (R Jain et al, Neurol Indian 2002; 50:470) and amphotericin B fluconazole and rifampin have also been used (J Vargas-Zepeda et al, Arch Med Research 2005;36:83). Case reports of other successful therapy have been published (FL Schuster and GS Visvesvara, Int J Parasitol 2004; 34:1001).
9. Several patients with granulomatous amebic encephalitis (GAE) have been successfully treated with combinations of pentamidine, sulfadiazine, flucytosine, and either fluconazole or itraconazole (GS Visvesvara et al, FEMS Immunol Med Microbiol 2007; 50:1, epub Apr 11). GAE in an AIDS patient was treated successfully with sulfadiazine, pyrimethamine and fluconazole combined with surgical resection of the CNS lesion (M Seijo Martinez et al, J Clin Microbiol 2000; 38:3892). Chronic *Acanthamoeba* meningitis was successfully treated in 2 children with a combination of oral trimethoprim/sulfamethoxazole, rifampin and ketoconazole (T Singhal et al, Pediatr Infect Dis J 2001; 20:623). Disseminated cutaneous infection in an immunocompromised patient was treated successfully with IV pentamidine, topical chlorhexidine and 2% ketoconazole cream, followed by oral itraconazole (CA Slater et al, N Engl J Med 1994; 331:85) and with voriconazole and amphotericin B lipid complex (R Walia et al, Transplant Infect Dis 2007; 9:51). Other reports of successful therapy have been described (FL Schuster and GS Visvesvara, Drug Resistance Updates 2004; 7:41). Susceptibility testing of *Acanthamoeba* isolates has shown differences in drug sensitivity between species and even among strains of a single species; antimicrobial susceptibility testing is advisable (FL Schuster and GS Visvesvara, Int J Parasitiol 2004; 34:1001).
10. *B. mandrillaris* is a free-living ameba that causes subacute to fatal granulomatous amebic encephalitis (GAE) and cutaneous disease. Two cases of *Balamuthia* encephalitis have been successfully treated with flucytosine, pentamidine, fluconazole and sulfadiazine plus either azithromycin or clarithromycin (phenothiazines were also used) combined with surgical resection of the CNS lesion (TR Deetz et al, Clin Infect Dis 2003; 37:1304). Another case was successfully treated following open biopsy with pentamidine, fluconazole, sulfadiazine and clarithromycin (S Jung et al, Arch Pathol Lab Med 2004; 128:466).
11. A free-living ameba once thought not to be pathogenic to humans. *S. diploidea* has been successfully treated with azithromycin, pentamidine, itraconazole and flucytosine combined with surgical resection of the CNS lesion (BB Gelman et al, J Neuropathol Exp Neurol 2003; 62:990).

12. Albendazole must be taken with food; a fatty meal increases oral bioavailability.
13. Pyrantel pamoate suspension can be mixed with milk or fruit juice.
14. *A. cantonensis* causes predominantly neurotropic disease. *A. costaricensis* causes gastrointestinal disease. Most patients infected with either species have a self-limited course and recover completely. Analgesics, corticosteroids and careful removal of CSF at frequent intervals can relieve symptoms from increased intracranial pressure (V Lo Re III and SJ Gluckman, Am J Med 2003; 114:217). Treatment of *A. cantonensis* is controversial and varies across endemic areas. No anthelminthic drug is proven to be effective and some patients have worsened with therapy (TJ Slom et al, N Engl J Med 2002; 346:668). Mebendazole and a corticosteroid, however, appear to shorten the course of infection (H-C Tsai et al, Am J Med 2001; 111:109; V Chotmongkol et al, Am J Trop Med Hyg 2006; 74:1122). Albendazole has also relieved symptoms of angiostrongyliasis (XG Chen et al, Emerg Infect Dis 2005; 11:1645).
15. A Repiso Ortega et al, Gastroenterol Hepatol 2003; 26:341. Successful treatment of *Anisakiasis* with albendazole 400 mg PO bid x 3-5d has been reported, but the diagnosis was presumptive (DA Moore et al, Lancet 2002; 360:54; E Pacios et al, Clin Infect Dis 2005; 41:1825).
16. Safety of ivermectin in young children (<15 kg) and pregnant women remains to be established. Ivermectin should be taken on an empty stomach with water.
17. Exchange transfusion has been used in severely ill patients and those with high (>10%) parasitemia (VI Powell and K Grima, Transfus Med Rev 2002; 16:239). In patients who were not severely ill, combination therapy with atovaquone and azithromycin was as effective as clindamycin and quinine and may have been better tolerated (PJ Krause et al, N Engl J Med 2000; 343:1454). Longer treatment courses may be needed in immunosuppressed patients and those with asplenia. Patients are commonly co-infected with Lyme disease (Med Lett Drugs Ther 2007; 49:49; AC Steere et al, Clin Infect Dis 2003; 36:1078).
18. Oral clindamycin should be taken with a full glass of water to minimize esophageal ulceration.
19. Quinine should be taken with or after a meal to decrease gastrointestinal adverse effects.
20. Atovaquone is available in an oral suspension that should be taken with a meal to increase absorption.
21. Use of tetracyclines is contraindicated in pregnancy and in children <8 years old. Tetracycline should be taken 1 hour before or 2 hours after meals and/or dairy products.
22. No drug has been demonstrated to be effective. Albendazole 25 mg/kg/d PO x 20 d started as soon as possible (up to 3 d after possible infection) might prevent clinical disease and is recommended for children with known exposure (ingestion of raccoon stool or contaminated soil) (WJ Murray and KR Kazacos, Clin Infect Dis 2004; 39:1484). Mebendazole, levamisole or ivermectin could be tried if albendazole is not available. Steroid therapy may be helpful, especially in eye and CNS infections (PJ Gavin et al, Clin Microbiol Rev 2005; 18:703). Ocular baylisascariasis has been treated successfully using laser photocoagulation therapy to destroy the intraretinal larvae (CA Garcia et al, Eye 2004; 18:624).
23. Clinical significance of these organisms is controversial; metronidazole 750 mg PO tid x 10 d, iodoquinol 650 mg PO tid x 20 d or trimethoprim/sulfamethoxazole 1 DS tab PO bid x 7d have been reported to be effective (DJ Stenzel and PFL Borenam, Clin Microbiol Rev 1996; 9:563; UZ Ok et al, Am J Gastroenterol 1999; 94:3245). Metronidazole resistance may be common in some areas (K Haresh et al, Trop Med Int Health 1999; 4:274). Nitazoxanide has been effective in clearing organism and improving symptoms (E Diaz et al, Am J Trop Med Hyg 2003; 68:384; JF Rossignol, Clin Gastroenterol Hepatol 2005; 18:703).
24. No drug has proven efficacy against cryptosporidiosis in advanced AIDS (I Abubakar et al, Cochrane Database Syst Rev 2007; 1:CD004932). Treatment with HAART is the mainstay of therapy. Nitazoxanide (JF Rossignol, Aliment Pharmacol Ther 2006; 24:807), paromomycin (P Maggi et al, Clin Infect Dis 2000; 33:1609), or a combination of paromomycin and azithromycin (NH Smith et al, J Infect Dis 1998; 178:900) may be

tried to decrease diarrhea and recalcitrant malabsorption of antimicrobial drugs, which can occur with chronic cryptosporidiosis.

25. G Albanese et al, Int J Dermatol 2001; 40:67; D Malvy et al, J Travel Med 2006; 13:244.
26. HIV-infected patients may need higher dosage and long-term maintenance. Successful use of nitazoxanide (see also footnote 5) has been reported in one patient with sulfa allergy (SM Zimmer et al, Clin Infect Dis 2007; 44:466).
27. A Norberg et al, Clin Microbiol Infect 2003; 9:65; O Vandenberg et al, Int J Infect Dis 2006; 10:255.
28. No drug is curative against *Dracunculus*. A program for monitoring local sources of drinking water to eliminate transmission has dramatically decreased the number of cases worldwide (M Barry, N Engl J Med 2007; 356:2561). The treatment of choice is slow extraction of worm combined with wound care and pain management (C Greenaway, CMAJ 2004; 170:495).
29. Since family members are usually infected, treatment of the entire household is recommended.
30. Antihistamines or corticosteroids may be required to decrease allergic reactions to components of disintegrating microfilariae that result from treatment, especially in infection caused by *Loa loa*. Endosymbiotic *Wolbachia* bacteria may have a role in filarial development and host response, and may represent a potential target for therapy. Addition of doxycycline 100 or 200 mg/d PO x 6-8 wks in lymphatic filariasis and onchocerciasis has resulted in substantial loss of *Wolbachia* and decrease in both micro- and macrofilariae (MJ Taylor et al, Lancet 2005; 365:2116; AY Debrah et al, Plos Pathog 2006; e92:0829); but use of tetracyclines is contraindicated in pregnancy and in children <8 yrs old.
31. Most symptoms are caused by adult worm. A single-dose combination of albendazole (400 mg PO) with either ivermectin (200 mcg/kg PO) or diethylcarbamazine (6 mg/kg PO) is effective for reduction or suppression of *W. bancrofti* microfilaria, but the albendazole/ivermectin combination does not kill all the adult worms (D Addiss et al, Cochrane Database Syst Rev 2004; CD003753).
32. For patients with microfilaria in the blood, Medical Letter consultants start with a lower dosage and scale up: d1: 50 mg; d2: 50 mg tid; d3: 100 mg tid; d4-14: 6 mg/kg in 3 doses (for *Loa loa* d4-14: 9 mg/kg in 3 doses). Multi-dose regimens have been shown to provide more rapid reduction in microfilaria than single-dose diethylcarbamazine, but microfilaria levels are similar 6-12 months after treatment (LD Andrade et al, Trans R Soc Trop Med Hyg 1995; 89:319; PE Simonsen et al, Am J Trop Med Hyg 1995; 53:267). A single dose of 6 mg/kg is used in endemic areas for mass treatment (J Figueredo-Silva et al, Trans R Soc Trop Med Hyg 1996; 90:192; J Noroes et al, Trans R Soc Trop Med Hyg 1997; 91:78).
33. Diethylcarbamazine should not be used for treatment of *Onchocerca volvulus* due to the risk of increased ocular side effects including blindness associated with rapid killing of the worms. It should be used cautiously in geographic regions where *O. volvulus* coexists with other filariae. Diethylcarbamazine is contraindicated during pregnancy. See also footnote 38.
34. In heavy infections with *Loa loa*, rapid killing of microfilariae can provoke encephalopathy. Apheresis has been reported to be effective in lowering microfilarial counts in patients heavily infected with Loa loa (EA Ottesen, Infect Dis Clin North Am 1993; 7:619). Albendazole may be useful for treatment of loiasis when diethylcarbamazine is ineffective or cannot be used, but repeated courses may be necessary (AD Klion et al, Clin Infect Dis 1999; 29:680; TE Tabi et al, Am J Trop Med Hyg 2004; 71:211). Ivermectin has also been used to reduce microfilaremia, but albendazole is preferred because of its slower onset of action and lower risk of precipitating encephalopathy (AD Klion et al, J Infect Dis 1993; 168:202; M Kombila et al, Am J Trop Med Hyg 1998; 58:458). Diethylcarbamazine, 300 mg PO once/wk, has been recommended for prevention of loiasis (TB Nutman et al, N Engl J Med 1988; 319:752).
35. Diethylcarbamazine has no effect. A single dose of ivermectin 200 mcg/kg PO reduces microfilaria densities and provides both short- and long-term reductions in *M. ozzardi* microfilaremia (AA Gonzalez et al, W Indian Med J 1999; 48:231).

36. Diethylcarbamazine is potentially curative due to activity against both adult worms and microfilariae. Ivermectin is active only against microfilariae.
37. AK Boggild et al, Clin Infect Dis 2004; 39:1123. Relapses occur and can be treated with a repeated course of diethylcarbamazine.
38. Diethylcarbamazine should not be used for treatment of this disease because rapid killing of the worms can lead to blindness. Periodic treatment with ivermectin (every 3-12 months), 150 mcg/kg PO, can prevent blindness due to ocular onchocerciasis (DN Udall, Clin Infect Dis 2007; 44:53). Skin reactions after ivermectin treatment are often reported in persons with high microfilarial skin densities. Ivermectin has been inadvertently given to pregnant women during mass treatment programs; the rates of congenital abnormalities were similar in treated and untreated women. Because of the high risk of blindness from onchocerciasis, the use of ivermectin after the first trimester is considered acceptable according to the WHO. Doxycycline (100 mg/day PO for 6 weeks), followed by a single 150 mcg/kg PO dose of ivermectin, resulted in up to 19 months of amicrofilaridermia and 100% elimination of *Wolbachia* species (A Hoerauf et al, Lancet 2001; 357:1415).
39. Praziquantel should be taken with liquids during a meal.
40. Unlike infections with other flukes, *Fasciola hepatica* infections may not respond to praziquantel. Triclabendazole (*Egaten* - Novartis) appears to be safe and effective, but data are limited (DY Aksoy et al, Clin Microbiol Infect 2005; 11:859). It is available from Victoria Pharmacy, Zurich, Switzerland (www.pharmaworld.com; 41-1-211-24-32) and should be given with food for better absorption. Nitazoxanide also appears to have efficacy in treating fascioliasis in adults and in children (L Favennec et al, Aliment Pharmacol Ther 2003; 17:265; JF Rossignol et al, Trans R Soc Trop Med Hyg 1998; 92:103; SM Kabil et al, Curr Ther Res 2000; 61:339).
41. J Keiser et al, Expert Opin Investig Drugs 2005; 14:1513.
42. Triclabendazole may be effective in a dosage of 5 mg/kg PO once/d x 3 d or 10 mg/kg PO bid x 1 d (M Calvopiña et al, Trans R Soc Trop Med Hyg 1998; 92:566). See footnote 40 for availability.
43. Another alternative is albendazole 400 mg/d PO x 5 d in adults and 10 mg/kg/d PO x 5 d in children (K Yereli et al, Clin Microbiol Infect 2004; 10:527; O Karabay et al, World J Gastroenterol 2004; 10:1215). Combination treatment with standard doses of metronidazole and quinacrine x 3 wks has been effective for a small number of refractory infections (TE Nash et al, Clin Infect Dis 2001; 33:22). In one study, nitazoxanide was used successfully in high doses to treat a case of *Giardia* resistant to metronidazole and albendazole (P Abboud et al, Clin Infect Dis 2001; 32:1792).
44. Poorly absorbed; may be useful for treatment of giardiasis in pregnancy.
45. Quinacrine should be taken with liquids after a meal.
46. P Nontasut et al, Southeast Asian J Trop Med Pub Health 2005; 36:650; M de Gorgolas et al, J Travel Med 2003; 10:358. All patients should be treated with medication whether surgery is attempted or not.
47. ME Wilson et al, Clin Infect Dis 2001; 32:1378; G Molavi et al, J Helminth 2006; 80:425.
48. Usually a self-limited illness in immunocompetent patients. Immunosuppressed patients may need higher doses, longer duration (TMP/SMX qid x 10 d, followed by bid x 3 wks) and long-term maintenance. In sulfonamide-sensitive patients, pyrimethamine 50-75 mg daily in divided doses (plus leucovorin 10-25 mg/d) has been effective.
49. To maximize effectiveness and minimize toxicity, the choice of drug, dosage, and duration of therapy should be individualized based on the region of disease acquisition, a likely infecting species, and host factors such as immune status (BL Herwaldt, Lancet 1999; 354:1191). Some of the listed drugs and regimens are effective only against certain *Leishmania* species/strains and only in certain areas of the world (J Arevalo et al, Clin Infect Dis 2007; 195:1846). Medical Letter consultants recommend consultation with physicians experienced in management of this disease.
50. Visceral infection is most commonly due to the Old World species *L. donovani* (kala-azar) and *L. infantum* and the New World species *L. chagasi*.
51. Liposomal amphotericin B (*AmBisome*) is the only lipid formulation of amphotericin B FDA-approved for treatment of visceral leishmania, largely based on clinical trials in

patients infected with *L. infantum* (A Meyerhoff, Clin Infect Dis 1999; 28:42). Two other amphotericin B lipid formulations, amphotericin B lipid complex (*Abelcet*) and amphotericin B cholesteryl sulfate (*Amphotec*) have been used, but are considered investigational for this condition and may not be as effective (C Bern et al, Clin Infect Dis 2006; 43:917).

52. The FDA-approved dosage regimen for immunocompromised patients (e.g., HIV infected) is 4 mg/kg/d IV on days 1-5, 10, 17, 24, 31 and 38. The relapse rate is high; maintenance therapy (secondary prevention) may be indicated, but there is no consensus as to dosage or duration.
53. Effective for both antimony-sensitive and -resistant *L. donovani* (Indian); miltefosine (*Impavido*) is manufactured in 10- or 50-mg capsules by Zentaris (Frankfurt, Germany at info@zentaris.com) and is available through consultation with the CDC. The drug is contraindicated in pregnancy; a negative pregnancy test before drug initiation and effective contraception during and for 2 months after treatment is recommended (H Murray et al, Lancet 2005; 366:1561). In a placebo-controlled trial in patients ≥12 years old, oral miltefosine 2.5 mg/kg/d x 28 d was also effective for treatment of cutaneous leishmaniasis due to *L.(V.) panamensis* in Colombia, but not *L.(V.) braziliensis* or *L. mexicana* in Guatemala (J Soto et al, Clin Infect Dis 2004; 38:1266). "Motion sickness," nausea, headache and increased creatinine are the most frequent adverse effects (J Soto and P Soto, Expert Rev Anti Infect Ther 2006; 4:177).
54. Paromomycin IM has been effective against leishmania in India; it has not yet been tested in South America or the Mediterranean and there is insufficient data to support its use in pregnancy (S Sundar et al, N Engl J Med 2007; 356:2371). Topical paromomycin should be used only in geographic regions where cutaneous leishmaniasis species have low potential for mucosal spread. A formulation of 15% paromomycin/12% methylbenzethonium chloride (*Leshcutan*) in soft white paraffin for topical use has been reported to be partially effective against cutaneous leishmaniasis due to *L. major* in Israel and *L. mexicana* and *L. (V.) braziliensis* in Guatemala, where mucosal spread is very rare (BA Arana et al, Am J Trop Med Hyg 2001; 65:466). The methylbenzethonium is irritating to the skin; lesions may worsen before they improve.
55. Cutaneous infection is most commonly due to the Old World species *L. major* and *L. tropica* and the New World species *L. mexicana*, *L. (Viannia) braziliensis*, and others.
56. Although azole drugs (fluconazole, ketoconazole, itraconazole) have been used to treat cutaneous disease, they are not reliably effective and have no efficacy against mucosal disease (AJ Magill, Infect Dis Clin North Am 2005; 19:241). For treatment of *L. major* cutaneous lesions, a study in Saudi Arabia found that oral fluconazole, 200 mg once/d x 6 wks appeared to speed healing (AA Alrajhi et al, N Engl J Med 2002; 346:891). Thermotherapy may be an option for cutaneous *L. tropica* infection (R Reithinger et al, Clin Infect Dis 2005; 40:1148). A device that generates focused and controlled heating of the skin has been approved by the FDA for this indication (ThermoMed—ThermoSurgery Technologies Inc., Phoenix, AZ, 602-264-7300; www.thermosurgery.com).
57. At this dosage pentamidine has been effective in Colombia predominantly against *L. (V.) panamensis* (J Soto-Mancipe et al, Clin Infect Dis 1993; 16:417; J Soto et al, Am J Trop Med Hyg 1994; 50:107). Activity against other species is not well established.
58. Mucosal infection is most commonly due to the New World species *L. (V.) braziliensis*, *L. (V.) panamensis*, or *L. (V.) guyanensis*.
59. Pediculocides should not be used for infestations of the eyelashes. Such infestations are treated with petrolatum ointment applied 2-4 x/d x 8-10 d. Oral TMP/SMX has also been used (TL Meinking and D Taplin, Curr Probl Dermatol 1996; 24:157). For pubic lice, treat with 5% permethrin or ivermectin as for scabies. TMP/SMX has also been effective when used together with permethrin for head lice (RB Hipolito et al, Pediatrics 2001; 107:E30).
60. Malathion is both ovicidal and pediculocidal; 2 applications at least 7 days apart are generally necessary to kill all lice and nits.
61. Permethrin and pyrethrin are pediculocidal; retreatment in 7-10 d is needed to eradicate the infestation. Some lice are resistant to pyrethrins and permethrin (TL Meinking et al, Arch Dermatol 2002; 138:220).

62. Ivermectin is pediculocidal, but more than one dose is generally necessary to eradicate the infestation (KN Jones and JC English 3rd, Clin Infect Dis 2003; 36:1355). The number of doses and interval between doses has not been established, but in one study of body lice, 3 doses administered at 7-day intervals were effective (C Fouault et al, J Infect Dis 2006; 193:474).
63. Chloroquine-resistant *P. falciparum* occurs in all malarious areas except Central America (including Panama north and west of the Canal Zone), Mexico, Haiti, the Dominican Republic, Paraguay, northern Argentina, North and South Korea, Georgia, Armenia, most of rural China and some countries in the Middle East (chloroquine resistance has been reported in Yemen, Oman, Saudi Arabia and Iran). For treatment of multiple-drug-resistant *P. falciparum* in Southeast Asia, especially Thailand, where mefloquine resistance is frequent, atovaquone/proguanil, quinine plus either doxycycline or clindamycin, or artemether/lumefantrine may be used.
64. *P. vivax* with decreased susceptibility to chloroquine is a significant problem in Papua-New Guinea and Indonesia. There are also a few reports of resistance from Myanmar, India, the Solomon Islands, Vanuatu, Guyana, Brazil, Colombia and Peru (JK Baird et al, Curr Infect Dis Rep 2007; 9:39).
65. Chloroquine-resistant *P. malariae* has been reported from Sumatra (JD Maguire et al, Lancet 2002; 360:58).
66. Uncomplicated or mild malaria may be treated with oral drugs. Severe malaria (e.g. impaired consciousness, parasitemia >5%, shock, etc.) should be treated with parenteral drugs (KS Griffin et al, JAMA 2007; 297:2264).
67. Primaquine is given for prevention of relapse after infection with *P. vivax* or *P. ovale*. Some experts also prescribe primaquine phosphate 30 mg base/d (0.6 mg base/kg/d for children) for 14 d after departure from areas where these species are endemic (Presumptive Anti-Relapse Therapy [PART], "terminal prophylaxis"). Since this is not always effective as prophylaxis (E Schwartz et al, N Engl J Med 2003; 349:1510), others prefer to rely on surveillance to detect cases when they occur, particularly when exposure was limited or doubtful. See also footnote 79.
68. Atovaquone/proguanil is available as a fixed-dose combination tablet: adult tablets (*Malarone*; 250 mg atovaquone/100 mg proguanil) and pediatric tablets (*Malarone Pediatric*; 62.5 mg atovaquone/25 mg proguanil). To enhance absorption and reduce nausea and vomiting, it should be taken with food or a milky drink. Safety in pregnancy is unknown; outcomes were normal in 24 women treated with the combination in the 2nd and 3rd trimester (R McGready et al, Eur J Clin Pharmacol 2003; 59:545). The drug should not be given to patients with severe renal impairment (creatinine clearance <30mL/min). There have been isolated case reports of resistance in *P. falciparum* in Africa, but Medical Letter consultants do not believe there is a high risk for acquisition of *Malarone*-resistant disease (E Schwartz et al, Clin Infect Dis 2003; 37:450; A Farnert et al, BMJ 2003; 326:628; S Kuhn et al, Am J Trop Med Hyg 2005; 72:407; CT Happi et al, Malaria Journal 2006; 5:82).
69. Although approved for once-daily dosing, Medical Letter consultants usually divide the dose in two to decrease nausea and vomiting.
70. Available in the US in a 324-mg capsule; 2 capsules suffice for adult dosage. In Southeast Asia, relative resistance to quinine has increased and treatment should be continued for 7 d. Quinine should be taken with or after meals to decrease gastrointestinal adverse effects.
71. Doxycycline should be taken with adequate water to avoid esophageal irritation. It can be taken with food to minimize gastrointestinal adverse effects.
72. For use in pregnancy and in children <8 yrs.
73. B Lell and PG Kremsner, Antimicrob Agents Chemother 2002; 46:2315; M Ramharter et al, Clin Infect Dis 2005; 40:1777.
74. At this dosage, adverse effects include nausea, vomiting, diarrhea and dizziness. Disturbed sense of balance, toxic psychosis and seizures can also occur. Mefloquine should not be used for treatment of malaria in pregnancy unless there is no other treatment option because of increased risk for stillbirth (F Nosten et al, Clin Infect Dis 1999; 28:808). It should be avoided for treatment of malaria in persons with active depression or with a history of psychosis or seizures and should be used with caution in persons with any psychiatric illness. Mefloquine can be given to patients taking β-blockers if they do not have

an underlying arrhythmia; it should not be used in patients with conduction abnormalities. Mefloquine should not be given together with quinine or quinidine, and caution is required in using quinine or quinidine to treat patients with malaria who have taken mefloquine for prophylaxis. Mefloquine should not be taken on an empty stomach; it should be taken with at least 8 oz of water.

75. *P. falciparum* with resistance to mefloquine is a significant problem in the malarious areas of Thailand and in areas of Myanmar and Cambodia that border on Thailand. It has also been reported on the borders between Myanmar and China, Laos and Myanmar, and in Southern Vietnam. In the US, a 250-mg tablet of mefloquine contains 228 mg mefloquine base. Outside the US, each 275-mg tablet contains 250 mg base.
76. The artemisinin-derivatives, artemether and artesunate, are both frequently used globally in combination regimens to treat malaria. Both are available in oral, parenteral and rectal formulations, but manufacturing standards are not consistent (HA Karunajeewa et al, JAMA 2007; 297:2381; EA Ashley and NJ White, Curr Opin Infect Dis 2005; 18:531). In the US, only the IV formulation of artesunate is available; it can be obtained through the CDC under an IND for patients with severe disease who do not have timely access, cannot tolerate, or fail to respond to IV quinidine (www.cdc.gov/malaria/features/artesunate_now_available.htm). To avoid development of resistance, monotherapy should be avoided (PE Duffy and CH Sibley, Lancet 2005; 366:1908). In animal studies artemisinins have been embryotoxic and caused a low incidence of teratogenicity; no adverse pregnancy outcome has been observed in limited studies in humans (S Dellicour et al, Malaria Journal 2007; 6:15).
77. Artemether/lumefantrine is available as a fixed-dose combination tablet (*Coartem* in countries with endemic malaria, *Riamet* in Europe and countries without endemic malaria); each tablet contains 20 mg artemether and 120 mg lumefantrine (M van Vugt et al, Am J Trop Med Hyg 1999; 60:936). It is contraindicated during the first trimester of pregnancy; safety during the second and third trimester is not known. The tablets should be taken with food. Artemether/lumefantrine should not be used in patients with cardiac arrhythmias, bradycardia, severe cardiac disease or QT prolongation. Concomitant use of drugs that prolong the QT interval or are metabolized by CYP2D6 is contraindicated.
78. Adults treated with artesunate should also receive oral treatment doses of either atovaquone/proguanil, doxycycline, clindamycin or mefloquine; children should take either atovaquone/proguanil, clindamycin or mefloquine (F Nosten et al, Lancet 2000; 356:297; M van Vugt, Clin Infect Dis 2002; 35:1498; F Smithuis et al, Trans R Soc Trop Med Hyg 2004; 98:182). If artesunate is given IV, oral medication should be started when the patient is able to tolerate it (SEAQUAMAT group, Lancet 2005; 366:717).
79. Primaquine phosphate can cause hemolytic anemia, especially in patients whose red cells are deficient in G-6-PD. This deficiency is most common in African, Asian and Mediterranean peoples. Patients should be screened for G-6-PD deficiency before treatment. Primaquine should not be used during pregnancy. It should be taken with food to minimize nausea and abdominal pain. Primaquine-tolerant *P. vivax* can be found globally. Relapses of primaquine-resistant strains may be retreated with 30 mg (base) x 28 d.
80. Chloroquine should be taken with food to decrease gastrointestinal adverse effects. If chloroquine phosphate is not available, hydroxychloroquine sulfate is as effective; 400 mg of hydroxychloroquine sulfate is equivalent to 500 mg of chloroquine phosphate.
81. Chloroquine combined with primaquine was effective in 85% of patients with *P. vivax* resistant to chloroquine and could be a reasonable choice in areas where other alternatives are not available (JK Baird et al, J Infect Dis 1995; 171:1678).
82. Exchange transfusion is controversial, but has been helpful for some patients with high-density (>10%) parasitemia, altered mental status, pulmonary edema or renal complications (VI Powell and K Grima, Transfus Med Rev 2002; 16:239; MS Riddle et al, Clin Infect Dis 2002; 34:1192).
83. Continuous EKG, blood pressure and glucose monitoring are recommended, especially in pregnant women and young children. For problems with quinidine availability, call the manufacturer (Eli Lilly, 800-821-0538) or the CDC Malaria Hotline (770-488-7788). Quinidine may have greater antimalarial activity than quinine. The loading dose should be decreased or omitted in patients who have received quinine or mefloquine. If more than 48 hours of parenteral treatment is required, the quinine or quinidine dose should be reduced by 30-50%.

84. No drug guarantees protection against malaria. Travelers should be advised to seek medical attention if fever develops after they return. Insect repellents, insecticide- impregnated bed nets and proper clothing are important adjuncts for malaria prophylaxis (Med Lett Drugs Ther 2005; 47:100). Malaria in pregnancy is particularly serious for both mother and fetus; prophylaxis is indicated if exposure cannot be avoided.
85. Alternatives for patients who are unable to take chloroquine include atovaquone/proguanil, mefloquine, doxycycline or primaquine dosed as for chloroquine-resistant areas.
86. Has been used extensively and safely for prophylaxis in pregnancy.
87. Beginning 1-2 wks before travel and continuing weekly for the duration of stay and for 4 wks after leaving.
88. Beginning 1-2 d before travel and continuing for the duration of stay and for 1 wk after leaving. In one study of malaria prophylaxis, atovaquone/proguanil was better tolerated than mefloquine in nonimmune travelers (D Overbosch et al, Clin Infect Dis 2001; 33:1015). The protective efficacy of *Malarone* against *P. vivax* is variable ranging from 84% in Indonesian New Guinea (J Ling et al, Clin Infect Dis 2002; 35:825) to 100% in Colombia (J Soto et al, Am J Trop Med Hyg 2006; 75:430). Some Medical Letter consultants prefer alternate drugs if traveling to areas where *P. vivax* predominates.
89. Beginning 1-2 d before travel and continuing for the duration of stay and for 4 wks after leaving. Use of tetracyclines is contraindicated in pregnancy and in children <8 years old. Doxycycline can cause gastrointestinal disturbances, vaginal moniliasis and photosensitivity reactions.
90. Mefloquine has not been approved for use during pregnancy. However, it has been reported to be safe for prophylactic use during the second and third trimester of pregnancy and possibly during early pregnancy as well (CDC Health Information for International Travel, 2008, page 228; BL Smoak et al, J Infect Dis 1997; 176:831). For pediatric doses <½ tablet, it is advisable to have a pharmacist crush the tablet, estimate doses by weighing, and package them in gelatin capsules.There is no data for use in children <5 kg, but based on dosages in other weight groups, a dose of 5 mg/kg can be used. Not recommended for use in travelers with active depression or with a history of psychosis or seizures and should be used with caution in persons with psychiatric illness. Mefloquine can be given to patients taking β-blockers if they do not have an underlying arrhythmia; it should not be used in patients with conduction abnormalities.
91. Beginning 1-2 wks before travel and continuing weekly for the duration of stay and for 4 wks after leaving. Most adverse events occur within 3 doses. Some Medical Letter consultants favor starting mefloquine 3 weeks prior to travel and monitoring the patient for adverse events, this allows time to change to an alternative regimen if mefloquine is not tolerated.
92. The combination of weekly chloroquine (300 mg base) and daily proguanil (200 mg) is recommended by the World Health Organization (www.WHO.int) for use in selected areas; this combination is no longer recommended by the CDC. Proguanil (Paludrine—AstraZeneca, United Kingdom) is not available alone in the US but is widely available in Canada and Europe. Prophylaxis is recommended during exposure and for 4 weeks afterwards. Proguanil has been used in pregnancy without evidence of toxicity (PA Phillips-Howard and D Wood, Drug Saf 1996; 14:131).
93. Studies have shown that daily primaquine beginning 1d before departure and continued until 3-7 d after leaving the malarious area provides effective prophylaxis against chloroquine-resistant *P. falciparum* (JK Baird et al, Clin Infect Dis 2003; 37:1659). Some studies have shown less efficacy against *P. vivax*. Nausea and abdominal pain can be diminished by taking with food.
94. A traveler can be given a course of medication for presumptive self-treatment of febrile illness. The drug given for self-treatment should be different from that used for prophylaxis. This approach should be used only in very rare circumstances when a traveler would not be able to get medical care promptly.
95. CM Chan et al, Ophthalmology 2003; 110:1420. Ocular lesions due to *E. hellem* in HIV-infected patients have responded to fumagillin eyedrops prepared from *Fumidil-B* (bicyclohexyl ammonium fumagillin) used to control a microsporidial disease of honey bees (MJ Garvey et al, Ann Pharmacother 1995; 29:872), available from Leiter's Park Avenue Pharmacy (see footnote 1). For lesions due to *V. corneae*, topical therapy is generally not effective and keratoplasty may be required (RM Davis et al, Ophthalmology 1990; 97:953).

96. Oral fumagillin (*Flisint*—Sanofi-Aventis, France) has been effective in treating *E. bieneusi* (J-M Molina et al, N Engl J Med 2002; 346:1963), but has been associated with thrombocytopenia and neutropenia. Highly active antiretroviral therapy (HAART) may lead to microbiologic and clinical response in HIV-infected patients with microsporidial diarrhea. Octreotide (*Sandostatin*) has provided symptomatic relief in some patients with large-volume diarrhea.
97. J-M Molina et al, J Infect Dis 1995; 171:245. There is no established treatment for *Pleistophora*. For disseminated disease due to*Trachipleistophora* or *Brachiola*, itraconazole 400 mg PO once/d plus albendazole may also be tried (CM Coyle et al, N Engl J Med 2004; 351:42).
98. Albendazole or pyrantel pamoate may be effective (JB Ziem et al, Ann Trop Med Parasitol 2004; 98:385).
99. Pneumocystis has been reclassified as a fungus. In severe disease with room air $PO_2 \leq$ 70 mmHg or Aa gradient ≥ 35 mmHg, prednisone should also be used (S Gagnon et al, N Engl J Med 1990; 323:1444; E Caumes et al, Clin Infect Dis 1994; 18:319).
100. Primary/secondary prophylaxis in patients with HIV can be discontinued after CD4 count increases to >200 x 10^6/L for >3 mos.
101. Plus leucovorin 25 mg with each dose of pyrimethamine. Pyrimethamine should be taken with food to minimize gastrointestinal adverse effects.
102. Treatment may need to be repeated in 10-14 days. A second ivermectin dose taken 2 weeks later increases the cure rate to 95%, which is equivalent to that of 5% permethrin (V Usha et al, J Am Acad Dermatol 2000; 42:236; O Chosidow, N Engl J Med 2006; 354:1718; J Heukelbach and H Feldmeier, Lancet 2006; 367:1767).
103. Lindane (γ-benzene hexachloride) should be reserved for treatment of patients who fail to respond to other drugs. The FDA has recommended it not be used for immunocompromised patients, young children, the elderly, pregnant and breast-feeding women, and patients weighing <50 kg.
104. Ivermectin, either alone or in combination with a topical scabicide, is the drug of choice for crusted scabies in immunocompromised patients (P del Giudice, Curr Opin Infect Dis 2004; 15:123).
105. Oxamniquine, which is not available in the US, is generally not as effective as praziquantel. It has been useful, however, in some areas in which praziquantel is less effective (ML Ferrari et al, Bull World Health Organ 2003; 81:190; A Harder, Parasitol Res 2002; 88:395). Oxamniquine is contraindicated in pregnancy. It should be taken after food.
106. In East Africa, the dose should be increased to 30 mg/kg, and in Egypt and South Africa to 30 mg/kg/d x 2 d. Some experts recommend 40-60 mg/kg over 2-3 d in all of Africa (KC Shekhar, Drugs 1991; 42:379).
107. In immunocompromised patients or disseminated disease, it may be necessary to prolong or repeat therapy, or to use other agents. Veterinary parenteral and enema formulations of ivermectin have been used in severely ill patients with hyperinfection who were unable to take or reliably absorb oral medications (J Orem et al, Clin Infect Dis 2003; 37:152; PE Tarr Am J Trop Med Hyg 2003; 68:453; FM Marty et al, Clin Infect Dis 2005; 41:e5). In disseminated strongyloidiasis, combination therapy with albendazole and ivermectin has been suggested (S Lim et al, CMAJ 2004; 171:479).
108. Niclosamide must be chewed thoroughly before swallowing and washed down with water.
109. JO Juan et al, Trans R Soc Trop Med Hyg 2002; 96:193; JC Chero et al, Trans R Soc Trop Med Hyg 2007; 101:203; E Diaz et al, Am J Trop Med Hyg 2003; 68:384.
110. Patients may benefit from surgical resection or percutaneous drainage of cysts. Praziquantel is useful preoperatively or in case of spillage of cyst contents during surgery. Percutaneous aspiration-injection-reaspiration (PAIR) with ultrasound guidance plus albendazole therapy has been effective for management of hepatic hydatid cyst disease (RA Smego, Jr. et al, Clin Infect Dis 2003; 37:1073; S Nepalia et al, J Assoc Physicians India 2006; 54:458; E Zerem and R Jusufovic Surg Endosc 2006; 20:1543).
111. Surgical excision is the only reliable means of cure. Reports have suggested that in nonresectable cases use of albendazole (400 mg bid) can stabilize and sometimes cure infection (P Craig, Curr Opin Infect Dis 2003; 16:437; O Lidove et al, Am J Med 2005; 118:195).

112. Initial therapy for patients with inflamed parenchymal cysticercosis should focus on symptomatic treatment with anti-seizure medication (LS Yancey et al, Curr Infect Dis Rep 2005; 7:39; AH del Brutto et al, Ann Intern Med 2006; 145:43). Patients with live parenchymal cysts who have seizures should be treated with albendazole together with steroids (dexamethasone 6 mg/d or prednisone 40-60 mg/d) and an anti-seizure medication (HH Garcia et al, N Engl J Med 2004; 350:249). Patients with subarachnoid cysts or giant cysts in the fissures should be treated for at least 30 d (JV Proaño et al, N Engl J Med 2001; 345:879). Surgical intervention (especially neuroendoscopic removal) or CSF diversion followed by albendazole and steroids is indicated for obstructive hydocephalus. Arachnoiditis, vasculitis or cerebral edema is treated with prednisone 60 mg/d or dexamethasone 4-6 mg/d together with albendazole or praziquantel (AC White, Jr., Annu Rev Med 2000; 51:187). Any cysticercocidal drug may cause irreparable damage when used to treat ocular or spinal cysts, even when corticosteroids are used. An ophthalmic exam should always precede treatment to rule out intraocular cysts.
113. To treat CNS toxoplasmosis in HIV-infected patients, some clinicians have used pyrimethamine 50-100 mg/d (after a loading dose of 200 mg) with sulfadiazine and, when sulfonamide sensitivity developed, have given clindamycin 1.8-2.4 g/d in divided doses instead of the sulfonamide. Treatment is usually given for at least 4-6 weeks. Atovaquone (1500 mg PO bid) plus pyrimethamine (200 mg loading dose, followed by 75 mg/d PO) for 6 weeks appears to be an effective alternative in sulfa-intolerant patients (K Chirgwin et al, Clin Infect Dis 2002; 34:1243). Atovaquone must be taken with a meal to enhance absorption. Treatment is followed by chronic suppression with lower dosage regimens of the same drugs. For primary prophylaxis in HIV patients with <100 x 106/L CD4 cells, either trimethoprim-sulfamethoxazole, pyrimethamine with dapsone, or atovaquone with or without pyrimethamine can be used. Primary or secondary prophylaxis may be discontinued when the CD4 count increases to >200 x 106/L for >3mos (MMWR Morb Mortal Wkly Rep 2004; 53 [RR15]:1). In ocular toxoplasmosis with macular involvement, corticosteroids are recommended in addition to antiparasitic therapy for an anti-inflammatory effect. In one randomized single-blind study, trimethoprim/sulfamethoxazole was reported to be as effective as pyrimethamine/sulfadiazine for treatment of ocular toxoplasmosis (M Soheilian et al, Ophthalmology 2005; 112:1876). Women who develop toxoplasmosis during the first trimester of pregnancy should be treated with spiramycin (3-4 g/d). After the first trimester, if there is no documented transmission to the fetus, spiramycin can be continued until term. If transmission has occurred in utero, therapy with pyrimethamine and sulfadiazine should be started (JG Montoya and O Liesenfeld, Lancet 2004; 363:1965). Pyrimethamine is a potential teratogen and should be used only after the first trimester.
114. Plus leucovorin 10-25 mg with each dose of pyrimethamine. Pyrimethamine should be taken with food to minimize gastrointestinal adverse effects.
115. Congenitally infected newborns should be treated with pyrimethamine every 2 or 3 days and a sulfonamide daily for about one year (JS Remington and G Desmonts in JS Remington and JO Klein, eds, *Infectious Disease of the Fetus and Newborn Infant*, 6th ed, Philadelphia: Saunders, 2006:1038).
116. Sulfadiazine should be taken on an empty stomach with adequate water.
117. Sexual partners should be treated simultaneously with same dosage. Metronidazole-resistant strains have been reported and can be treated with higher doses of metronidazole (2-4 g/d x 7-14 d) or with tinidazole (MMWR Morb Mortal Wkly Rep 2006; 55 [RR11]:1).
118. MP Barrett et al, Lancet 2003; 362:1469. Treatment of chronic or indeterminate Chagas' disease with benznidazole has been associated with reduced progression and increased negative seroconversion (R Viotti et al, Ann Intern Med 2006; 144:724).
119. Benznidazole should be taken with meals to minimize gastrointestinal adverse effects. It is contraindicated during pregnancy.
120. Pentamidine and suramin have equal efficacy, but pentamidine is better tolerated.
121. Eflornithine is highly effective in *T.b. gambiense*, but not in *T.b. rhodesiense* infections. In one study of treatment of CNS disease due to *T.b. gambiense*, there were fewer serious complications with eflornithine than with melarsoprol (F Chappuis et al, Clin Infect Dis 2005; 41:748). Eflornithine is available in limited supply only from the WHO. It is contraindicated during pregnancy.

122. E Schmid et al, J Infect Dis 2005; 191:1922. Corticosteroids have been used to prevent arsenical encephalopathy (J Pepin et al, Trans R Soc Trop Med Hyg 1995; 89:92). Up to 20% of patients with *T.b.gambiense* fail to respond to melarsoprol (MP Barrett, Lancet 1999; 353:1113). In one study, a combination of low-dose melarsoprol (1.2 mg/kg/d IV) and nifurtimox (7.5 mg/kg PO bid) x 10d was more effective than standard-dose melarsoprol alone (S Bisser et al, J Infect Dis 2007; 195:322).
123. Optimum duration of therapy is not known; some Medical Letter consultants would treat x 20 d. For severe symptoms or eye involvement, corticosteroids can be used in addition (D Despommier, Clin Microbiol Rev 2003; 16:265).

Safety of Antiparasitic Drugs in Pregnancy

Drug	Toxicity in Pregnancy	Recommendations
Albendazole (*Albenza*)	Teratogenic and embryotoxic in animals	Caution*
Amphotericin B (*Fungizone*, and others)	None known	Caution*
Amphotericin B liposomal (*AmBisome*)	None known	Caution*
Artemether/lumefantrine (*Coartem, Riamet*)[1]	Embryocidal and teratogenic in animals	Caution*
Artesunate[1]	Embryocidal and teratogenic in animals	Caution*
Atovaquone (*Mepron*)	Maternal and fetal toxicity in animals	Caution*
Atovaquone/proguanil (*Malarone*)[2]	Maternal and fetal toxicity in animals	Caution*
Azithromycin (*Zithromax*, and others)	None known	Probably safe
Benznidazole (*Rochagan*)	Unknown	Contraindicated
Chloroquine (*Aralen*, and others)	None known with doses recommended for malaria prophylaxis	Probably safe in low doses
Clarithromycin (*Biaxin*, and others)	Teratogenic in animals	Contraindicated
Clindamycin (*Cleocin*, and others)	None known	Caution*
Crotamiton (*Eurax*)	Unknown	Caution*
Dapsone	None known; carcinogenic in rats and mice; hemolytic reactions in neonates	Caution*, especially at term
Diethylcarbamazine (DEC; *Hetrazan*)	Not known; abortifacient in one study in rabbits	Contraindicated
Diloxanide (*Furamide*)	Safety not established	Caution*
Doxycycline (*Vibramycin*, and others)	Tooth discoloration and dysplasia, inhibition of bone growth in fetus; hepatic toxicity and azotemia with IV use in pregnant patients with decreased renal function or with overdosage	Contraindicated
Eflornithine (*Ornidyl*)	Embryocidal in animals	Contraindicated
Fluconazole (*Diflucan*, and others)	Teratogenic	Contraindicated for high dose; caution* for single dose
Flucytosine (*Ancoban*)	Teratogenic in rats	Contraindicated
Furazolidone (*Furoxone*)	None known; carcinogenic in rodents; hemolysis with G-6-PD deficiency in newborn	Caution*; contraindicated at term

Drug	Toxicity in Pregnancy	Recommendations
Hydroxychloroquine (*Plaquenil*)	None known with doses recommended for malaria prophylaxis	Probably safe in low doses
Itraconazole (*Sporanox*, and others)	Teratogenic and embryotoxic in rats	Caution*
Iodoquinol (*Yodoxin*, and others)	Unknown	Caution*
Ivermectin (*Stromectol*)	Teratogenic in animals	Contraindicated
Ketoconazole (*Nizoral*, and others)	Teratogenic and embryotoxic in rats	Contraindicated; topical probably safe
Lindane	Absorbed from the skin; potential CNS toxicity in fetus	Contraindicated
Malathion, topical (*Ovide*)	None known	Probably safe
Mebendazole (*Vermox*)	Teratogenic and embryotoxic in rats	Caution*
Mefloquine (*Lariam*)[3]	Teratogenic in animals	Caution*
Meglumine (*Glucantine*)	Not known	Caution*
Metronidazole (*Flagyl*, and others)	None known—carcinogenic in rats and mice	Caution*
Miconazole (*Monistat* i.v.)	None known	Caution*
Miltefosine (*Impavido*)	Teratogenic in rats and induces abortions in animals	Contraindicated; effective contraception must be used for 2 months after the last dose
Niclosamide (*Niclocide*)	Not absorbed; no known toxicity in fetus	Probably safe
Nitazoxanide (*Alinia*)	None known	Caution*
Oxamniquine (*Vansil*)	Embryocidal in animals	Contraindicated
Paromomycin (*Humatin*)	Poorly absorbed; toxicity in fetus unknown	Oral capsules probably safe
Pentamidine (*Pentam 300*, *NebuPent*, and others)	Safety not established	Caution*
Permethrin (*Nix*, and others)	Poorly absorbed; no known toxicity in fetus	Probably safe
Praziquantel (*Biltricide*)	Not known	Probably safe
Primaquine	Hemolysis in G-6-PD deficiency	Contraindicated
Pyrantel pamoate (*Antiminth*, and others)	Absorbed in small amounts; no known toxicity in fetus	Probably safe
Pyrethrins and piperonyl butoxide (*RID*, and others)	Poorly absorbed; no known toxicity in fetus	Probably safe
Pyrimethamine (*Daraprim*)[4]	Teratogenic in animals	Caution*; contraindicated during 1st trimester
Quinacrine (*Atabrine*)	Safety not established	Caution*
Quinidine	Large doses can cause abortion	Probably safe
Quinine (*Qualaquin*)	Large doses can cause abortion; auditory nerve hypoplasia, deafness in fetus; visual changes, limb anomalies, visceral defects also reported	Caution*

Appendix

Drug	Toxicity in Pregnancy	Recommendations
Sodium stibogluconate (*Pentostam*)	Not known	Caution*
Sulfonamides	Teratogenic in some animal studies; hemolysis in newborn with G-6-PD deficiency; increased risk of kernicterus in newborn	Caution*; contraindicated at term
Suramin sodium (*Germanin*)	Teratogenic in mice	Caution*
Tetracycline (*Sumycin*, and others)	Tooth discoloration and dysplasia, inhibition of bone growth in fetus; hepatic toxicity and azotemia with IV use in pregnant patients with decreased renal function or with overdosage	Contraindicated
Tinidazole (*Tindamax*)	Increased fetal mortality in rats	Caution*
Trimethoprim (*Proloprim*, and others)	Folate antagonism; teratogenic in rats	Caution*
Trimethoprim-sulfamethoxazole (*Bactrim*, and others)	Same as sulfonamides and trimethoprim	Caution*; contraindicated at term

* Use only for strong clinical indication in absence of suitable alternative.
1. See also footnote 76 in previous table.
2. See also footnote 68 in previous table.
3. See also footnotes 74 and 90 in previous table.
4. See also footnote 113 in previous table.

Manufacturers of Drugs Used to Treat Parasitic Infections

albendazole—Albenza (GlaxoSmithKline)
Albenza (GlaxoSmithKline)—albendazole
Alinia (Romark)—nitazoxanide
AmBisome (Gilead)—amphotericin B, liposomal
amphotericin B—*Fungizone* (Apothecon), others
amphotericin B, liposomal—*AmBisome* (Gilead)
Ancobon (Valeant)—flucytosine
§ *Antiminth* (Pfizer)—pyrantel pamoate
• *Aralen* (Sanofi)—chloroquine HCl and chloroquine phosphate
§ artemether—*Artenam* (Arenco, Belgium)
§ artemether/lumefantrine—*Coartem, Riamet* (Novartis)
§ *Artenam* (Arenco, Belgium)—artemether
§ artesunate—(Guilin No. 1 Factory, People's Republic of China)
atovaquone—*Mepron* (GlaxoSmithKline)
atovaquone/proguanil—*Malarone* (GlaxoSmithKline)
azithromycin—*Zithromax* (Pfizer), others
• *Bactrim* (Roche)—TMP/Sulfa
§ benznidazole—*Rochagan* (Brazil)
• *Biaxin* (Abbott)—clarithromycin
§ *Biltricide* (Bayer)—praziquantel
† bithionol—*Bitin* (Tanabe, Japan)
† *Bitin* (Tanabe, Japan)—bithionol
§ *Brolene* (Aventis, Canada)—propamidine isethionate chloroquine HCl
and chloroquine phosphate—*Aralen* (Sanofi), others
clarithromycin—*Biaxin* (Abbott), others
• *Cleocin* (Pfizer)—clindamycin
clindamycin—*Cleocin* (Pfizer), others
Coartem (Novartis)—artemether/lumefantrine
crotamiton—*Eurax* (Westwood-Squibb)
dapsone—(Jacobus)
§ *Daraprim* (GlaxoSmithKline)—pyrimethamine USP
† diethylcarbamazine citrate (DEC)—*Hetrazan*
• *Diflucan* (Pfizer)—fluconazole
§ diloxanide furoate—*Furamide* (Boots, United Kingdom)
doxycycline—*Vibramycin* (Pfizer), others
eflornithine (Difluoromethylornithine, DFMO)—*Ornidyl* (Aventis)
§ *Egaten* (Novartis)—triclabendazole
Elimite (Allergan)—permethrin
Ergamisol (Janssen)—levamisole
Eurax (Westwood-Squibb)—crotamiton
• *Flagyl* (Pfizer)—metronidazole
§ *Flisint* (Sanofi-Aventis, France)—fumagillin
fluconazole—*Diflucan* (Pfizer), others
flucytosine—*Ancobon* (Valeant)
§ fumagillin—*Flisint* (Sanofi-Aventis, France)
• *Fungizone* (Apothecon)—amphotericin
§ *Furamide* (Boots, United Kingdom)—diloxanide furoate

§ furazolidone—*Furozone* (Roberts)
§ *Furozone* (Roberts)—furazolidone
† *Germanin* (Bayer, Germany)—suramin sodium
§ *Glucantime* (Aventis, France)—meglumine antimonate
† *Hetrazan*—diethylcarbamazine citrate (DEC)
Humatin (Monarch)—paromomycin
§ *Impavido* (Zentaris, Germany)—miltefosine
iodoquinol—*Yodoxin* (Glenwood), others
itraconazole—*Sporanox* (Janssen-Ortho), others
ivermectin—*Stromectol* (Merck)
ketoconazole—*Nizoral* (Janssen), others
† *Lampit* (Bayer, Germany)—nifurtimox
L ariam (Roche)—mefloquine
§ *Leshcutan* (Teva, Israel)—topical paromomycin
levamisole—*Ergamisol* (Janssen)
lumefantrine/artemether—*Coartem, Riamet* (Novartis)
Malarone (GlaxoSmithKline)—atovaquone/proguanil
malathion—*Ovide* (Medicis)
mebendazole—*Vermox* (McNeil), others
mefloquine—*Lariam* (Roche)
§ meglumine antimonate—*Glucantime* (Aventis, France)
† melarsoprol—*Mel-B*
† *Mel-B*—melarsoprol
Mepron (GlaxoSmithKline)—atovaquone
metronidazole—*Flagyl* (Pfizer), others
§ miconazole—*Monistat i.v.*
§ miltefosine—*Impavido* (Zentaris, Germany)
§ *Monistat i.v.*—miconazole
NebuPent (Fujisawa)—pentamidine isethionate
Neutrexin (US Bioscience)—trimetrexate
§ niclosamide—*Yomesan* (Bayer, Germany)
† nifurtimox—*Lampit* (Bayer, Germany)
nitazoxanide—*Alinia* (Romark)
• *Nizoral* (Janssen)—ketoconazole
Nix (GlaxoSmithKline)—permethrin
§ ornidazole—*Tiberal* (Roche, France)
Ornidyl (Aventis)—eflornithine (Difluoromethylornithine, DFMO)
Ovide (Medicis)—malathion
§ oxamniquine—*Vansil* (Pfizer)
§ *Paludrine* (AstraZeneca, United Kingdom)—proguanil
paromomycin—*Humatin* (Monarch); *Leshcutan* (Teva, Israel; (topical formulation not available in US)
Pentam 300 (Fujisawa)—pentamidine isethionate
pentamidine isethionate—*Pentam 300* (Fujisawa), *NebuPent* (Fujisawa)
† *Pentostam* (GlaxoSmithKline, United Kingdom)—sodium stibogluconate
permethrin—*Nix* (GlaxoSmithKline), *Elimite* (Allergan)
§ praziquantel—*Biltricide* (Bayer)
primaquine phosphate USP
§ proguanil—*Paludrine* (AstraZeneca, United Kingdom)
proguanil/atovaquone—*Malarone* (GlaxoSmithKline)
§ propamidine isethionate—*Brolene* (Aventis, Canada)
§ pyrantel pamoate—*Antiminth* (Pfizer)

pyrethrins and piperonyl butoxide—*RID* (Pfizer), others
§ pyrimethamine USP—*Daraprim* (GlaxoSmithKline)
Qualaquin—quinine sulfate (Mutual Pharmaceutical Co/AR Scientific)
* quinidine gluconate (Eli Lilly)
§ quinine dihydrochloride
quinine sulfate—*Qualaquin* (Mutual Pharmaceutical Co/AR Scientific)
Riamet (Novartis)—artemether/lumefantrine
• *RID* (Pfizer)—pyrethrins and piperonyl butoxide
• *Rifadin* (Aventis)—rifampin
rifampin—*Rifadin* (Aventis), others
§ *Rochagan* (Brazil)—benznidazole
* *Rovamycine* (Aventis)—spiramycin
† sodium stibogluconate—Pentostam (GlaxoSmithKline, United Kingdom)
* spiramycin—*Rovamycine* (Aventis)
• *Sporanox* (Janssen-Ortho)—itraconazole
Stromectol (Merck)—ivermectin
sulfadiazine—(Eon)
† suramin sodium—*Germanin* (Bayer, Germany)
§ *Tiberal* (Roche, France)—ornidazole
Tindamax (Mission)—tinidazole
tinidazole—*Tindamax* (Mission)
TMP/Sulfa—*Bactrim* (Roche), others
§ triclabendazole—*Egaten* (Novartis)
trimetrexate—*Neutrexin* (US Bioscience)
§ *Vansil* (Pfizer)—oxamniquine
• *Vermox* (McNeil)—mebendazole
• *Vibramycin* (Pfizer)—doxycycline
• *Yodoxin* (Glenwood)—iodoquinol
§ *Yomesan* (Bayer, Germany)—niclosamide
• *Zithromax* (Pfizer)—azithromycin

* Available in the US only from the manufacturer.
§ Not available in the US; may be available through a compounding pharmacy (see footnote 4 in previous *Drugs for Parasitic Infections* table).
† Available from the CDC Drug Service, Centers for Disease Control and Prevention, Atlanta, Georgia 30333; 404-639-3670 (evenings, weekends, or holidays: 770-488-7100).
• Also available generically.

Index

D

E

F

G

H

I

K

L

M

R

S

T

U

V

W